TIMESCALES OF MAGMATIC PROCESSES

Timescales of Magmatic Processes
From Core to Atmosphere

EDITED BY

**Anthony Dosseto, Simon P. Turner
and James A. Van Orman**

A John Wiley & Sons, Ltd., Publication

Registered Office
John Wiley & Sons Ltd, The Atrium, Southern Gate, Chichester, West Sussex, PO19 8SQ, UK

Editorial Offices
9600 Garsington Road, Oxford, OX4 2DQ, UK
The Atrium, Southern Gate, Chichester, West Sussex, PO19 8SQ, UK
111 River Street, Hoboken, NJ 07030-5774, USA

For details of our global editorial offices, for customer services and for information about how to apply for permission to reuse the copyright material in this book please see our website at www.wiley.com/wiley-blackwell.

Library of Congress Cataloguing-in-Publication Data
Timescales of magmatic processes : from core to atmosphere / edited by Anthony Dosseto, Simon P. Turner, James A. Van Orman.
       p. cm.
    Includes index.
    ISBN 978-1-4443-3260-5 (cloth) – ISBN 978-1-4443-3261-2 (pbk.)   1. Magmatism.   2. Magmas.   3. Volcanic ash, tuff, etc.   4. Geological time.   I. Dosseto, Anthony.   II. Turner, Simon P.   III. Van Orman, James A.
    QE461.T526 2010
    551.1'3–dc22
                                        2010023309
A catalogue record for this book is available from the British Library.

This book is published in the following electronic formats:
eBook [9781444328516]; Wiley Online Library [9781444328509]

Set in 9/11.5pt TrumpMediaeval by SPi Publisher Services, Pondicherry, India
Printed and bound in Singapore by Markono Print Media Pte Ltd.

1   2011

Contents

List of Contributors

OLIVIER BACHMANN *Department of Earth and Space Sciences, University of Washington, Johnson Hall, room 435, Mailstop 351310, Seattle, WA 98195-1310, USA*

KIM BERLO *Department of Earth and Planetary Sciences, McGill University, 3450 University Street, Montreal, QC H3A 2A7, Canada*

JONATHAN D. BLUNDY *Department of Earth Sciences, University of Bristol, Wills Memorial Building, Queen's Road, Bristol BS8 1RJ, UK*

BERNARD BOURDON *Institute of Geochemistry and Petrology, ETH Zurich, NW D 81.4, 8092 clausiusstrasse 25, Switzerland*

G. CARO *CRPG-CNRS, Nancy Université, 15 rue Notre-Dame-des-Pauvres, Vandoeuvre-les-Nancy 54501 cedex, France*

FIDEL COSTA *Departamento de Geofísica y Georiesgos, Institut de Ciencies de la Terra 'Jaume Almera', C/Lluís Solé i Sabarís s/n 08028, CSIC, Barcelona, Spain and Earth Observatory of Singapore, Nanyang Technological University, Singapore 637879*

ANTHONY DOSSETO *GeoQuEST Research Centre, School of Earth and Environmental Sciences, University of Wollongong, Northfields Avenue, Wollongong, NSW 2522, Australia*

TIM ELLIOTT *Bristol Isotope Group, University of Bristol, Wills Memorial Building, Queen's Road, Bristol BS8 1RJ, UK*

JAMES E. GARDNER *Department of Geological Sciences, University of Texas at Austin, 1 University Station C1100, Austin, TX 78712, USA*

W. L. GRIFFIN *GEMOC ARC National Key Centre, Department of Earth and Planetary Sciences, Macquarie University, Talavera Road, Sydney, NSW 2109, Australia*

T. KLEINE *Institut für Planetologie, Westfälische Wilhelms-Universität Münster, Wilhelm-Klemm-Str. 10, 48149 Münster, Germany*

KURT KNESEL *Department of Earth Sciences, University of Queensland, Brisbane, Qld 4072, Australia*

DANIEL MORGAN *School of Earth and Environment, University of Leeds, LS2 9JT, UK*

CRAIG O'NEILL *GEMOC ARC National Key Centre, Department of Earth and Planetary Sciences, Macquarie University, Talavera Road, Sydney, NSW 2109, Australia*

SUZANNE Y. O'REILLY *GEMOC ARC National Key Centre, Department of Earth and Planetary Sciences, Macquarie University, Talavera Road, Sydney, NSW 2109, Australia*

TRACY RUSHMER *GEMOC ARC National Key Centre, Department of Earth and Planetary Sciences, Macquarie University, Talavera Road, Sydney, NSW 2109, Australia*

ALBERTO E. SAAL *Department of Geological Sciences, Brown University, 324 Brook Street, Box 1846, Providence, RI 02912, USA*

MARC SPIEGELMAN *Lamont-Doherty Earth Observatory of Columbia University, PO Box 1000, Palisades, NY 10964, USA*

SIMON P. TURNER *GEMOC ARC National Key Centre, Department of Earth and Planetary Sciences, Macquarie University, Talavera Road, Sydney, NSW 2109, Australia*

JAMES A. VAN ORMAN *Department of Geological Sciences, 112 A.W. Smith Building, Case Western Reserve University, 10900 Euclid Avenue, Cleveland, OH 44106-7216, USA*

Introduction to the Timescales of Magmatic Processes

ANTHONY DOSSETO[1], SIMON P. TURNER[2], FIDEL COSTA[3,4] AND JAMES A. VAN ORMAN[5]

[1]GeoQuEST Research Centre, School of Earth and Environmental Sciences, University of Wollongong, Wollongong, Australia
[2]GEMOC ARC National Key Centre, Department of Earth and Planetary Sciences, Macquarie University, Sydney, NSW, Australia
[3]Institut de Ciencies de la Terra 'Jaume Almera', CSIC, Barcelona, Spain
[4]Earth Observatory of Singapore, Nanyang Technological University, Singapore
[5]Department of Geological Sciences, 112 Case Western Reserve University, Cleveland, OH, USA

The publication in 1928 of *The Evolution of Igneous Rocks* by Norman L. Bowen laid the foundation to understand the formation and evolution of igneous rocks. Most of the magmatic processes that he proposed are still used and discussed today, and his search for the physical parameters to understand these processes are major themes of current research. At that time, radioactivity had been known for 30 years, but there were as yet no applications to dating of magmatic rocks. Thus any quantification of the timescales of the processes that he proposed was not possible. Determining accurate isotope ratios of natural rocks for the purposes of dating and understanding their formation would come decades later (Paterson, 1956). An example of how critical the quantification of timescales is, and how radiometric dating revolutionized our understanding of the world around us, is the determination of the age of the Earth. Early scientific estimates varied widely (Dalrymple, 1994), but

by measuring the trace abundance of isotopes produced by radioactive decay of uranium and thorium, it was possible to demonstrate that the Earth is actually 4.55 billion years old.

Great progress in understanding magmatic processes has been made by a combination of fieldwork, geochemical analysis, experiments and numerical models (Young, 2003). On the one hand, there has been an avalanche of geochemical data (e.g. databases PETDB, GEOROC and NAVDAT) that have allowed the relationship between plate tectonics and magma genesis and allowed the formation and differentiation of the different reservoirs of the Earth to be explored (Wilson, 1989). The detailed petrological and geochemical studies of individual rocks and crystals have shown that magmas can carry a mixture of components from various sources and with various ages (Davidson et al., 2007). On the other hand, experiments, theory and numerical models have brought a robust understanding of the conditions and intensive variables under which magmas can be generated, differentiated, and may erupt (Carmichael & Eusgter, 1987; MELTS algorithm of Ghiorso & Sack, 1995). Numerical models built upon robust physical properties of multicomponent silicate melts (viscosity, density and diffusivities; Stebbins *et al.*, 1995) can be used to investigate the dynamics of magmatic processes (Nicholls & Russell, 1990; Spera & Bohrson, 2001).

Early dating of rock and mineral suites by various radioactive decay systems (e.g.^{14}C, Rb-Sr, K-Ar, Sm-Nd, U-Pb) provided a calibrated sequence of magmatic and volcanic events (eruption or emplacement of plutonic bodies) that could be used for the first time to estimate rates of magma production and evolution. However, the errors resulting from the large amount of sample required and the precisions of the measurements did not allow distinction between eruption and crystallization ages, or for dating a series of eruptions from the same vent, which may record the magmatic differentiation (Faure, 1986).

A turning point in understanding the dynamics of magmatic processes has occurred in the last 10 to 20 years, with technological advances in mass spectrometry and the accurate analysis of short-lived radioactive isotopes of the U and Th decay series (Bourdon *et al.*, 2003), as well as the possibility of *in-situ* dating of extremely small amounts of minerals (diameter of several tens of μm (Hanchar & Hoskin, 2003)). Furthermore, a large amount of data on the kinetics of element migration in minerals together with natural observations has led to the realization that the chemical heterogeneities in crystals can also be used to extract time information on magmatic processes. Thus, it is now possible to study magmatic processes over timescales ranging from a few minutes to millions of years. This allows us to tackle critical questions about the evolution of Earth systems: How long is a magma stored in a reservoir prior to eruption? How long does it take to evolve from mafic to felsic magma compositions? How long does it take for a magma generated in the mantle to reach the surface? How long after the Earth was created did a metallic core form?

In this book we bring together syntheses of work aimed at tackling these questions. Although the age of the Earth is now well known, until recently little was known about how the

seemingly homogeneous proto-Earth differentiated into different envelopes (core, mantle, crust). In the first chapter, Caro and Kleine show how extinct radionuclides can be used to demonstrate that differentiation of the Earth's core and the initiation of plate tectonics all occurred very early in the Earth's history. The differentiation of the Earth into various reservoirs is still a work in progress. Material is continuously transferred from the mantle towards the surface and surface material is recycled back into the mantle at subduction zones. To understand the mechanisms that allow melting of mantle rocks and the transfer of produced magmas to the surface, it is necessary to quantify the timescales of production and transport of magmas below mid-ocean ridges and at hotspots. Two complementary approaches can be used to constrain the timescales of magma production in the mantle: uranium-series disequilibrium and the diffusion of trace elements in minerals.

Uranium and thorium-series isotopes compose the decay chains that start with ^{238}U, ^{235}U and ^{232}Th, and end with stable isotopes ^{206}Pb, ^{207}Pb and ^{208}Pb, respectively (Figure 0.1). These decay chains are composed of radioactive systems where a parent nuclide decays into the daughter by alpha or beta emission. For instance, ^{230}Th decays into ^{226}Ra by alpha emission. For systems closed for more than 1 Myr, all radioactive systems in the decay chains are in *secular equilibrium*: the activities (i.e. rates of decay) of parent and daughter nuclides are equal. For instance, $(^{230}$Th$)$ = $(^{226}$Ra$)$, or $(^{226}$Ra$/^{230}$Th$)$ = 1 (where parentheses denote activities). This is the case for mantle rocks prior to melting. During geological processes, such as partial melting or fractional crystallization, the different radionuclides behave differently inducing *radioactive disequilibrium*. Thus, because ^{226}Ra is usually more incompatible than ^{230}Th (i.e. it partitions preferentially into the magma relative to the residue during partial melting), magmas are often characterized by $(^{226}$Ra$/^{230}$Th$)$ >1. Once radioactive disequilibrium is produced, the system returns to secular equilibrium by radioactive decay over a timescale that is about five times the half-life of the

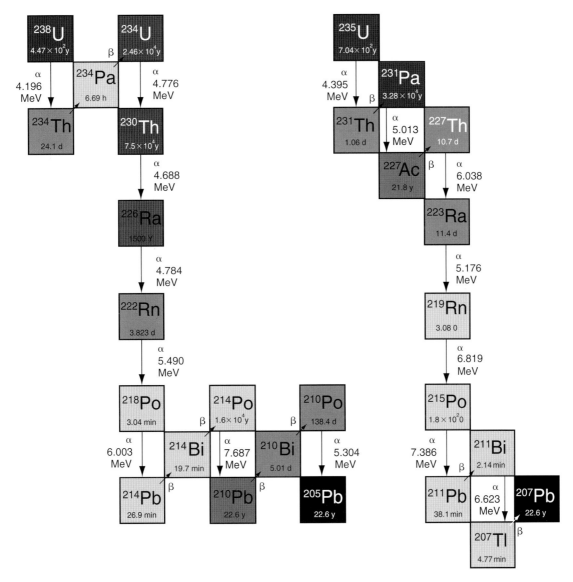

Fig. 0.1 ^{238}U and ^{235}U decay chains. See Plate 0.1 for a colour version of this image.

shorter-lived daughter nuclide. For example, if a magma with (^{226}Ra/^{230}Th) >1 behaves as a closed system once the disequilibrium is acquired, secular equilibrium will be attained after ~8,000 years (^{226}Ra half-life is 1,602 years (Bourdon *et al.* 2003)) (Figure 0.2). Thus, because U-series isotopes fractionate during magmatic processes and their

ratios are time-dependent, they provide a critical tool for constraining the timescales of magmatic processes.

In Chapter 3, Bourdon and Elliott show that the study of the U-series isotope composition of basaltic rocks can be used to constrain the conditions and timescales of mantle melting. It is

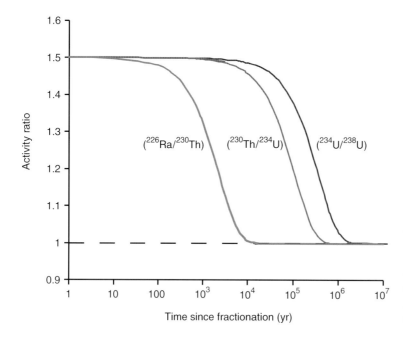

Fig. 0.2 Decay of radioactive disequilibria for different systems. The timescale of return to secular equilibrium is directly a function of the daughter nuclide half-life in the system considered: ~8,000 years for ^{226}Ra-^{230}Th, 300,000 years for ^{230}Th-^{238}U. See Plate 0.2 for a colour version of this image.

shown that the porosity of the matrix undergoing melting needs to be as low as a few per mil, in order to account for observed U-series isotope compositions. Moreover, the rate of melting beneath oceanic islands (e.g. Hawaii, Reunion) is inferred to be related to the upwelling rate of the hotspot (i.e. mantle instability). They show that melt production is greater at large hotspots, and that melting beneath oceanic islands occurs faster than beneath mid-ocean ridges. During partial melting of mantle rocks, trace elements partition between minerals and the newly formed magma and it is generally assumed that a partitioning equilibrium is reached between minerals and magma, as trace elements are believed to diffuse rapidly into the magma. Nevertheless, in Chapter 2, Van Orman and Saal show that diffusion of trace elements can be very slow, leading to an incomplete equilibrium between minerals and magma. They also show that different rates of diffusion for ^{238}U, ^{230}Th, ^{226}Ra and ^{210}Pb can explain most of the observed radioactive disequilibria in basaltic rocks. Thus, diffusion of trace elements (including

radionuclides) needs to be taken into account when inferring rates of magma production.

Mantle melting operates at variable rates between different tectonic environments (hotspots, mid-ocean ridges, arcs). In all cases, once a significant amount of melt is produced, it will rise towards the surface, because it is less dense than the surrounding rocks. In Chapter 4, O'Neill and Spiegelman show how numerical modeling can provide some understanding of the mechanisms and timescales involved in the transfer of magmas from their source region towards the surface. They suggest that melt transport can occur over a wide range of timescales: from several millimeters per year when a pluton intrudes into the brittle crust, up to several tens of kilometers per hour during dyke propagation and elastic fracturing.

Because radioactive disequilibrium between U-series isotopes is created during mantle melting and any disequilibrium decreases will disappear within up to 300,000 years after production, U-series isotopes can also be used to constrain the timescales of magma transport. For instance, in Chapter 5, Turner and Bourdon show that magma ascent rates must be at least ~1 to 20 m/yr in order to preserve radioactive disequilibria produced in the mantle and observed in many basaltic rocks. Furthermore, because any ^{226}Ra-^{230}Th disequilibrium vanishes after ~8000 years, the observation of ^{226}Ra-^{230}Th disequilibrium in island arc basalts implies magma ascent rates in subduction zones as high as 70 m/yr. Magma transport in channels and/or rapidly propagating fractures is required to account for such large ascent rates. The study of xenoliths (i.e. fragments of rock detached from the deep lithosphere and transported in a magma) can also provide information on the timescales of magma ascent. Because xenoliths are denser than their host magma, the magma is required to have a minimum velocity in order to transport them to the surface and prevent their settling. Moreover, the chemical composition of xenoliths is not in equilibrium with that of the host magma, and this disequilibrium disappears with time. Thus, the study of xenoliths and host magma compositions can be used to determine magma ascent rates. In Chapter 6, O'Reilly and Griffin show that ascent rates for (volatile-rich) alkali basalts range from 0.2 to 2 m/s, i.e. 6 orders of magnitude faster than that of calc-alkaline basalts, inferred from U-series isotopes. Furthermore, kimberlites are believed to erupt at near-supersonic speeds (≥300 m/s). To summarize, melt transport can occur over a wide range of timescales that reflect the variety of mechanisms involved (e.g. slow porous vs. fast channel flow). Moreover, magma transport can be very rapid, with transfer from the Earth's mantle to the surface in only a few hours in the case of alkali basalts.

Magmas do not always reach the surface directly from the region of their production, but frequently stall in the crust where they differentiate, leading to the observed diversity of igneous rock compositions. To understand how magma differentiation produces such a wide range of compositions, it is important to constrain the rates of magma differentiation to discriminate between physical models of magma emplacement in the crust, cooling and interaction with the country rock. In Chapter 7, Costa and Morgan show how the study of elemental and isotopic composition profiles in minerals can be used to infer rates of magma cooling. The approach is based on several concepts:
• A gradient of concentrations for a given element exists between crystals and the magma they are derived from;
• Concentrations between the crystal and the melt will equilibrate with time by diffusion: and
• Diffusion rates are different for each element.
Thus, because chemical equilibration between a melt and a crystal takes time, concentration and isotope composition profiles can be used to infer the timescales of magma evolution. Diffusion studies suggest that characteristic timescales of magma evolution in the crust (crystal fractionation, crust assimilation, magma recharge, etc.) are typically up to a few hundred years.

Similarly to previous chapters presenting the use of U-series isotopes to investigate rates of magma production and transfer, Chapter 8 by Dosseto and Turner shows how the U-series isotope composition of co-genetic volcanic rocks can be used to infer timescales of magma differentiation. Two approaches are presented: the first one is based on a study of the U-series isotope composition of minerals. It is shown that depending on the degree of differentiation of the volcanic rock hosting the minerals (i.e. more or less silica-rich), timescales of differentiation can vary from less than one thousand years to several hundred thousand years. However, minerals from the same rock can also yield contrasting timescales. This is most likely explained by the complex history of minerals during magmatic evolution; for instance, minerals carried by a volcanic rock may not be in equilibrium with the host magma and have formed from an earlier magmatic batch. To circumvent this problem, it would be necessary to analyse profiles of the U-series isotope compositions in minerals. This represents a major analytical challenge, although recent advances in *in-situ* geochemical techniques promise more detailed investigations of mineral isotope composition in the near future. Another approach is to study the U-series isotope composition of whole

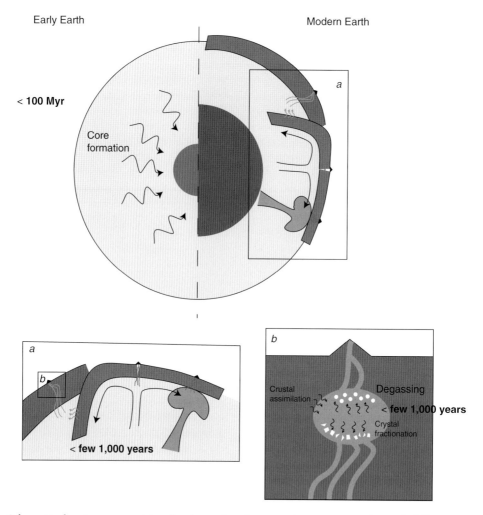

Fig. 0.3 Schematic drawing summarizing the timescales of magmatic processes. Early Earth differentiation occurs shortly after the beginning of the solar system (less than a few 100 million years). Inset (a) shows melt production at different tectonic settings (from left to right: subduction, divergent plates and hotspot). Timescales of magma production and transfer towards the surface are inferred to be very short, of the order of a few 1,000 years or less. Inset (b) shows magma differentiation in the crust (regardless the tectonic setting). Crystal fractionation, crustal assimilation or magma degassing occur over short timescales, typically a few 1,000 years or less. Note however that some magma bodies can exist for several 100,000 years, but the actual process of differentiation of the magma into a wide range of composition is very short. Moreover, it takes 100,000's of years to heat up the crust to allow crustal assimilation, but once the required thermal regime is reached, production and assimilation of crustal melts into mantle-derived magmas is very rapid. See Plate 0.3 for a colour version of these images.

rock samples. It is shown in Dosseto and Turner's chapter that changes in U-series isotope composition of volcanic rocks with differentiation indices yield timescales of differentiation of the order of a few thousand years or less, even when crustal

assimilation and frequent recharge of the magma body are taken into account.

Silicic magma bodies are often characterized by timescales of differentiation of the order of several hundreds of thousands of years, as indicated by U-series isotopes in crystals. In Chapter 10, Bachmann explores the timescales related to large eruptions of silicic magma bodies such as Yellowstone Caldera. Large silicic eruptions can have dramatic consequences for life and there is a clear inverse relationship between the frequency of these eruptions and their size. Although it is crucial to be able to predict such events, our understanding of the mechanisms involved is still incomplete. The techniques presented in this volume carry the hope of further advances in the near future.

As mantle-derived magmas are emplaced into the crust, they interact with the host rock and can induce significant melting of the surrounding rock. In Chapter 9, Rushmer and Knesel show how melting experiments and the geochemistry of melts and restites (the residual solid from partial melting) shed light on the mechanisms and timescales of crustal melting. It is shown that the composition of melts produced from the crust depends not only on the composition of the starting material (i.e. the crustal rock that undergoes melting) but also on the conditions of melting (pressure, temperature, presence or not of deformation). Rushmer and Knesel show that crustal melting and 'contamination' of mantle-derived magmas occur over timescales of a few hundred years.

Finally, when magmas erupt, they release a wide range of gases in variable abundances. This phenomenon is of primary importance since it is linked to the origin and evolution of the Earth's atmosphere. It can also represent a significant hazard, as major degassing is generally associated with the most destructive eruptions. In Chapter 11, Berlo *et al.* discuss how the timescales of degassing can be quantified, in particular using short-lived radionuclides such as ^{210}Pb. It is shown that degassing is a complex process where some magma bodies can accumulate significant volumes of gas derived from the degassing of other volumes of magma. Timescales of magma degassing are very short compared to timescales of differentiation and are generally on the order of a few years.

In conclusion, magmatism on Earth encompasses a variety of processes – from partial melting, to transport, to crystallization and degassing

in magma chambers – that operate on vastly different timescales (Figure 0.3). Our understanding of these processes and their timescales, in various tectonic settings, relies on integration of observational, theoretical and experimental constraints from many sub-disciplines in geochemistry and geophysics. This book provides an introduction to the various approaches that are used to study magmatic timescales, and the key constraints that have been derived from each. By introducing the reader to all of the aspects that should be treated to gain a better understanding of magmatic processes, we hope that this book opens new doors onto the future.

REFERENCES

Bourdon B, Henderson GM, *et al.* 2003. *Uranium-series Geochemistry*. Washington Geochemical Society – Mineralogical Society of America, Washington DC.

Bowen NL. 1921. Diffusion in silicate melts. *Journal of Geology* **30**: 295–317.

Bowen NL. 1928. *The Evolution of Igneous Rocks*. Princeton University Press, Princeton, 332 pp.

Carmichael ISE, Eugster HP. 1987. Thermodynamic modeling of geological materials: minerals, fluids and melts. *Reviews in Mineralogy and Geochemistry* **17**: 500 pp.

Dalrymple GB. 1994. *The Age of the Earth*. Stanford University Press, Stanford, 492 pp.

Davidson JP, Morgan DJ *et al.* 2007. Microsampling and isotopic analysis of igneous rocks: implications for the study of magmatic systems. *Annual Reviews of Earth Planetary Science* **35**: 273–311.

Faure G. 1986. *Principles of Isotope Geology*. John Wiley & Sons, New York, Chichester, Brisbane, Toronto, Singapore.

Ghiorso MS, Sack RO. 1995. Chemical mass transfer in magmatic processes IV. A revised and internally consistent thermodynamic model for the interpolation and extrapolation of liquid-solid equilibria in magmatic systems at elevated temperatures and pressures. *Contributions to Mineral Petrology* **119**: 197–212.

Hanchar JM, Hoskin PWO. 2003. Zircon. *Reviews in Mineralogy and Geochemistry* **53**: 500 pp.

Nicholls J, Russell JK. 1990. Modern methods of igneous petrology: understanding magmatic processes. *Reviews in Mineralogy and Geochemistry* **24**: 314 pp.

Paterson C. 1956. Age of meteorites and the Earth. *Geochimica et Cosmochimica Acta* **10**: 230–237.

Spera FJ, Bohrson WA. 2001. Energy-Constrained Open-System Magmatic Processes I: General Model and Energy-Constrained Assimilation and Fractional Crystallization (EC-AFC) Formulation. *Journal of Petrology* **42**: 999–1018.

Stebbins JF, McMillan PF, Dingwell DB. 1995. Structure, dynamics and properties of silicate melts. *Reviews in Mineralogy and Geochemistry* **32**: 616 pp.

Wilson M. 1989. *Igneous Petrogenesis: A Global Tectonic Approach.* Springer-Verlag, Berlin, 466 pp.

Young DA. 2003. *Mind over Magma.* Princeton University Press, Princeton NJ, 686 pp.

1 Extinct Radionuclides and the Earliest Differentiation of the Earth and Moon

G. CARO[1] AND T. KLEINE[2]

[1]CRPG-CNRS, Nancy Université, Vandoeuvre-les-Nancy, France
[2]Institut für Planetologie, Westfälische Wilhelms-Universität Münster, Münster, Germany

SUMMARY

The extinct ^{182}Hf-^{182}W and ^{146}Sm-^{142}Nd systems provide key chronological constraints on the major episodes of planetary differentiation. Both chronometers can be considered extinct after approximately 5–6 half-lives (i.e., after 50 Myr and 400 Myr respectively) and are therefore selectively sensitive to early events. Application of ^{182}Hf-^{182}W chronometry shows that segregation of the Earth's core may have been complete no earlier than 30 Myr after formation of the solar system and probably involved at least partial re-equilibration of newly accreted metallic cores within the terrestrial magma ocean. The current best estimate for the termination of the major stage of Earth's accretion and segregation of its core is provided by the age of the Moon, which formed as a result of a giant collision of a Mars-sized embryo with the proto-Earth. According to ^{182}Hf-^{182}W systematics this event occurred 50–150 Myr after CAI formation. As the Earth cooled down following the giant impact, crystallization of the magma ocean resulted in the formation of the earliest terrestrial crust. While virtually no remnant of this proto-crust survived in the present-day rock record, the isotopic fingerprint of this early event is recorded in the form of small ^{142}Nd anomalies in early Archean rocks. These anomalies show that magma ocean solidification must have taken place 30–100 Myr after formation of the solar system. In contrast, ^{146}Sm-^{142}Nd systematics of lunar samples show that the lunar mantle may have remained partially molten until 300 Myr after CAI formation. Therefore, extinct chronometers indicate that accretion and differentiation of the Earth proceeded rapidly. The core, mantle and crust were completely differentiated less than 100 Myr after formation of the solar system.

INTRODUCTION

The accretion and earliest history of the Earth was an episode of major differentiation of a magnitude that probably will never be repeated. Frequent and highly energetic impacts during the Earth's growth caused widespread melting, permitting separation of a metallic core from a silicate mantle in a magma ocean. Soon after the major stages of the Earth's growth were complete, the magma ocean solidified and a first proto-crust formed. Earth is an active planet, however, concealing most of the evidence of its earliest evolutionary history by frequent rejuvenation of its crust. Consequently,

Timescales of Magmatic Processes: From Core to Atmosphere, 1st edition. Edited by Anthony Dosseto, Simon P. Turner and James A. Van Orman.
© 2011 by Blackwell Publishing Ltd.

there is no direct rock record of the Earth's origin and earliest evolution, but fortunately witnesses of the Earth's earliest evolution have been preserved as small isotope anomalies in the terrestrial rock record. The short-lived ^{182}Hf-^{182}W and ^{146}Sm-^{142}Nd isotope systems provide key constraints for understanding the Earth's accretion and earliest differentiation, and in this chapter, the basic theory of these isotope systems and their application to models of the Earth's formation and differentiation will be discussed.

The starting place for the accretion of the Earth is the solar nebula, a flattened disk of gas and dust orbiting the young Sun. Within the inner regions of the solar nebula, dust grains collided and stuck together to form a large population of meter- to kilometer-sized objects. Gravity and gas drag caused these planetesimals to collide and form increasingly larger bodies in a period of runaway growth, the products of which include numerous Moon- to Mars-sized planetary embryos. Collisions among these bodies mark the late stages of accretion, culminating in the formation of a few terrestrial planets that sweep up all the other bodies. The Moon probably formed during this period and involved a 'giant impact' of a Mars-sized body with Earth at the very end of Earth's accretion (Chambers, 2004).

Planetary accretion is intimately linked with heating and subsequent melting of the planetary interiors. The decay of short-lived radioactive isotopes, especially ^{26}Al ($t_{1/2} = 0.73$ Myr), and collisions among the planetary embryos, caused the planetary interiors to heat up. At some critical size, melting will have started within the planetary bodies, causing separation and segregation of a metallic core (Stevenson, 1990; Rubie et al., 2007). As a consequence, all major bodies of the inner solar system and also many smaller bodies, are chemically differentiated into a metallic core and a silicate mantle. However, some planetary bodies, the chondrite parent bodies, escaped differentiation. They are too small for impact heating to be significant and formed after most ^{26}Al

had decayed away. Chondrites thus provide information on what planetary bodies looked like before differentiation began. As such, the chondrites are invaluable archives for investigating planetary differentiation.

Melting in the interior of a planetary object permits the denser components to migrate towards the center, thereby forming a core. Metallic iron melts at lower temperatures than silicates, such that core formation can occur either by migration of molten metal through solid silicate matrix or by separation of metal droplets from molten silicate. The latter process is probably appropriate for core formation in the Earth, where giant impacts caused the formation of widespread magma oceans. Once differentiation began, it proceeded rapidly. The downward motion of dense metal melts result in the release of potential energy and hence further heating, which further triggers differentiation (Stevenson, 1990; Rubie et al., 2007).

As the Earth's mantle cooled following the last giant impact, the terrestrial magma ocean started to crystallize, ultimately resulting in the formation of the earliest terrestrial crust by migration and crystallization of residual melts near the surface. This process probably took place over timescales of the order of 10,000 to 100,000 years, depending on the blanketing effect of the early atmosphere (Abe, 1997; Solomatov, 2000). As demonstrated by the presence of detrital zircon in an Archean sedimentary formation from Western Australia (Wilde et al., 2001), the earliest terrestrial crust must have solidified <150 Myr after formation of the solar system. Little is known, however, about this ancient proto-crust, as subsequent mantle-crust exchanges led to a complete rejuvenation of the Earth's surface, leaving virtually no remnant older than 3.8 Gyr. Clues on the age, lifetime and composition of the Earth's crust can thus only be obtained through the study of early Archean rocks, which sampled the mantle at a time when chemical and isotopic fingerprints of the earliest differentiation processes had not been completely erased by mantle mixing.

This chapter will be divided into three sections. The first section will introduce the main concepts and the reference parameters used for constraining the chronology of core-mantle and mantle-crust differentiation using the ^{182}Hf-^{182}W and ^{146}Sm-^{142}Nd chronometers, respectively. The second section will be dedicated to the ^{182}Hf-^{182}W chronology of the Earth's accretion and core formation on the Earth and the Moon. The last section will examine the chronology and mechanisms of mantle-crust differentiation on the Earth, Mars and the Moon, as obtained from application of the ^{146}Sm-^{142}Nd system to meteorites and planetary material.

SYSTEMATICS AND REFERENCE PARAMETERS FOR SHORT-LIVED RADIONUCLIDES

The two-stage model for short-lived nuclide systems: ^{182}Hf-^{182}W

Tungsten has five stable isotopes, all non-radiogenic with the exception of ^{182}W, which was produced by β-decay of the short-lived isotope ^{182}Hf ($t_{1/2}$ = 9 Myr). Because W is a moderately siderophile (iron-loving) element, whilst Hf is lithophile (rock-loving), the Hf/W ratio is fractionated by processes involving segregation of metal from silicates during formation of planetary cores. The ^{182}Hf-^{182}W system has thus been extensively studied in order to derive chronological constraints on terrestrial accretion and core formation. The development of the Hf-W system as a chronometer of core formation goes back to the pioneering work of Lee & Halliday (1995) and Harper & Jacobsen (1996).

The siderophile behavior of W depends on several parameters including pressure, temperature and in particular oxygen fugacity. As a consequence, partitioning of W in planetary cores varies drastically among the terrestrial planets. The Hf/W ratio of the bulk silicate Earth is estimated to be ~17, which is significantly higher than the chondritic Hf/W ratio of ~1. As Hf and W are both refractory elements, this difference cannot be accounted for by cosmochemical fractionation in the solar nebula and is thus attributed to W partitioning into the core. Hence, assuming that core formation occurred at a time when ^{182}Hf was still extant, the bulk silicate Earth should have evolved towards a W isotopic composition that is more radiogenic in ^{182}W compared to chondrites.

The two-stage model represents the simplest conceptual framework for calculating ages of differentiation using radiogenic systems. The formalism described below applies to core-mantle segregation but a similar development can be used for mantle-crust differentiation using the ^{146}Sm-^{142}Nd chronometer. In this simple model, an undifferentiated primitive reservoir experiences instantaneous differentiation at time t_d. The core and mantle reservoirs are characterized by Hf/W ratios lower and higher than the bulk Earth value, respectively, and subsequently evolve as closed systems during the second stage. The basic decay equation for the short-lived ^{182}Hf-^{182}W chronometer can then be obtained from the following mass balance relationship:

$$^{182}Hf(t_0) + {}^{182}W(t_0) = {}^{182}Hf(t) + {}^{182}W(t) \quad (1)$$

where t_0 is the origin of the solar system and t is time running forward from $t_0 = 0$ to present day (t_p = 4.567 Gyr). Dividing both sides of Equation (1) using a stable and non-radiogenic isotope of the daughter element (e.g. ^{184}W), we obtain, for the bulk Earth (BE):

$$\left(\frac{^{182}W}{^{184}W}\right)_t^{BE} = \left(\frac{^{182}W}{^{184}W}\right)_{t_0}^{} + \left(\frac{^{182}Hf}{^{184}W}\right)_{t_0}^{BE} - \left(\frac{^{182}Hf}{^{184}W}\right)_{t_0}^{BE} \quad (2)$$

where $(^{182}W/^{184}W)_{t0}$ is the initial W isotopic composition of the solar system.

Radioactive decay for ^{182}Hf can be expressed as a function of time:

$$\left(\frac{^{182}Hf}{^{184}W}\right)_t^{BE} = \left(\frac{^{182}Hf}{^{184}W}\right)_{t_0}^{BE} e^{-\lambda_{182}t} \quad (3)$$

where λ_{182} is the decay constant of ^{182}Hf of 0.079 Myr^{-1}. The second term on the right-hand side of Equation (2) can then be written as:

$$\left(\frac{^{182}Hf}{^{184}W}\right)_{t_0}^{BE} = \left(\frac{^{182}Hf}{^{180}Hf}\right)_{t_0} \left(\frac{^{180}Hf}{^{184}W}\right)^{BE} \quad (4)$$

where $(^{182}Hf/^{180}Hf)_{t_0}$ represents the initial abundance of the short-lived radionuclide at the time of formation of the solar system. The decay equation for the bulk Earth finally reads:

$$\left(\frac{^{182}W}{^{184}W}\right)_{t}^{BE} = \left(\frac{^{182}W}{^{184}W}\right)_{t_0} + \left(\frac{^{182}Hf}{^{180}Hf}\right)_{t_0} \left(\frac{^{180}Hf}{^{184}W}\right)^{BE} \\ \times \left[1 - e^{-\lambda_{182}t}\right] \quad (5)$$

Isotopic heterogeneities due to radioactive decay are usually normalized to the composition of the bulk silicate Earth and expressed as deviations in parts per 10^4 using the epsilon notation:

$$\varepsilon^{182}W = \left(\frac{\left(^{182}W/^{184}W\right)_R}{\left(^{182}W/^{184}W\right)_{BSE}} - 1\right) \times 10^4 \quad (6)$$

Using this notation, the present-day chondritic $\varepsilon^{182}W$ is -1.9 ε-units, whilst the bulk silicate Earth (BSE), by definition, has $\varepsilon^{182}W = 0$.

In the two-stage model, differentiation occurs as an instantaneous event at $t_d > t_0$. The core incorporates no Hf, so that its isotopic composition during the second stage remains constant and identical to that of the bulk Earth at time t_d:

$$\left(\frac{^{182}W}{^{184}W}\right)_{t}^{Core} = \left(\frac{^{182}W}{^{184}W}\right)_{t_d}^{BE} = \left(\frac{^{182}W}{^{184}W}\right)_{t_0} \\ + \left(\frac{^{182}Hf}{^{180}Hf}\right)_{t_0} \left(\frac{^{180}Hf}{^{184}W}\right)^{BE} \times \left[1 - e^{-\lambda_{182}t_d}\right] \quad (7)$$

The residual silicate reservoir (i.e. the bulk silicate Earth) is depleted in W and its isotopic composition evolves towards a more radiogenic $^{182}W/^{184}W$ signature according to the following equation:

$$\left(\frac{^{182}W}{^{184}W}\right)_{t}^{BSE} = \left(\frac{^{182}W}{^{184}W}\right)_{t_d}^{BE} + \left(\frac{^{182}Hf}{^{180}Hf}\right)_{t_0} \left(\frac{^{180}Hf}{^{184}W}\right)^{BSE} \\ \times \left[e^{-\lambda_{182}t_d} - e^{-\lambda_{182}t}\right] \quad (8)$$

Combining Equations (7) and (8), we obtain the final two-stage model equation for short-lived radionuclides:

$$\left(\frac{^{182}W}{^{184}W}\right)_{t}^{BSE} = \left(\frac{^{182}W}{^{184}W}\right)_{t_0} + \left(\frac{^{182}Hf}{^{180}Hf}\right)_{t_0} \\ \begin{Bmatrix} \left(\frac{^{180}Hf}{^{184}W}\right)^{BE} \times \left[1 - e^{-\lambda_{182}t_d}\right] \\ + \left(\frac{^{180}Hf}{^{184}W}\right)^{BSE} \times \left[e^{-\lambda_{182}t_d} - e^{-\lambda_{182}t}\right] \end{Bmatrix} \quad (9)$$

A model age for core formation can thus be directly derived from the present-day W isotopic composition and Hf/W ratio of the BSE (i.e. $(^{182}W/^{184}W)^{BSE}$ and $(^{180}Hf/^{184}W)^{BSE}$). The results of this equation for various values of t_d between 0 and 50 Myr are illustrated in Figure 1.1. This figure reveals that core-mantle segregation can only affect the $^{182}W/^{184}W$ composition of the mantle if $t_d < 50$ Myr. After this point in time, ^{182}Hf was effectively extinct and any later Hf-W fractionation could not have led to variations in the $^{182}W/^{184}W$ ratio between the mantle and the core.

Calculating two-stage model ages requires knowledge of the W isotopic evolution of chondrites, which is assumed to represent that of the bulk, undifferentiated planet. To calculate the Hf-W isotopic evolution of chondrites, their initial ^{182}Hf/^{180}Hf and $^{182}W/^{184}W$ compositions, ^{180}Hf/^{184}W ratio and present-day W isotopic composition needs to be known (Equation (5)). The most direct approach for determining the initial Hf and W isotopic compositions of chondrites is to obtain internal Hf-W isochrons for the oldest objects formed in the solar system, the Ca,Al-rich inclusions (CAIs) found in carbonaceous chondrites. The slope of the CAI isochron does not directly provide an age, as is the case for

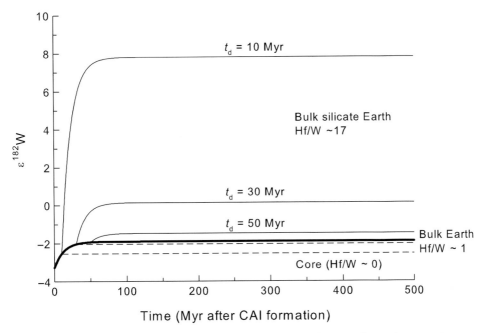

Fig. 1.1 A simple two-stage model illustrating the isotopic evolution of core and mantle reservoirs segregated at various times (t_d) after formation of the solar system.

long-lived chronometers, but instead yields the abundance of the parent nuclide (i.e. $^{182}Hf/^{180}Hf$) at the time of CAI formation. Thus, the chronology obtained from Hf-W on planetary differentiation is not absolute but is established relative to the formation of CAIs at $t_0 = 0$. In order to determine an absolute chronology, it is necessary to estimate independently the age of CAIs using a long-lived radioactive system. The U-Pb chronometer gives ages of ~4.568 Gyr for CAI formation (Amelin *et al.*, 2002; Bouvier *et al.*, 2007).

An internal Hf-W isochron for mineral separates from several CAIs from the Allende and NWA 2364 CV chondrites gives the following initial values: initial $^{182}Hf/^{180}Hf = (9.72 \pm 0.44) \times 10^{-5}$ and initial $\varepsilon^{182}W = -3.28 \pm 0.12$ (Figure 1.2). Bulk carbonaceous chondrites have a more radiogenic W isotopic composition of $\varepsilon^{182}W = -1.9 \pm 0.1$ (Kleine *et al.*, 2002, 2004a; Schoenberg *et al.*, 2002; Yin *et al.*, 2002b). Note that the first

W isotope data for chondrites, obtained by Lee & Halliday (1995), yielded $\varepsilon^{182}W \sim 0$, but later studies showed that this initial measurement was inaccurate. The elevated $^{182}W/^{184}W$ of chondrites compared to the initial value of the solar system reflects the decay of ^{182}Hf to ^{182}W over the age of the solar system, and from this difference the time-integrated $^{180}Hf/^{184}W$ of chondrites can be calculated using Equation (5). This approach results in a time-integrated $^{180}Hf/^{184}W$ of chondrites of 1.23, identical to the results of high-precision concentration measurements that yield an $^{180}Hf/^{184}W$ ratio of 1.23 (Kleine *et al.*, 2004a).

Once the Hf-W isotopic evolution of chondrites is defined, two-stage model ages of core formation can be calculated from the Hf/W ratio and $\varepsilon^{182}W$ value of the mantle or core of a differentiated planetary body (Equation (9)). We will discuss these two-stage model ages for several planetary bodies in detail in the section on 'Hf-W

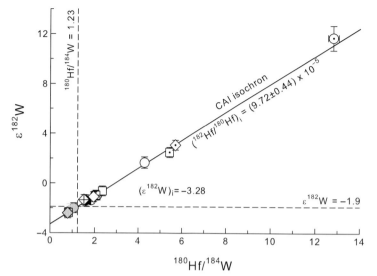

Fig. 1.2 Hf-W isochron of Ca,Al-rich inclusions from CV chondrites. Data points are for different mineral separates from CAIs. Those with the highest $^{180}Hf/^{184}W$ and $\varepsilon^{182}W$ are for fassaites. The solar system initial Hf and W isotopic compositions are defined by the CAI isochron. Also shown with dashed lines are the Hf-W parameters for average chondrites. Modified from Burkhardt *et al.* (2008).

chronology of the accretion and early differentiation of the Earth and Moon.'

However, when considering the short half-life of ^{182}Hf (9 Myr) and the characteristic timescale of accretion and core formation (10–100 Myr), it is evident that core formation cannot be treated as an instantaneous event and more complex, continuous segregation models are required. These also will be discussed in more detail in that section. The two-stage model, however, is extremely useful as it provides a maximum age for the completion of core formation.

Because both Hf and W are refractory, it has been customary to assume that their relative abundances remained more or less constant throughout the solar system. Consequently, for refractory elements, the composition of the bulk Earth must be close to that of chondrites (i.e. $(^{180}Hf/^{184}W)_{BE} = 1.23$). In detail, however, the composition of the bulk Earth may deviate from that of chondrites due to collisional erosion of early-formed crust (O'Neil & Palme, 2008; Warren, 2008). Due to the higher incompatibility of W compared to Hf, planetary crusts have sub-chondritic Hf/W ratios, such that impact-induced erosion of such crustal material would tend to increase the Hf/W ratio of the bulk Earth com-

pared to chondrites. Obviously this would affect the chronology derived from Hf-W systematics. This aspect and its consequences on chronological constraints derived from ^{182}Hf-^{182}W will be briefly discussed in later sections.

Coupled ^{146}Sm-^{142}Nd and ^{147}Sm-^{143}Nd chronometry

The coupled Sm-Nd system is a chronometer composed of an extinct and an extant radioactivity: ^{143}Nd is produced by α-decay of the long-lived ^{147}Sm ($T_{1/2} = 106$ Gyr), and ^{142}Nd was produced by α-decay of now-extinct ^{146}Sm ($T_{1/2} = 103$ Myr). The domain of application of these chronometers is thus conditioned by their respective half-life and by the chemical properties of the Rare Earth Elements (REE). As with most REE, Sm and Nd are characterized by high condensation temperatures (Boynton, 1975; Davis & Grossman, 1979), moderate incompatibility during magmatic processes and an extremely lithophile behavior during metal-silicate segregation. Core formation is thus unlikely to fractionate the REE significantly, even under highly reducing conditions, whilst volatility-controlled fractionation has only been observed in high-T

Table 1.1 ^{147}Sm-^{143}Nd systematics in the major terrestrial reservoirs (upper crust, depleted mantle, and bulk silicate Earth)

	^{143}Nd/^{144}Nd	^{147}Sm/^{144}Nd	Source
Upper crust	0.5116	0.117	Taylor & McLennan (1985)
Depleted mantle	0.5131	0.215–0.235	Salters & Stracke (2004)
BSE (chondritic model)	0.512638	0.1966	Jacobsen & Wasserburg (1984)
BSE (non-chondritic model)	0.51299	0.2082	Caro *et al.* (2008)

condensates such as CAIs (Ireland *et al.*, 1988). At the exception of extremely reducing nebular environments, where Samarium can be reduced to a divalent state (Lodders & Fegley, 1993; Pack *et al.*, 2004), Sm and Nd are trivalent and do not experience redox-controlled fractionation. As a consequence, the relative abundances of Sm and Nd in planetary reservoirs are essentially controlled by partial melting and fractional crystallization processes involved in the formation of planetary crusts (Table 1.1). As Nd is normally more incompatible than Sm, crustal reservoirs formed by partial melting of the mantle, or by fractional crystallization of a magma ocean, will have low Sm/Nd ratios and high Nd concentration, whilst the residual mantle reservoirs are characterized by a higher Sm/Nd (Table 1.1). Differentiated crust or mantle are thus expected to develop distinct ^{143}Nd/^{144}Nd signatures, usually noted ε^{143}Nd:

$$\varepsilon^{143}Nd = \left(\frac{\left(^{143}Nd/^{144}Nd\right)}{\left(^{143}Nd/^{144}Nd\right)_{CHUR}} - 1 \right) \times 10^4 \quad (10)$$

where CHUR (Chondritic Uniform Reservoir) represents the average chondritic composition. A similar notation is used for the ^{146}Sm-^{142}Nd system, with $\left(^{142}Nd/^{144}Nd\right)_{BSE}$ as the normalizing ratio. Long-lived differentiated crustal reservoirs are thus expected to develop unradiogenic ε^{143}Nd, whilst an elevated ε^{143}Nd indicates derivation from a mantle previously depleted in incompatible elements. If differentiation takes place prior to extinction of ^{146}Sm (i.e. >4.2 Gyr ago), then crustal and mantle reservoirs will also develop distinct ε^{142}Nd signatures, respectively lower and higher than the BSE composition.

The existence of distinct ε^{143}Nd signatures between terrestrial silicate reservoirs (Figure 1.3) has permitted the establishment of constraints on the chronology of mantle-crust differentiation (Jacobsen & Wasserburg, 1979; Allegre *et al.*, 1979; DePaolo & Wasserburg, 1979; DePaolo, 1980), the exchange fluxes between crust and mantle (Albarede & Rouxel, 1987; Albarede, 2001) and the fate of recycled crustal material in the mantle (Zindler & Hart, 1986). However, the constraints derived from ^{147}Sm-^{143}Nd systematics only depict a relatively recent (<2–3 Gyr) history of the mantle-crust system. Events that took place earlier in Earth's history are far more difficult to constrain, because their isotopic fingerprint on mantle composition has been erased by later crustal formation and recycling. In contrast, the ^{146}Sm-^{142}Nd chronometer is only sensitive to events that took place during the very early history of the Earth. Given the current analytical uncertainties on Nd isotopic measurements (2σ = 2–5 ppm), radiogenic ingrowth in silicate reservoirs becomes negligible after approximately 4 periods. Distinct ^{142}Nd signatures can thus only develop in silicate reservoirs differentiated prior to 4.2 Gyr. Thus, whilst the ^{143}Nd/^{144}Nd composition of modern silicate reservoirs is mainly inherited from continuous crustal formation during the past 2 to 3 Gyr, their ^{142}Nd signature can only be the result of much earlier REE fractionation, thereby making the ^{146}Sm-^{142}Nd system a selective tool for dating the formation of the earliest crusts on Earth and other terrestrial planets.

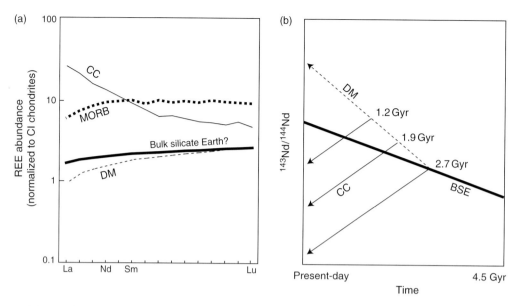

Fig. 1.3 (a) REE abundance patterns of the major terrestrial silicate reservoirs normalized to CI chondrites. The CI-normalized pattern of the bulk silicate Earth is not directly constrained but has been conservatively assumed to be perfectly flat, with $[REE]_{BSE} = 2.75 \times [REE]_{CHUR}$ (McDonough & Sun, 1995). However, a slightly non-chondritic composition, perhaps similar to the hypothetical pattern shown here, is suggested by the non-chondritic ^{142}Nd composition of terrestrial samples. A large fraction (30–50%) of the LREE budget of the silicate Earth is concentrated in the continental crust, which shows an enrichment by 1–2 orders of magnitude compared with bulk silicate Earth values. In contrast, the uppermost mantle (as sampled by mid-ocean ridge basalts) is characterized by a depleted REE pattern (LREE<HREE) and a radiogenic ^{143}Nd/^{144}Nd signature indicative of long-term evolution with high Sm/Nd ratio. (b) Schematic illustration showing the ^{143}Nd/^{144}Nd evolution of major silicate reservoirs. The present-day continental mass was formed during three major episodes of crustal growth at 2.7, 1.9 and 1.2 Gyr (Condie, 2000). This process extracted about one-third of the total Nd budget of the bulk Earth, creating an enriched reservoir with low ^{147}Sm/^{144}Nd ratio (0.12), which subsequently evolved towards unradiogenic ϵ^{143}Nd (−15 ϵ-units), and a complementary depleted mantle characterized by high ^{147}Sm/^{144}Nd (0.23–0.25) and radiogenic ϵ^{143}Nd signature (ϵ^{143}Nd = +9) (Allegre *et al.*, 1979; DePaolo & Wasserburg, 1979; Jacobsen & Wasserburg, 1979). Recycling of oceanic crust in the deep mantle may also contribute to depleting the uppermost mantle in incompatible elements (Christensen & Hofmann, 1994).

Two-stage mantle-crust differentiation model

The development of ^{146}Sm-^{142}Nd systematics has long been hampered by the difficulty of measuring with sufficient precision the small radiogenic effects on ^{142}Nd (Sharma *et al.*, 1996), a problem essentially due to the scarcity of ^{146}Sm in the early solar system. The presence of this p-process radionuclide was first demonstrated by Lugmair & Marti (1977), who reported radiogenic ^{142}Nd effects in mineral separates from the achondrite Angra Dos Reis. This observation was then confirmed by

the discovery of highly radiogenic ^{142}Nd signatures in carbon-chromite fractions from the chondrite Allende, which were attributed to α-recoil effects from ^{146}Sm in carbon films surrounding Sm-bearing grains (Lugmair *et al.*, 1983). In the early 1990s, accurate determinations of the initial abundance of ^{146}Sm were obtained using a new generation of mass spectrometers. Lugmair & Galer (1992) reported an initial ^{146}Sm/^{144}Sm ratio of 0.0071 ± 17 for Angra Dos Reis and Prinzhofer *et al.* (1989, 1992) suggested a value of 0.008 ± 1 from mineral separates of the meteorites Ibitira and Morristown (Figure 1.4). These estimates are in general agree-

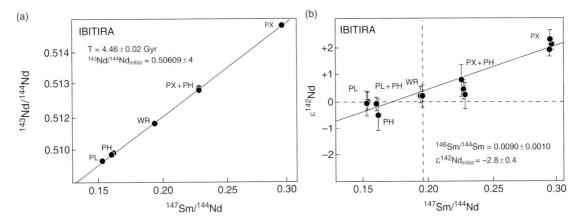

Fig. 1.4 (a)^{147}Sm-^{143}Nd and (b) ^{146}Sm-^{142}Nd mineral isochrons from the Eucrite Ibitira (Prinzhofer *et al.*, 1992).^{142}Nd heterogeneities in mineral separates are correlated with their respective Sm/Nd ratio, demonstrating the presence of live ^{146}Sm at the time of crystallization of the rock. In contrast with the long-lived ^{147}Sm-^{143}Nd system, the slope of the ^{146}Sm-^{142}Nd isochron does not yield an absolute age but is proportional to the abundance of the parent nuclide (i.e.^{146}Sm/^{144}Sm) at the time of crystallization. Mineral isochrons established in chondrites and achondrites provide a best estimate for the initial ^{146}Sm/^{144}Sm ratio of 0.008 ± 1. Reprinted from *"Geochimica et Cosmochimica Acta"*, **56**:2, A Prinzhofer, D.A Papanastassiou, G.J Wasserburg, Samarium–neodymium evolution of meteorites, pp. 1–2, 1992, with permission from Elsevier.

ment with more recent results obtained by Nyquist *et al.* (1994) and Amelin & Rotenberg (2004) and were widely applied as reference parameters in later studies (Harper & Jacobsen, 1992; Caro *et al.*, 2003; Boyet & Carlson, 2005).

The two-stage model described for the ^{182}Hf-^{182}W chronometer can be easily transposed to mantle-crust differentiation. In this case, the two-stage model equation describing the evolution of a depleted mantle reservoir reads:

$$
\begin{aligned}
\left(\frac{^{142}Nd}{^{144}Nd}\right)^{EDM}_{t} &= \left(\frac{^{142}Nd}{^{144}Nd}\right)^{BSE}_{t_p} + \frac{\left(^{146}Sm/^{144}Sm\right)_{t_0}}{\left(^{147}Sm/^{144}Sm\right)_{t_p}} \\
&\times \left[\begin{array}{l} \left(\frac{^{147}Sm}{^{144}Nd}\right)^{BSE}_{t_p} \times e^{-\lambda_{146}t_d} \\ +\left(\frac{^{147}Sm}{^{144}Nd}\right)^{EDM}_{t_p} \times \left[e^{-\lambda_{146}t_d} - e^{-\lambda_{146}t}\right] \end{array} \right]
\end{aligned} \quad (11)
$$

where the acronyms EDM and BSE stand for Early Depleted Mantle and Bulk Silicate Earth, respectively. The term EDM is used here to avoid confusion with the modern depleted mantle, whose chemical composition reflects continuous crustal extraction and is thus not directly relevant to early differentiation processes. Note also that in order to homogenize notations with the ^{147}Sm-^{143}Nd decay equation, the parent-daughter ratio in Equation (11) is often expressed as ^{147}Sm/^{144}Nd.

As can be seen from Equation (11), a model age (t_d) of mantle-crust differentiation can be derived from ^{146}Sm-^{142}Nd systematics, assuming that $(^{142}Nd/^{144}Nd)^{EDM}$ and $(^{147}Sm/^{144}Nd)^{EDM}$ are known. In practice, the latter cannot be estimated directly, as mantle-crust exchanges during the Archean and Proterozoic have overprinted the geochemical signature of the earlier magmatic events. However, this problem can be circumvented by combining the ^{147}Sm-^{143}Nd and ^{146}Sm-^{142}Nd systems using a simple two-stage model (Figure 1.5). The ^{143}Nd/^{144}Nd evolution of the EDM can then be expressed as:

$$
\begin{aligned}
\left(\frac{^{143}Nd}{^{144}Nd}\right)^{EDM}_{t} &= \left(\frac{^{143}Nd}{^{144}Nd}\right)^{BSE}_{t_p} + \left(\frac{^{147}Sm}{^{144}Nd}\right)^{BSE}_{t_p} \\
& \left[1 - e^{\lambda_{147}(t_p - t_d)}\right] + \left(\frac{^{147}Sm}{^{144}Nd}\right)^{EDM}_{t_p} \left[e^{\lambda_{147}(t_p - t_d)} - e^{\lambda_{147}(t_p - t)}\right]
\end{aligned}
$$

$$(12)$$

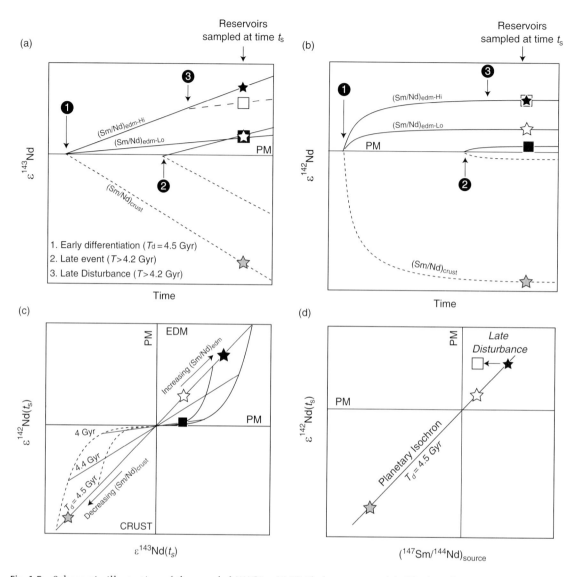

Fig. 1.5 Schematic illustration of the coupled ^{146,147}Sm-^{142,143}Nd chronometer. (a) ϵ^{143}Nd evolution of a proto-crust (labeled *crust*) and complementary mantle reservoirs (labeled *edm-Hi* and *edm-Lo*) differentiated at 4.5 Gyr (1) and 4 Gyr (2). The Archean ^{143}Nd record is consistent with either a moderate Sm/Nd fractionation at a very early stage of the Earth's history (1), or more substantial fractionation at a later stage (2). (b) The ^{146}Sm-^{142}Nd chronometer provides more accurate chronological constraints. Here only the early event (1) can generate significant ^{142}Nd excesses. In contrast, any Sm/Nd fractionation occurring after ~4.2 Gyr would not generate significant ^{142}Nd excess (2). Note that later depletion/enrichment of the EDM would modify the ^{143}Nd/^{144}Nd ratio of the EDM without necessarily affecting its ^{142}Nd/^{144}Nd composition (3). (c) Using coupled ^{147}Sm-^{143}Nd and ^{146}Sm-^{142}Nd chronometry, one can determine the age of differentiation of the mantle-crust system and the magnitude of Sm/Nd fractionation in the differentiated mantle (or crustal) reservoir. (d) When plotted in a Sm/Nd$_{srce}$ vs. ϵ^{142}Nd$_{initial}$ diagram, cogenetic reservoirs plot along a planetary isochron whose slope yields the age of differentiation of the mantle-crust system. Disturbance of this isochron relationship can occur during subsequent mantle depletion or enrichment (e.g. remelting of a magma ocean cumulate, metasomatism (3)). Note that the bulk composition of the primitive reservoir should also plot along the isochron.

where λ_{147} is the decay constant of ^{147}Sm. By combining Equations (11) and (12), we obtain a system of two chronometric equations from which both the ^{147}Sm/^{144}Nd ratio and the age of differentiation of the EDM can be calculated (Figure 1.5c). Using this coupled ^{146,147}Sm-^{142,143}Nd chronometer, the age of the crust can be estimated solely from its isotopic fingerprint in the mantle.

Planetary isochrons

The two-stage model described above implies that the EDM evolved as a closed-system between t_d and t_s. However, any magmatic event affecting the EDM after 4.2 Gyr would impact the long-lived ^{147}Sm-^{143}Nd system without necessarily affecting the ^{146}Sm-^{142}Nd system (Figure 1.5a) and this could result in spurious age estimates if considered in the context of a two-stage model. A possible way of identifying such decoupling is to establish a planetary isochron (Figure 1.5d). These are constructed from a series of samples derived from cogenetic crustal and/or mantle reservoirs such as magma ocean cumulates. Although the scarcity of terrestrial samples with ^{142}Nd anomalies has so far made this method inapplicable to early Earth differentiation, planetary isochrons have been used for determining ages of differentiation for the lunar and Martian mantle (see the section on '^{146}Sm-^{142}Nd constraints on the evolution of the Hadean crust'). In an isochron diagram, the ^{142}Nd/^{144}Nd composition of samples extracted at time t_s from differentiated reservoirs are plotted against the ^{147}Sm/^{144}Nd ratio of their source (Figure 1.5d). The latter are estimated using a two-stage model based on the initial ^{143}Nd/^{144}Nd composition of each sample:

$$\left(\frac{^{147}Sm}{^{144}Nd}\right)_{t_p}^{source} = $$
$$\frac{\left(^{143}Nd/^{144}Nd\right)_{t_d}^{PM} - \left(^{143}Nd/^{144}Nd\right)_{t_s}^{source}}{e^{\lambda t_s} - e^{\lambda t_d}} \quad (13)$$

where $\left(^{143}Nd/^{144}Nd\right)_{t_s}^{source}$ is the initial isotopic composition of a sample extracted at time t_s and $\left(^{143}Nd/^{144}Nd\right)_{t_d}^{PM}$ is the isotopic composition of the primitive mantle at t_d. An age of mantle differentiation is then obtained from the slope of the isochron (S) in an ε^{142}Nd vs ^{147}Sm/^{144}Nd diagram:

$$T_d = T_0 - \frac{1}{\lambda}$$
$$\ln\left[S \times \frac{\left(^{147}Sm/^{144}Sm\right)_{t_p}}{\left(^{146}Sm/^{144}Sm\right)_{t_0}} \left(\frac{^{142}Nd}{^{144}Nd}\right)_{t_p}^{Std} 10^{-4} \right] \quad (14)$$

where $(^{142}Nd/^{144}Nd)^{Std}$ is the composition of the terrestrial standard used for normalizing ε^{142}Nd values. The age of differentiation is thus estimated using an iterative scheme from the slope of the regression line passing through all samples, but is not forced to intersect a hypothetical primitive mantle composition. Planetary isochrons thus provide more than simple chronological constraints. As shown in Figure 1.5d, non-cogenetic reservoirs would not plot along the same isochron, and this can be used to identify decoupling of the ^{147}Sm-^{143}Nd and ^{146}Sm-^{142}Nd systems. In addition, the position of the isochron compared with the chondritic reference provides constraints on the composition of the primitive reservoir from which planetary crusts differentiated (see the section on '^{146}Sm-^{142}Nd constraints on the evolution of the Hadean crust'). Ideally, cogenetic reservoirs differentiated from a primitive chondritic mantle should plot along an isochron passing through the chondritic composition (i.e. CHUR, Table 1.1). Failure to do so would indicate that the source reservoir experienced at least one additional episode of REE fractionation prior to the differentiation event recorded by the isochron. This aspect and its consequences on the composition of the terrestrial planets will be further discussed in the section on '^{146}Sm-^{142}Nd constraints on the evolution of the Hadean crust.'

HF-W CHRONOLOGY OF THE ACCRETION AND EARLY DIFFERENTIATION OF THE EARTH AND MOON

Hf-W systematics of planetary reservoirs and two-stage model ages of core formation

The two main parameters that need to be known to calculate Hf-W ages of core formation are the Hf/W ratio and W isotope composition of the bulk mantle or core of a planetary body. Since we do not have samples from the core and mantle of an individual planetary object, the ages are calculated relative to the composition of the bulk, undifferentiated body, which we assume to be chondritic (see section on 'Systematics and reference parameters for short-lived radionuclides'). In the simplest model of core formation, it is assumed that the core formed instantaneously and an age for core formation can then be calculated by assuming a two-stage model (Lee & Halliday, 1995; Harper & Jacobsen, 1996; Kleine et al., 2002). The underlying approach and the equation for calculating a two-stage Hf-W model age of core formation is described in detail in the section on 'Systematics and reference parameters for short-lived radionuclides.'

The only samples available from the metallic core of differentiated planetary bodies are the magmatic iron meteorites. They contain virtually no Hf and their present-day W isotope composition is identical to that at the time of core formation (Horan et al., 1998). All iron meteorites exhibit a deficit in ^{182}W compared to chondrites and their ε^{182}W values are among the lowest yet measured for solar system materials. Some iron meteorites even have ε^{182}W values lower than the solar system initial value (as determined from Ca,Al-rich inclusions, see the section on 'Systematics and reference parameters for short-lived radionuclides') but these low values reflect the interaction with thermal neutrons produced during the extended exposure of iron meteoroids to cosmic rays (Kleine et al., 2005a; Markowski et al., 2006; Schérsten et al., 2006). If these effects

Table 1.2 Hf-W systematics of planetary reservoirs (Kleine et al., 2009)

	^{180}Hf/^{184}W	ε^{182}W
Chondrites	~0.6–1.8	−1.9 ± 0.1
Basaltic eucrites	24–42	21–33
Angrites	4–7	1–5
Bulk silicate Mars	~3–4	0.4 ± 0.2
Bulk silicate Moon	~26	0.09 ± 0.1
Bulk silicate Earth	~17	≡ 0

are fully taken into account, all iron meteorites have ε^{182}W values indistinguishable from the initial ε^{182}W of CAIs (Kleine et al., 2005a; Burkhardt et al., 2008). The unradiogenic ε^{182}W values of iron meteorites indicate that their parent bodies differentiated within <1 Myr after formation of CAIs. Thus iron meteorites are samples from some of the oldest planetesimals that had formed in the solar system.

Hafnium-tungsten data are available for a variety of samples that derive from the silicate part of differentiated planetary bodies, including the parent bodies of some basaltic achondrites (eucrites, angrites) and Mars as well as the Earth and Moon (Table 1.2). All samples exhibit elevated Hf/W ratios and ε^{182}W values compared to chondrites, indicating that Hf/W fractionation by core formation in these planetary bodies occurred during the lifetime of ^{182}Hf. An important observation is that the Hf-W data for samples derived from the silicate and metal parts of differentiated planetary bodies show the pattern that is expected for Hf/W fractionation during core formation and subsequent decay of ^{182}Hf to ^{182}W. As such, these data confirm the basic theory of the Hf-W system as a chronometer of core formation.

Whilst determining an Hf-W age of core formation based on iron meteorite data is straightforward, because the W isotope composition of iron meteorites solely reflects Hf-W fractionation during core formation, the interpretation of Hf-W data for silicate rocks in terms of core formation timescales is more complicated. Owing to the different behavior of Hf and W during mantle

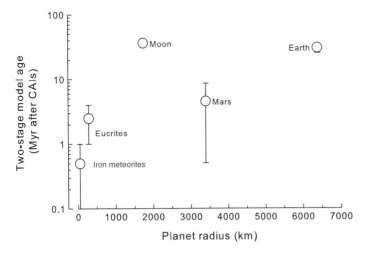

Fig. 1.6 Two-stage model age of core formation vs planet radius. After Kleine *et al.* (2002) with updated age.

melting, the Hf/W ratio of a rock is different from that of the bulk mantle of its parent body. Similarly, the W isotope composition of a mantle-derived rock may be different from that of the bulk mantle, if mantle melting occurred during the lifetime of ^{182}Hf. Thus, the Hf/W ratio and W isotope composition characteristic for the entire mantle of a differentiated planetary body can often not be measured directly and the effects of Hf-W fractionation during mantle melting need to be taken into account.

Eucrites and angrites are basaltic rocks that crystallized early in solar system history near the surface of small, differentiated planetary bodies. Their Hf-W systematics were established during the early differentiation of their parent bodies and thus can be used to constrain the timing of this early differentiation. Basaltic eucrites have ^{180}Hf/^{184}W ~20 to 30 and ε^{182}W ~21 to 33 (Kleine *et al.*, 2004a; Touboul *et al.*, 2008), whilst angrites have ^{180}Hf/^{184}W ~4 to 6 and ε^{182}W ~1 to 5 (Markowski *et al.*, 2007; Kleine *et al.*, 2009). The two-stage model ages of both basaltic eucrites and angrites are similar and range from ~1 to ~4 Myr after CAI formation (Figure 1.6). This range in model ages is unlikely to reflect a real spread in the timing of core formation but probably is due to additional Hf-W fractionations during mantle-crust differentiation, partial melting

and, in the case of the basaltic eucrites, also thermal metamorphism.

In contrast to the basaltic achondrites, samples from Mars as well as the Earth and Moon are relatively young igneous rocks and their Hf/W ratios, which are the product of a long history of multiple melting events, are unrelated to their W isotope composition, which only reflect the earliest differentiation history of their parent bodies. The Hf/W ratio that is characteristic of the entire mantle of Mars, the Earth and the Moon can thus not be measured directly but must be determined by comparing the W concentrations in the samples with another element that behaves similarly during silicate melting (i.e. has a similar incompatibility), tends to stay in the mantle and whose abundance relative to Hf is known. The latter two conditions are met by refractory lithophile elements because their relative abundances in bulk planetary mantles should be chondritic. This is because they are neither fractionated by core formation (because they are lithophile) nor by volatile element depletion (because they are refractory). Trace element studies on Martian, lunar and terrestrial basalts show that U, Th and W have similar incompatibilities, such that the U/W or Th/W ratios of these rocks can be used to obtain an Hf/W ratio of the bulk mantle (Newsom *et al.*, 1996).

Using this approach, the following estimates for the bulk mantle Hf/W ratios are obtained: 17 ± 5 (Earth), 26 ± 2 (Moon) and 3.5 ± 0.5 (Mars) (Kleine et al., 2009).

The $^{182}W/^{184}W$ ratios of Martian meteorites fall into two groups (Lee & Halliday, 1997; Kleine et al., 2004a; Foley et al., 2005b): the basaltic shergottites having $\varepsilon^{182}W$ ~0.3 to 0.6 and the nakhlites and Chassigny having $\varepsilon^{182}W$ ~2 to 3. This ^{182}W heterogeneity of the Martian mantle makes it difficult to estimate the $\varepsilon^{182}W$ that is characteristic for the bulk Martian mantle. Most studies have used the $\varepsilon^{182}W$ value of shergottites for the bulk Martian mantle because some shergottites have near-chondritic $^{142}Nd/^{144}Nd$ ratios, indicating only limited trace element fractionation during early melting of their source. Using $\varepsilon^{182}W = 0.4 \pm 0.2$ and Hf/W $= 3.5 \pm 0.5$ for the bulk Martian mantle, results in two-stage model ages of core formation of 4.6 ± 4.1 Myr after CAI formation. The large uncertainty of this age largely reflects uncertainties in the mantle Hf/W ratio and the fact that this ratio is only slightly higher than the chondritic value (Nimmo & Kleine, 2007). As will be discussed in more detail below for core formation in the Earth, for larger bodies the two-stage model age does not necessarily provide a good estimate of the timescale of core formation and the two-stage model age may underestimate the time taken to accrete and differentiate Mars by a factor of ~3 (Nimmo & Kleine, 2007). Thus, the Martian core may not have formed before ~20 Myr after the beginning of the solar system.

In contrast to the parent bodies of basaltic achondrites and Mars, there are no ^{182}W variations in the lunar and terrestrial mantles, such that the $\varepsilon^{182}W$ values of the bulk lunar mantle and the bulk silicate Earth can be measured directly and are known precisely. Determining the indigenous W isotope composition of the Moon and the extent to which $^{182}W/^{184}W$ variations exist that reflect ^{182}Hf decay within the Moon has proven to be a difficult task. There are large variations in the $^{182}W/^{184}W$ ratios of lunar whole-rocks, which initially were interpreted to reflect an early formation of the Moon and ^{182}Hf-decay within the Moon (Lee et al., 1997). More recent studies, however, showed that elevated $^{182}W/^{184}W$ ratio in lunar whole-rocks are entirely due to the production of ^{182}Ta by neutron-capture of ^{181}Ta and subsequent decay of ^{182}Ta to ^{182}W (Kleine et al., 2005b; Touboul et al., 2007), and that all lunar samples have indistinguishable $\varepsilon^{182}W$ values averaging at 0.09 ± 0.10 (Touboul et al., 2007, 2009b). Given that lunar rocks derive from sources with highly variable Hf/W, the lack of ^{182}W variations among lunar samples indicates that differentiation of the lunar mantle occurred after extinction of ^{182}Hf, that is >~60 Myr after the beginning of the solar system. The Hf/W ratio and W isotope composition of the lunar mantle correspond to a two-stage model age of 36 ± 2 Myr after CAI formation, slightly younger than the two-stage model age of the Earth's core of 30 ± 5 Myr (Touboul et al., 2007; Kleine et al., 2009).

Figure 1.6 reveals that there is a correlation between two-stage model ages of core formation and planet size (Kleine et al., 2002): the larger a planetary body, the younger is its two-stage model age. Such a pattern is unexpected if core formation did not occur until some critical stage of planetary accretion. Whilst the small meteorite parent bodies accreted early and underwent differentiation rapidly, larger bodies such as the Earth grew and differentiated over a much longer timescale. Thus, the basic assumption of the two-stage model – that is that the entire planet differentiates in one instant – is appropriate for the small meteorite parent bodies because their accretion rate is short compared to the ^{182}Hf half-life. However, the two-stage model age for the Earth of ~30 Myr indicates that accretion of the Earth occurred over a timescale that is long compared to the ^{182}Hf half-life. Thus, for the Earth the assumption of instantaneous core formation is not justified, and the two-stage model time of ~30 Myr would only date core formation if during this event the *entire* core was first remixed and homogenized with the *entire* mantle before final segregation of metal to form the present core. This is physically implausible.

The two-stage model age, therefore, is not a reasonable estimate of the exact age of the Earth's core but nevertheless provides an important constraint on the timescale of core formation. Because Hf is lithophile and W is siderophile, the bulk silicate Earth cannot have an Hf/W ratio and, hence a $^{182}W/^{184}W$ ratio, lower than chondritic at any time, provided that the bulk Earth has chondritic relative abundances of refractory elements. Since the two-stage model assumes a chondritic W isotopic composition at the time of core formation, the calculated model age corresponds to the earliest time from which the Earth's mantle could subsequently have evolved to its present-day ^{182}W excess (Halliday *et al.*, 1996; Kleine *et al.*, 2004b).

Hf-W isotopic evolution during continuous core formation and protracted accretion

The accretion of the Earth involved multiple collisions between smaller proto-planets. These collisions delivered the energy required for melting and core formation, such that the timescale of core formation is equivalent to the timescale of the major stage of accretion. Since the core did not segregate in a single instant at the end of accretion but formed continuously during accretion, there is no single, well-defined 'age' of core formation. An age rather corresponds to a certain growth stage of the core. A widely used model for the Earth's accretion is one that assumes an exponentially decreasing rate, such that:

$$\frac{m}{M_E}(t) = 1 - e^{-\alpha t} \qquad (15)$$

where m/M_E is the cumulative fractional mass of the Earth at time t and α is the time constant of accretion (Harper & Jacobsen, 1996; Jacobsen, 2005). This model provides a reasonable approximation to the Wetherill (1990) accretion model for the formation of the Earth, which is among the first of its kind. It is important to note however, that the exponential model does not mean that accretion occurred by the incremental

growth of small mass fractions. Dynamical modeling suggests that the main mass of the Earth was delivered by multiple and stochastic impacts that bring in large core masses at once (Agnor *et al.*, 1999). From the stochastic nature of these large, late-stage impacts, it follows that growth of the Earth's core occurred episodically rather than being a continuous process during accretion. The exponential model, therefore, assumes that these collisions occurred at an exponentially decreasing rate. In this model, the end of accretion is poorly constrained because accretion probably terminated with one large impact (the Moon-forming impact), the timing of which is not well defined in the exponential model. It is thus more useful to characterize the timescale in terms of a mean age of accretion, τ, which is the inverse of the time constant α and corresponds to the time taken to achieve ~63% growth (cf. Harper & Jacobsen, 1996). It is also useful to characterize the timescale of the Earth's accretion by the time taken to accrete 90% of the Earth, because current models of lunar origin predict that the Moon formed during the collision of a Mars-sized impactor with a proto-Earth that was 90% accreted (Halliday, 2004, 2008; Kleine *et al.*, 2004b, 2009). Thus, t_{90} for the Earth is equivalent to the timing of the Moon-forming impact and the end of the major stage of the Earth's accretion and core formation. Using Equation (15), we show that $t_{90} = 2.3\tau$.

Several lines of evidence indicate that a large fraction of the Earth was accreted from objects that had already undergone core-mantle differentiation. Hafnium-tungsten data for iron meteorites and basaltic achondrites demonstrate that differentiation of their parent bodies occurred very early, within <1 Myr after CAI formation (Kleine *et al.*, 2009). These data suggest that at least some of the planetesimals that accreted to the Earth were already differentiated. However, a significant amount of the Earth's mass was delivered by much larger Moon- to Mars-sized embryos. It seems inevitable that these were already differentiated prior to their collision with the proto-Earth, because models of planetary accretion predict that at 1 AU, planetary embryos had formed within ~1 Myr after CAI formation. In this case, they will have melted owing to heating by the decay of then abundant ^{26}Al. Furthermore, planetary embryos are large enough that collisions between two embryos release sufficient energy for melting and differentiation.

A question of considerable interest for interpreting the W isotope composition of the Earth's mantle in terms of core formation timescales is the mechanism by which metal is transported to the core during the collision of a pre-differentiated object with the Earth (Harper & Jacobsen, 1996; Kleine et al., 2004b, 2009; Nimmo & Agnor, 2006; Rubie et al., 2007; Halliday, 2008). The key issue is if the incoming impactor metal core directly merged with the Earth's core' or finely dispersed as small droplets in the terrestrial magma ocean. In the case of *core merging*, no equilibration between the impactor core and the Earth's mantle will have taken place and the chemical and isotopic composition of the Earth's mantle would largely reflect the conditions of core formation in Earth's building blocks. In contrast, small metal droplets will have easily equilibrated with the surrounding silicate material in a magma ocean. This model, therefore, is often referred to as *magma ocean differentiation*. In this model any information on the differentiation of the impactor is lost and the chemical and isotopic composition of Earth's mantle reflects the conditions of core formation within the Earth.

Whether core merging or magma ocean differentiation is more appropriate for a given collision depends on the relative size of the two colliding bodies. Collisions in which the impactor is much smaller than the target result in vaporization of the impactor, in which case the impactor material can efficiently mix and homogenize within the magma ocean of the target. What happens in detail during larger collisions, however, is less well understood. Hydrocode simulations of giant impacts (Canup & Asphaug, 2001) show that the cores of target and impactor merge rapidly, although more recent simulations indicate that some re-equili-

bration might occur (Canup, 2004). The problem is that these simulations currently provide a resolution in the order of 100 km, whereas the length-scale on which chemical and isotopic re-equilibration occurs is probably in the order of centimeters (Stevenson, 1990; Rubie et al., 2007). Thus, the extent to which metal-silicate equilibration occurred during large impacts is currently not well understood.

The W isotope evolution of the Earth's mantle during accretion from pre-differentiated objects can be followed using a three-box model (Harper & Jacobsen, 1996; Halliday, 2004; Kleine et al., 2004a, 2009; Jacobsen, 2005; Nimmo & Agnor, 2006). In this model, the ^{182}W/^{184}W ratio of Earth's mantle immediately after addition of a new object reflects the contribution of the following three components: Earth's mantle prior to the impact (component 1); the impactor mantle (component 2); and the impactor core (component 3). We will assume that a fraction $1-k$ of the impactor core is added directly to the target core without any prior re-equilibration with the Earth's mantle and will term k the equilibration factor ($k = 1$ indicates complete re-equilibration). Let R_i be the ^{182}W/^{184}W ratio of component i ($i = 1,2,3$), then R_1', the ^{182}W/^{184}W ratio of the Earth's mantle immediately after the collision, is given by:

$$R'_1 = \frac{y_1 R_1 + y_2 R_2 + k y_3 R_3}{y_1 + y_2 + k y_3} \qquad (16)$$

where $y_i = m_i[\text{W}]_i$ is the total mass of W, m_i is the mass and $[\text{W}]_i$ is the W concentration in component i. R_i for the three components at the time of the collision can be calculated using the decay equation of extinct short-lived nuclides and for the time interval between two collisions at t_1 and t_2 is given by:

$$R_i(t_2) = R_i(t_1) + \left(\frac{^{180}\text{Hf}}{^{184}\text{W}}\right)_i \left(\frac{^{182}\text{Hf}}{^{180}\text{Hf}}\right)_{t_1} \left(1 - e^{-\lambda(t_2 - t_1)}\right) \quad (17)$$

The total mass of W in any of these three components depends on the mass of this component and its W concentration. The latter is given by the metal-silicate partition coefficient of *W*, which is defined as:

$$D^{m/s} = \frac{[W]_m}{[W]_s} \qquad (18)$$

For the concentration of W in the mantle of a differentiated object, mass balance considerations and the definition of *D* give:

$$[W]_m = \frac{[W]_0}{D + \gamma(1 - \gamma)} \qquad (19)$$

where $[W]_0$ is the W concentration of the bulk Earth (assumed to be chondritic) and γ is the silicate mass fraction (in case of the Earth $\gamma = 0.675$). For the concentration of W in the core we have:

$$[W]_c = \frac{[W]_0}{1 + \gamma(D - 1)} \qquad (20)$$

From the equations summarized above it is clear that a number of parameters need to be known to calculate the W isotope evolution of Earth's mantle during protracted accretion with concomitant core formation. Important parameters include:

• Equilibration factor *k*;
• D for W in both the Earth and the differentiated objects that are accreted to form the Earth; and
• The $^{182}W/^{184}W$ ratios in impactor mantle and core at the time of their collision with proto-Earth.

All these parameters are unknown and for the sake of simplicity, we will first assume that the metal-silicate partition coefficient *D* for W remained constant throughout accretion and was the same in all bodies that accreted to the Earth. We will further assume that all the bodies the collided with the Earth underwent core-mantle differentiation at t_0, the start of the solar system.

Figure 1.7 schematically illustrates the W isotope evolution of Earth's mantle in the two end-member models of metal-silicate equilibration during core formation. In the core-merging model (*k* = 0), metal-silicate re-equilibration does not occur and the W isotope composition of the target's mantle immediately after the impact results from addition of impactor mantle material to the target's mantle. The resulting $^{182}W/^{184}W$ ratio will always be higher than chondritic because no core material (with subchondritic $^{182}W/^{184}W$) is involved in these mixing processes. In the case of core merging, no isotopic record of the collision was generated and the W isotope effect in the Earth's mantle would largely reflect the timing of core formation in the pre-merged objects. In contrast, small metal droplets in a magma ocean could have equilibrated efficiently with the surrounding molten silicates. If re-equilibration was complete (i.e. *k* = 1), this is equivalent to adding an undifferentiated object to the Earth's mantle, followed by metal segregation. In this case, the resulting W isotope effect reflects the rate of accretion and timing of core formation.

A variety of growth curves for the Earth's accretion, calculated by assuming an exponentially decaying accretion rate and mean times of accretion, τ of 11, 15, and 25 Myr, are shown in Figure 1.8a. Using these growth curves and Equations (16) and (17), the W isotope evolution of the Earth's mantle can be calculated for each value of τ and an assumed equilibration factor *k*. In Figure 1.8b, the W isotope evolution of Earth's mantle in the core merging and magma ocean differentiation models are compared for an assumed τ of 11 Myr. Obviously these two models result in very different $\varepsilon^{182}W$ values for the bulk silicate Earth. The core merging model predicts highly radiogenic ^{182}W in the Earth's mantle, which would reflect the timescale of core formation in the objects that later were added to the Earth (assumed to be t_0). The magma ocean differentiation model predicts much lower $\varepsilon^{182}W$ values for the Earth's mantle, a reflection of the high degree of metal-silicate equilibration and protracted timescale of accretion in this model. Obviously, the magma ocean differentiation model is consistent with the observed W isotope composition of Earth's mantle, whilst the core merging model is not. Thus, some metal-silicate re-equilibration during accretion of the Earth from pre-differentiated objects clearly is required (Halliday *et al.*,

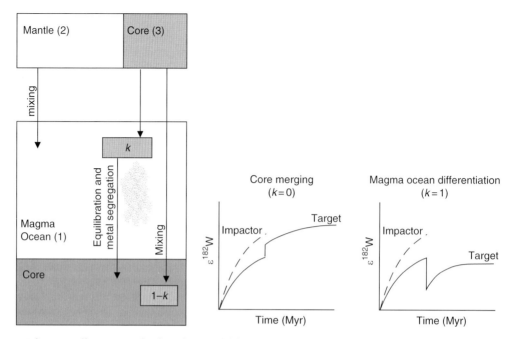

Fig. 1.7 Schematic illustration of a three-box model for W isotope evolution during protracted accretion with concomitant core formation. The components that contribute to the W isotope composition of Earth's mantle are labeled with 1, 2 and 3. The W isotope effects in the two end-member models of core formation – core merging and magma ocean differentiation – are shown schematically. Modified from Kleine *et al.* (2009).

1996; Halliday, 2004; Kleine *et al.*, 2004a, b; Jacobsen, 2005; Nimmo & Agnor, 2006).

In Figure 1.8c, W isotope evolution curves of the Earth's mantle are shown for three different combinations of τ and k. In all three models, the present-day W isotopic composition of the Earth's mantle ($\varepsilon^{182}W = 0$) can be reproduced, demonstrating that determining the accretion rate of Earth (τ) requires knowledge of the degree to which metal-silicate re-equilibration occurred (k). In Figure 1.9a, the calculated ages of core formation are plotted as a function of the equilibration factor k. As is evident from this figure, the calculated core formation ages become younger with decreasing values of k (Halliday, 2004; Kleine *et al.*, 2004b, 2009). Metal-silicate re-equilibration during core formation results in a decrease of the ^{182}W excess in Earth's mantle that had previously accumulated due to ^{182}Hf decay, such that, for a given accretion rate, decreasing k values will result in an increasingly radiogenic W isotope composition of the Earth's mantle. Thus, to match the present-day W isotope composition of Earth's mantle, a decreasing degree of metal-silicate re-equilibration must be accompanied by longer accretion timescales, as shown in Figure 1.9a.

In all models discussed above, it was assumed that the Hf/W ratio in the mantle of the Earth, as well all other bodies added to the Earth, was constant and identical to the present-day Hf/W ratio of the Earth's mantle. This assumption is unlikely to be valid because the metal-silicate distribution coefficient for W depends on several parameters such as pressure, temperature, and particularly oxygen fugacity (Cottrell *et al.*, 2009). These parameters are likely to have varied during accretion (Wade & Wood, 2005; Wood *et al.*, 2008). Unfortunately, there is currently little

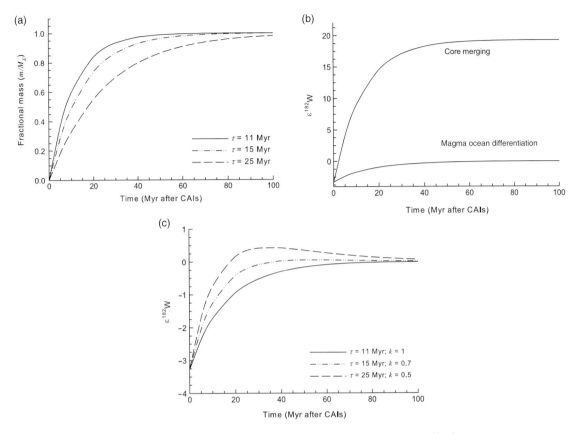

Fig. 1.8 (a) Fractional mass of the Earth as a function of time assuming an exponentially decreasing accretion rate and a mean time of accretion of 11, 15 and 25 Myr. (b) W isotope evolution of Earth's mantle calculated for the core merging and magma ocean differentiation models, assuming a mean time of accretion of 11 Myr. (c) W isotope evolution of Earth's mantle calculated for the growth curves shown in (a) and assuming different equilibration factors k. The present-day W isotope composition of Earth's mantle ($\varepsilon^{182}W = 0$) can be reproduced in all three scenarios, indicating that τ cannot be constrained independent of k.

understanding of how W partitioning into the Earth's core evolved as accretion proceeded. Most models predict that conditions were highly reducing during the early stages of accretion and became more oxidizing towards the end of accretion (Wade & Wood, 2005; Wood et al., 2008). However, other models argue for the opposite (Halliday, 2004; Rubie et al., 2004). In any event, the W partitioning into the core was clearly different in different planetary bodies, as is evident from the large range in Hf/W ratios in planetary mantles.

In Figure 1.9b, the calculated core formation ages are plotted against the Hf/W ratios in the impactor's mantles. In these calculations, a constant equilibration factor was assumed. This figure reveals that for a given equilibration factor (in this case $k = 0.5$) decreasing Hf/W ratios in the impactor's mantles result in shorter calculated accretion rates of the Earth. The reason for this is that a mantle with a relatively low Hf/W ratio will never develop a large ^{182}W excess, even if differentiation occurred very early. Consequently, for a given value of k, decreasing Hf/W ratios in

(a)

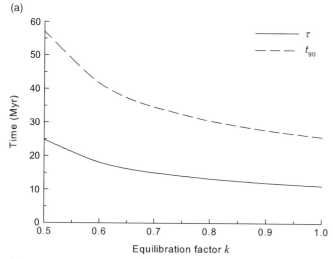

(b)

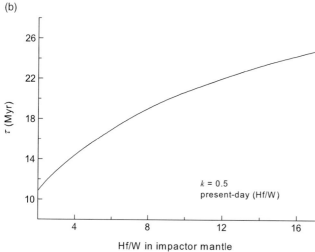

Fig. 1.9 (a) Effect of incomplete metal-silicate equilibration on calculated core formation ages. Shown are results for the calculated value of τ and for the time at which 90% growth was achieved (b) Effect of variable Hf/W ratios in the impactor's mantles on calculated core formation ages. In these calculations, it is assumed that the equilibration factor k was 0.5 throughout. After Kleine et al. (2004b; 2009).

the impactor's mantles led to less radiogenic $^{182}W/^{184}W$ in the Earth's mantle. Thus, in such a model, the present-day W isotope composition of the Earth's mantle is reproduced by assuming a shorter accretion timescale.

In summary, determining the age of the Earth's core using the Hf-W system requires knowledge of the degree of isotopic equilibrium during metal-silicate separation, on the conditions of core formation in the growing Earth and its building blocks, and on the timescale of core formation in the planetary embryos that were accreted to form the Earth. However, these aspects are currently not well understood, making determining the exact age of the Earth's core difficult. In spite of these uncertainties, the Hf-W data allow some important and robust conclusions to be drawn:

• The relatively small ^{182}W excess found in the Earth's mantle compared to ^{182}W anomalies found in eucrites and iron meteorites indicates a much more protracted timescale of accretion and core formation in the Earth compared to the smaller meteorite parent bodies;

• Core formation in the Earth can have terminated no earlier than ~30 Myr after solar system formation, as shown by its two-stage Hf-W model age;

• Core formation in the Earth must have involved some re-equilibration of newly accreted metal within a terrestrial magma ocean, otherwise the ^{182}W excess of the Earth's mantle should be one order of magnitude higher than observed.

However, as discussed above, the degree to which this re-equilibration occurred is not known.

Age of the Moon and accretion of the Earth

The models discussed in the preceding section use the present-day Hf/W ratio and W isotope composition of the Earth's mantle as the sole criterion to infer the rate at which the Earth and its core formed. We have seen, however, that the outcome of these models depends on a number of assumptions regarding the process of accretion and the conditions of core formation. Obviously, additional constraints are required and these are provided from the age and isotopic composition of the Moon. The current paradigm for the formation of the Moon is that a roughly Mars-sized body struck the proto-Earth at an oblique angle. Some fraction of the hot mantle of the impactor spun off and re-accreted outside the Roche limit to form the present-day Moon. It is widely believed that the Moon-forming impact was the last major event in the Earth's accretion, occurring when the Earth was ~90% accreted (Canup & Asphaug, 2001). This giant impact caused widespread melting of the Earth's mantle and led to the final stage of core formation in the Earth (Tonks & Melosh, 1993). Thus, dating the Moon provides an independent constraint on the timescale of Earth's accretion and core formation and can be used to test the models discussed in the previous section.

Lunar samples exhibit variable and large excess in ^{182}W but these W isotope variations are the result of cosmogenic ^{182}W production (see above). Once these cosmogenic effects are eliminated, all lunar samples have a W isotopic composition indistinguishable from that of the Earth's mantle (Touboul *et al.*, 2007, 2009b). This similarity is unexpected and the currently favored explanation for this observation is the Earth and Moon isotopically equilibrated in the aftermath of the giant impact via a shared silicate atmosphere (cf. Pahlevan & Stevenson, 2007). As to whether this model is viable for equilibrating W isotopes is unclear and the alternative model to account for the similar W isotopes in the lunar and terrestrial mantles is that large parts of the Moon are derived from Earth's rather than from the impactor's mantle. This, however, is difficult to reconcile with results from dynamical models of lunar origin, all of which predict that the Moon predominantly consists of impactor mantle material (Canup, 2004). Whatever the correct interpretation, the indistinguishable W isotopic compositions of the lunar and terrestrial mantles indicate that the Earth and Moon formed in isotopic equilibrium, because it is highly unlikely that the identical W isotopic compositions evolved by coincidence (Touboul *et al.*, 2007; Kleine *et al.*, 2009).

Since the lunar and terrestrial mantles have different Hf/W ratios, they should have different $\varepsilon^{182}W$ values if formation of the Moon and the final stage of core formation in the Earth occurred whilst ^{182}Hf was still extant. However, the bulk silicate Moon and Earth have indistinguishable W isotopic compositions, indicating that the Moon must have formed after ^{182}Hf extinction (Touboul *et al.*, 2007, 2009b). This is quantified in Figure 1.10, where the expected $\varepsilon^{182}W$ difference between the lunar and terrestrial mantles is plotted as a function of time. The current best estimates indicate that the bulk silicate Moon has an Hf/W ratio that is ~50 % higher than that of the bulk silicate Earth (i.e. $f_{Hf/W}$ ~ 0.5 in Figure 1.10). In this case, the Moon must have formed >~50 Myr after CAI formation. However, the uncertainties on the estimated Hf/W ratios are large and their more precise determination will be an important future task for better determining the age of the Moon.

The indistinguishable $\varepsilon^{182}W$ values of the lunar and terrestrial mantles only provide a maximum age for the formation of the Moon. Its minimum

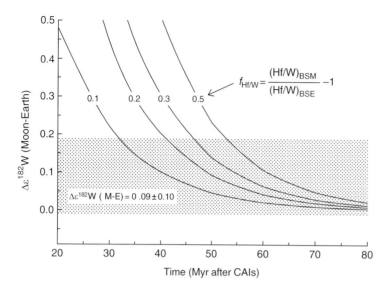

Fig. 1.10 Difference in ε^{182}W values between the lunar and terrestrial mantles as a function of time (after Kleine *et al.* 2009).

age is given by the oldest lunar rocks, which are the ferroan anorthosites that have Sm-Nd ages of 4.46 ± 0.04 Gyr (Carlson & Lugmair, 1988; Norman *et al.*, 2003). These two age constraints combined indicate that the Moon formed between 50 and 150 Myr after CAI formation (Touboul *et al.*, 2007, 2009b).

The age of the Moon can be used to test some of the accretion and core formation models discussed in the previous section. Two selected models, which resemble those of Halliday (2008), are shown in Figure 1.11. These were calculated by assuming that 90% of the Earth accreted at an exponentially decreasing rate and that accretion terminated by a giant Moon-forming impact at ~100 Myr, adding the remaining 10% of the Earth's mass. The main difference between the two growth curves is that in the first model it is assumed that the Earth accreted at an exponentially decreasing rate until the giant impact at 100 Myr (Figure 1.11a), whereas in the second model 90% of the Earth was accreted rapidly within the first ~23 Myr ($\tau \sim 10$ Myr), followed by an accretion hiatus until the final giant impact at 100 Myr (Figure 1.11c). In both models, the present-day W isotopic composition of the bulk silicate Earth (ε^{182}W = 0) can be reproduced, but in the first model a low degree of metal-silicate

equilibration is required ($k = 0.4$), whereas in the latter full equilibration occurred during core formation ($k = 1$).

These two exemplary models illustrate that variable scenarios can reproduce the observed present-day Hf-W systematics of the bulk silicate Earth and the age of the Moon. An important difference between these two models is the degree to which metal-silicate equilibration occurred. Thus, constraining the degree of which equilibrium was achieved during core formation may help to distinguish among different scenarios of the Earth's accretion. The abundances of siderophile elements in the Earth's mantle may be brought to bear on this issue. They are commonly explained by metal-silicate re-equilibration under high pressures and temperatures in a deep terrestrial magma ocean (Rubie *et al.*, 2007; Wood *et al.*, 2008), in which case the model shown in Figures 1.11 c and d might best represent the Earth's accretion and core formation. However, the effects of incomplete metal-silicate equilibration on siderophile element abundances in the growing Earth have yet not been investigated and it is currently unclear as to whether they require full equilibration during core formation. Clearly, investigation of this issue will provide essential information on how the Earth accreted.

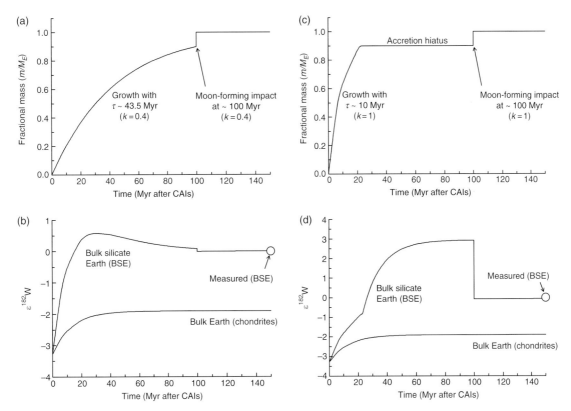

Fig. 1.11 (a) Growth curves for the Earth assuming an exponentially decreasing accretion rate for the first 90% of growth, which was terminated by a final Moon-forming impact at ~100 Myr that added the remaining 10% of Earth's mass. (b) W isotope evolution of the bulk silicate Earth for the growth curve shown in (a). The present-day $\varepsilon^{182}W$ of the BSE is reproduced by assuming that only 40% of the incoming impactor cores equilibrated with the Earth's mantle ($k = 0.4$). (c) Growth curve for the Earth assuming an initially rapid growth with a mean life $\tau \sim 10$ Myr, followed by an accretion hiatus and a final Moon-forming impact at ~100 Myr. (d) W isotope evolution of the bulk silicate Earth for the growth curve shown in (c) and complete metal-silicate equilibration during core formation ($k = 1$). Modified after Halliday (2008).

[146,147]SM-[142,143]ND CHRONOLOGY OF MANTLE-CRUST DIFFERENTIATION ON THE EARTH, MARS, AND THE MOON

The [147]Sm-[143]Nd record of Archean terranes

Estimating the [143]Nd evolution of the early Earth's mantle requires back-calculating the present-day [143]Nd/[144]Nd ratio of the most ancient Archean rocks at the time of their formation, in order to obtain their initial $\varepsilon^{143}Nd$. If the rocks are juvenile (i.e. crystallized from mantle-derived melts), then their $\varepsilon^{143}Nd_{initial}$ gives the isotopic composition of the mantle at the time of their extraction. A depleted mantle reservoir having evolved for a few hundred million years with high Sm/Nd will be characterized by positive $\varepsilon^{143}Nd$ (Figure 1.5a), whilst a primitive mantle, assuming a chondritic composition, should have $\varepsilon^{143}Nd = 0$. Therefore, the initial isotopic composition of juvenile Archean rocks should reflect the degree of depletion of the early Earth's mantle, and provide constraints on crustal evolution in the early Earth.

The expression used for back-calculating the present-day $^{143}Nd/^{144}Nd$ composition of a rock sample (S) is given by:

$$\left(\frac{^{143}Nd}{^{144}Nd}\right)_{T_s} = \left(\frac{^{143}Nd}{^{144}Nd}\right)_{T_p} + \left(\frac{^{147}Sm}{^{144}Nd}\right)_{T_p} \left(1 - e^{-\lambda_{147}T_S}\right) \quad (21)$$

where T_s is age of the rock, T_p is present day, and T is time running backward from the origin of the solar system ($T_p = 4.56\,Gyr$) to the present day. Estimating initial $^{143}Nd/^{144}Nd$ ratios thus requires the knowledge of the present-day isotopic $(^{143}Nd/^{144}Nd)_{Tp}$ and chemical $(^{147}Sm/^{144}Nd)_{Tp}$ composition of the rock, but also a precise estimate of its age (T_s). The latter parameter is the most difficult to obtain, as long-lived chronometric systems in early Archean rocks are often disturbed by tectono-metamorphic events. A classic method for simultaneously estimating T_s and $(^{143}Nd/^{144}Nd)_{Ts}$ is to establish a whole-rock ^{147}Sm-^{143}Nd isochron from a series of cogenetic samples (Moorbath *et al.*, 1997). In many cases, however, this method yields imprecise results due to insufficient spread in Sm/Nd, minor disturbance of the Sm-Nd system due to metamorphism, or mixing of the juvenile magmas with pre-existing crustal components. An alternative method is to independently estimate T_s using U-Pb dating of zircon grains (Bennett *et al.*, 1993). This has the advantage that U-Pb zircon ages are usually precise within a few million years, whilst whole-rock Sm-Nd isochrons have uncertainties of 50 Myr or more. This method, however, is problematic because the U-Pb age of the zircons and the Sm-Nd age of their host rock are not always identical. For example, the Acasta (Northern Canada) and Amitsôq gneisses (Western Greenland), dated at 3.6 to 4.0 Gyr and 3.73 to 3.87 Gyr using U-Pb in zircon, have younger Sm-Nd ages of 3.37 and 3.65 Gyr, respectively. Back-calculating initial $^{143}Nd/^{144}Nd$ compositions at the zircon age thus yields highly heterogeneous $\varepsilon^{143}Nd_{initial}$, which can be incorrectly interpreted as reflecting mantle heterogeneity (Bowring & Housh, 1995; Bennett *et al.*, 1993).

Despite the difficulties in reconstructing a reliable ^{147}Sm-^{143}Nd record for the oldest terrestrial rocks, it is now well established that Archean terranes have initial $^{143}Nd/^{144}Nd$ ratios more radiogenic than predicted for a primitive mantle of chondritic composition (Shirey & Hanson, 1986; Bennett, 2003) (Figure 1.12). These radiogenic signatures require that the most ancient crustal rocks were extracted from a mantle reservoir having evolved for several hundred million years with an Sm/Nd ratio 5 to 10% higher than chondritic, which has been considered as evidence that a long-lived crust formed very early (>3.8 Gyr) in the history of the Earth (Armstrong, 1981). Ancient cratons, however, represent <5% of the total mass of the present-day continents (Condie, 2000) (Figure 1.12) and this is by far insufficient to account for large-scale mantle depletion in the early Archean. It was thus suggested that Hadean continents had been recycled over shorter timescales (Armstrong, 1981; Jacobsen, 1988), or that the earliest crust was a mafic/ultramafic reservoir which, by analogy to the modern oceanic crust, did not contribute significantly to the Archean sedimentary mass (Chase & Patchett, 1988; Galer & Goldstein, 1991).

Extensive efforts to model the ^{143}Nd evolution of the Archean mantle showed that the Archean ^{147}Sm-^{143}Nd record is consistent with a wide range of evolution scenarios for the mantle-crust system (Armstrong, 1981; Chase & Patchett, 1988; Jacobsen, 1988; Galer & Goldstein, 1991; Boyet & Carlson, 2006; Caro *et al.*, 2006). The radiogenic $\varepsilon^{143}Nd$ signature of +1 to +3 ε-units characterizing the early Archean mantle could, for example, result from very early (4.5 Gyr) depletion, due to magma ocean crystallization (Caro *et al.*, 2005), or from later Sm/Nd fractionation resulting from continuous crustal formation during the Hadean (Jacobsen, 1988). This ambiguity is due to the very long half-life of ^{147}Sm, which limits the chronological resolution of this system (Figure 1.5a), and also to the lack of an isotopic record prior to 3.8 Gyr. In addition, the interpretation of ^{147}Sm-^{143}Nd data is tied to the assumption that the bulk silicate Earth has a

(a)

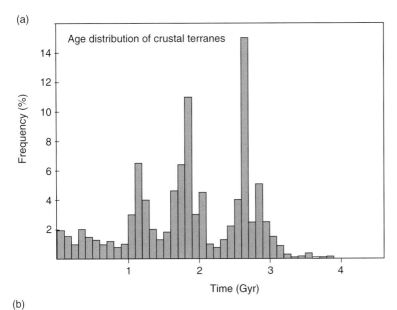

(b)

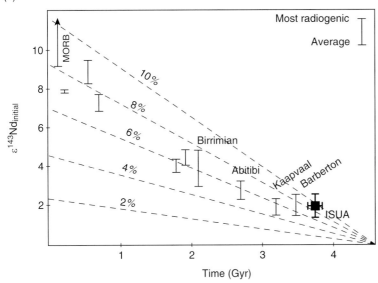

Fig. 1.12 (a) Histogram showing the age distribution of continental terranes. This distribution is characterized by three peaks of growth centered at 2.7, 1.9 and 1.2 Gyr and by a negligible amount of pre-3 Gyr crust. Reprinted from *"Tectonophysics"*, **322**: 1–2, Kent C. Condie, Episodic continental growth models: afterthoughts and extension, 2000, with permission from Elsevier. (b) $\varepsilon^{143}Nd$ evolution of the upper mantle estimated from the isotopic composition of mafic and ultramafic rocks through time. Positive $\varepsilon^{143}Nd_{initial}$ compositions are observed as far back as 3.8 Gyr, indicating that the mantle has evolved for most of its history with an Sm/Nd ratio at least 4 to 6% higher than chondritic, which cannot be explained by extraction of the present-day continents. The more radiogenic $\varepsilon^{143}Nd_{initial}$ recorded by Proterozoic juvenile rocks and modern mid-ocean ridge basalts indicates that the Proterozoic upper mantle was more depleted than the Archean mantle. This additional depletion can be explained by extraction and stabilization of large continental masses in the late Archean and Proterozoic.

perfectly chondritic composition. However, positive $\varepsilon^{143}Nd_{initial}$ may also reflect a slightly non-chondritic Sm/Nd ratio for the bulk Earth, in which case no early crustal extraction would be needed to account for the seemingly depleted signature of the Archean mantle. Thus, whilst a realistic model of early mantle-crust evolution should always be consistent with the ^{147}Sm-^{143}Nd record, the chronology and mechanisms of early mantle depletion cannot be unambiguously constrained from the observed $\varepsilon^{143}Nd_{initial}$ in the Archean mantle.

The terrestrial ^{146}Sm-^{142}Nd record

Because of its short half-life, the ^{146}Sm-^{142}Nd chronometer was expected to provide more precise chronological constraints on early mantle-crust evolution. It was also anticipated that the ^{142}Nd signature of early Archean rocks would be less sensitive to metamorphic disturbance than their ^{143}Nd/^{144}Nd signature, owing to the negligible production rate of ^{142}Nd after 4.2 Gyr. Consequently, the first searches for ^{142}Nd anomalies focused on early Archean rocks from West Greenland characterized by large positive ε^{143}Nd$_{initial}$ (Goldstein & Galer, 1992; Harper & Jacobsen, 1992; Sharma et al., 1996). However, these first attempts were seriously hampered by the limited precision achievable using thermal-ionization mass spectrometry (i.e. $2\sigma\sim20$ ppm). Harper & Jacobsen (1992) first reported a +33 ppm anomaly in a metasediment sample from the 3.8-Gyr-old Isua Greenstone Belt (West Greenland) but this discovery was met with skepticism as an in-depth investigation of the precision limits of thermal ionization mass spectrometers suggested that the precision quoted by the authors was perhaps overestimated (Sharma et al., 1996). In addition, several studies failed to detect ^{142}Nd anomalies in early Archean rocks, including samples from the Isua Greenstone Belt (Goldstein & Galer, 1992; McCulloch & Bennett, 1993; Regelous & Collerson, 1996).

With the advent of more precise thermal-ionization mass spectrometers in the last decade, it became possible to achieve precisions better than 5 ppm (2σ) on the ^{142}Nd/^{144}Nd ratio (Caro et al., 2003, 2006). This led to the discovery of ubiquitous ^{142}Nd anomalies in a series of 3.75-Gyr-old metasediments from the Isua Greenstone belt (Caro et al., 2003). Similar effects were also found in Isua metabasalts and in virtually all 3.6- to 3.8-Gyr-old gneisses from the West Greenland craton (Caro et al., 2006). These positive anomalies (Figure 1.13) typically ranging from 5 to 20 ppm are thus significantly smaller than the 30 to 40 ppm effects initially reported by Harper & Jacobsen (1992) and later on by Boyet et al. (2004) using ICP-MS technique. However, the magni-

tude of the effects reported by Caro et al. (2003, 2006) was confirmed by Boyet & Carlson (2005) and Bennett et al. (2007). In addition, Bennett et al. (2007) reported hints of a positive ^{142}Nd anomaly in Early Archean rocks from the Narryer gneiss complex (Western Australia).

There is no ambiguity in the presence of positive ^{142}Nd effects in West Greenland rocks. This depleted signature is consistent with the radiogenic ε^{143}Nd$_{initial}$ characterizing these rocks (Figure 1.12b), and indicates that this ancient craton was extracted from a reservoir depleted by crustal extraction processes >4.2 Gyr ago. As illustrated in Figure 1.14, this chronology can be further refined using coupled ^{146,147}Sm-^{142,143}Nd systematics (see the section on 'Coupled ^{146}Sm-^{142}Nd and ^{147}Sm-^{143}Nd chronometry'). ^{147}Sm-^{143}Nd investigations of the metasedimentary samples studied by Caro et al. (2003) yielded a whole-rock isochron age of 3.75 Gyr (Moorbath et al., 1997), consistent with U-Pb zircon ages from associated sedimentary formations (3.7–3.8 Gyr). This was considered as evidence that ^{147}Sm-^{143}Nd systematics in these samples had not been significantly disturbed by metamorphism on a whole-rock scale. The ε^{143}Nd$_{initial}$ of +1.9 ± 0.6 ε-units derived from this isochron was thus considered as a reliable estimate of the signature of mantle depletion characterizing the early Archean mantle. Application of coupled ^{146,147}Sm-^{142,143}Nd chronometry using ε^{143}Nd = +1.9 ε-units and ε^{142}Nd = 7 to 15 ppm yield an age of differentiation of <100 Myr (i.e. T_d <4.45 Gyr; Figure 1.14) and this age range can be narrowed to 30 to 100 Myr if we assume that the extraction of a long-lived crust must postdate core formation. The presence of positive ^{142}Nd effects in the early Archean mantle thus represents strong evidence that a crustal reservoir formed at a very early stage of the Earth's history, perhaps immediately following the end of terrestrial accretion and solidification of the terrestrial magma ocean (Caro et al., 2005). Together with constraints from Hf-W (see the section on 'Hf-W chronology of the accretion and early differentiation of the Earth and Moon') and Pu-I-Xe chronometers (Staudacher & Allegre, 1982; Ozima & Podosek,

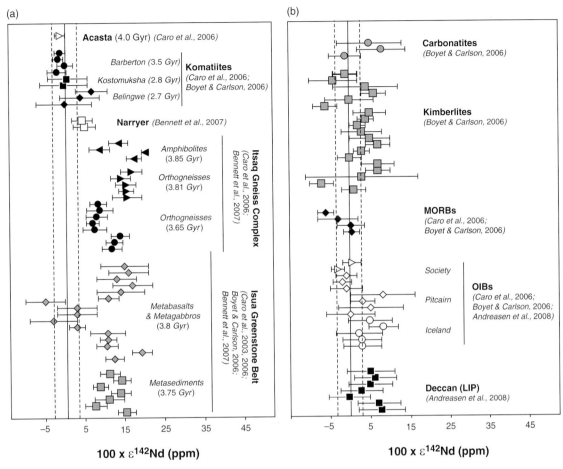

Fig. 1.13 Compilation of high-precision ^{142}Nd measurements in (a) Archean and (b) modern or recent mantle-derived rocks. The homogeneity of the modern mantle suggests that heterogeneities generated in the Hadean mantle have been efficiently remixed by mantle convection. A remnant of this ancient crust may have been preserved within early Archean cratonic roots, as suggested by negative ^{142}Nd effects observed in the NSB greenstone belt and in younger rocks from the Khariar craton (O'Neil *et al.*, 2008; Upadhyay *et al.*, 2009).

1999), these results indicate that the Earth was completely differentiated with a core, an atmosphere and long-lived silicate reservoirs <100 Myr after formation of the solar system.

^{146}Sm-^{142}Nd constraints on the evolution of the Hadean crust

A proto-crust formed >4.45 Gyr ago would have developed a negative ^{142}Nd anomaly of 0.5 to 1

ε-units, depending on its exact age and composition. Thus, even a moderate contribution of this ancient crustal material to modern magmatism would result in negative ^{142}Nd effects, which should be detectable at the level of precision currently reached by mass spectrometers (Bourdon *et al.*, 2008). However, negative ^{142}Nd anomalies have so far only been identified in early Archean rocks from the Nuvvuagittuq Supracrustal Belt (3.8 Gyr, Northern Canada) and in Proterozoic

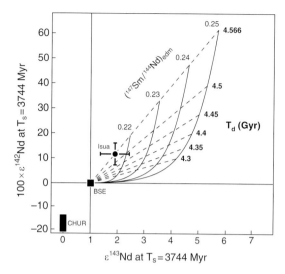

Fig. 1.14 Calculation of a model age of crustal extraction using coupled ^{142}Nd-^{143}Nd systematics in early Archean rocks from West Greenland. This model age was estimated by assuming a super-chondritic ^{147}Sm/^{144}Nd = 0.2082 for the BSE (see the section on '^{146}Sm-^{142}Nd constraints on the evolution of the Hadean crust'). A slightly younger model age (50–200 Myr) would be obtained by assuming differentiation from a perfectly chondritic bulk silicate Earth (Caro *et al.*, 2006).

syenites intruding the early Archean Khariar craton (<3.85 Gyr; India) (O'Neil *et al.*, 2008; Upadhyay *et al.*, 2009). If confirmed, these preliminary results would indicate that >4.2 Gyr old crust was reworked within early Archean terranes, where it was locally preserved from erosion and recycling. There is currently no evidence that such ancient crustal components were also incorporated within younger cratons, or that Hadean terranes represent a significant reservoir on a global scale. Analyses of modern mantle-derived rocks (OIBs, kimberlites, MORBs, LIPs) have also failed to provide evidence for a preserved Hadean crustal component in the mantle (Figure 1.13a), indicating that mantle heterogeneity is not inherited from early differentiation processes. Overall, these observations suggest that most of the Hadean crust was recycled and remixed within the mantle prior to the major epi-

sodes of crustal growth, making it unlikely that a large Hadean crustal reservoir was preserved until the present day.

As ^{142}Nd anomalies can only be generated prior to 4.2 Gyr, their preservation in the early Archean mantle depends on the rate at which the Hadean crust was recycled and remixed in the EDM. Based on this observation, Caro *et al.* (2006) presented a simplified mantle-crust model aimed at providing a first-order estimate of the lifetime of the Hadean crust. The central postulate of this model is that the positive anomalies measured in West Greenland rocks are representative of the early Archean mantle and represent the diluted signature of a reservoir differentiated 4.5 Gyr ago. In contrast to the two-stage model presented in previous sections, the proto-crust is progressively recycled and continuously replaced by an equal volume of juvenile crust (Figure 1.15). Radiogenic mantle material is thus remixed with non-radiogenic crustal components at a rate that depends on the residence time of Nd in the crustal reservoir (i.e. the lifetime of the crust). For the sake of simplicity, it was assumed that no significant crustal growth occurs during this period, so that the mass of crust and its chemical composition remain constant. The mass transport equation for this model thus reads:

$$\frac{dM_c}{dt} = \frac{dM_m}{dt} = J_{m \to c} - J_{c \to m} = 0 \qquad (22)$$

where the subscripts m and c stand for mantle and crust, respectively. M_i is the mass of the reservoir i and $J_{i \to j}$ is the mass flux from reservoir i towards reservoir j. Equations describing the isotopic evolution of the mantle-crust system in this simple two-box model can be found in Albarede (1995) and Caro *et al.* (2006).

Figure 1.16 shows the possible isotopic evolution curves for the early crust and its complementary depleted mantle as a function of the lifetime of the crustal reservoir R_c. The results are dependant on crustal composition, and we have therefore contrasted two different scenarios (basaltic vs continental crust), whose parameters are summarized in Table 1.3. Several constraints

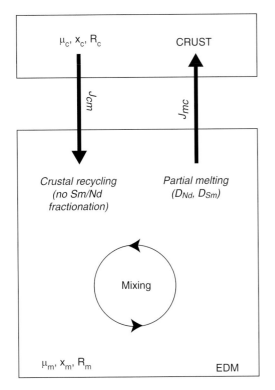

Fig. 1.15 A simplified two-box model describing a continuously interacting mantle-crust system.

on Hadean crustal dynamics can be obtained from Figure 1.16. First, it is unlikely that the Hadean crust had a lifetime shorter than 400 Myr. In this case, ^{142}Nd anomalies in the mantle would be erased prior to the formation of early Archean cratons, leaving virtually no residual imprint of the earliest crust in the 3.6 to 3.8 Gyr mantle. As shown in Figure 1.16, both the basaltic and continental crust models provide a satisfying fit to the ε^{142}Nd composition of the Archean mantle for significantly longer crustal residence times. In the basaltic crust model, a best fit to the early Archean mantle is obtained for values of R_c between 700 and 1,500 Myr. This value is essentially unchanged if we consider the continental crust model (R_c = 1–2 Gyr). A major difference between these two case scenarios, however, lies in the preservation of ^{142}Nd heterogeneities in the mantle-crust sys-

tem. In the basaltic crust model, ^{142}Nd anomalies in the mantle and crust become negligible after 3.5 Gyr. Only the early Archean terranes (3.6–3.8 Gyr) would then be expected to show ^{142}Nd effects. In the continental crust model, ^{142}Nd anomalies are preserved over a significantly longer period of time and could be found in rocks as young as 2.5 Gyr. There is currently not enough data to decide the matter. However, the study of late Archean Greenstone Belts could provide important information regarding the lifetime and composition of the early Earth's crust.

There is currently a very limited ^{142}Nd database for late Archean rocks, only including two komatiites and one carbonatite (Figure 1.13), which do not show any ^{142}Nd effect at the ±10 ppm level of precision (Boyet & Carlson, 2006). However, further examination of ^{146}Sm-^{142}Nd systematics in Archean samples is needed in order to better constrain the evolution and composition of the terrestrial protocrust. If confirmed, the lack of ^{142}Nd heterogeneities in the late Archean mantle-crust system could indicate that the Hadean proto-crust was more mafic than the present-day continents, and was characterized by a shorter lifetime (Figure 1.16a). This Hadean proto-crust would have been extensively recycled prior 3 Gyr, thereby making only a marginal contribution to the oldest continental cratons.

^{146}Sm-^{142}Nd systematics of chondrites

Early studies of ^{142,143}Nd systematics in planetary material have relied on two important assumptions regarding the bulk composition of the terrestrial planets. It was first assumed that these planets have perfectly chondritic Sm/Nd and ^{143}Nd/^{144}Nd ratios (i.e. the CHUR reference parameters (Table 1.1)). This general paradigm follows from the observation that chondrites have a homogeneous Sm/Nd ratio (Jacobsen & Wasserburg, 1984; Patchett *et al.*, 2004; Carlson *et al.*, 2007), which is also indistinguishable (within ± 10%) from the composition of the solar photosphere (Asplund *et al.*, 2006). This, in turn, suggests that high-T processes that took place in the solar nebula had a negligible effect on the REE

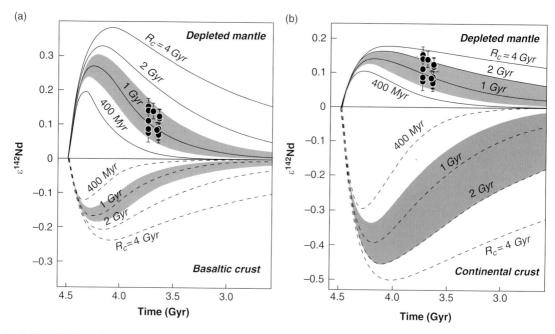

Fig. 1.16 ε^{142}Nd evolution of crustal and mantle reservoirs corresponding to the two-box model schematized in Figure 1.15. The curves are parameterized as a function of the crustal residence time (R_c). The model was investigated for different crustal composition models: (a) basaltic crust and (b) continental crust. Data from Caro *et al.* (2006) and Bennett *et al.* (2007)

Table 1.3 Parameters used for modeling the ^{146}Sm-^{142}Nd evolution of the early depleted mantle and crust. The Aggregate melt fraction (F) and partition coefficients (D_i) for the continental crust model are from Hofmann (1988) and those for the basaltic crust model are from Chase & Patchett (1988). Using these parameters, one can determine the mass fraction of Nd contained in the crustal reservoir (x_c) and the concentration of Sm and Nd in the crust and depleted mantle. The corresponding isotopic evolution of ^{142}Nd/^{144}Nd in the mantle-crust system are shown in Figure 1.16.

	Continental crust model	Basaltic crust model
F	1.6%	5%
D^{Sm}	0.09	0.05
D^{Nd}	0.045	0.03
x_c	27%	64%
$[Nd]_c$ (ppm)	20	15
$[Nd]_m$ (ppm)	0.9	0.5
$(^{147}Sm/^{144}Nd)_c$	0.117	0.1583

composition of the chondrite parent bodies and, by extension, on the composition of the terrestrial planets (Allegre *et al.*, 1979; DePaolo & Wasserburg, 1979; DePaolo, 1980). A second assumption is that the homogeneous ^{142}Nd composition of the modern mantle and crust is representative of that of the bulk silicate Earth, from which it follows that early differentiated crustal and mantle reservoirs must have been extensively remixed over the past 4.5 Gyr (see previous section). These assumptions, however, were recently challenged by high precision studies of meteorites, showing that the chondritic ^{142}Nd/^{144}Nd composition differ from that of all terrestrial samples by ~20 ppm (Figure 1.17a) (Boyet & Carlson, 2005; Andreasen & Sharma, 2006; Carlson *et al.*, 2007). In detail, ordinary chondrites are characterized by a homogeneous composition of −18 ± 5 ppm (2σ, 1 data excluded), whilst carbonaceous chondrites have more negative and also more heterogeneous

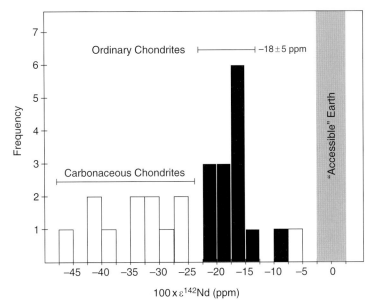

Fig. 1.17 Histogram showing the ^{142}Nd composition of ordinary and carbonaceous chondrites, normalized to a terrestrial standard. Ordinary chondrites display a homogeneous composition of -18 ± 5 ppm (2σ), whilst carbonaceous chondrites have more negative and also more heterogeneous signatures. Data from Boyet & Carlson (2005), Andreasen & Sharma (2006) and Carlson *et al.* (2007).

compositions ranging between -20 and -45 ppm. The difference between O- and C-chondrites can be explained by a heterogeneous distribution of ^{142}Nd and ^{146}Sm in the early solar system, resulting from incomplete mixing of nucleosynthetic products in the solar nebula (Andreasen & Sharma, 2006; Ranen & Jacobsen, 2006; Carlson *et al.*, 2007). This is best illustrated by the anomalous isotopic abundances observed for heavy elements such as Ba, Os, and Sm in C-chondrites (Carlson *et al.*, 2007; Dauphas *et al.*, 2002; Yin *et al.*, 2002a; Brandon *et al.*, 2005; Andreasen & Sharma, 2006; Ranen & Jacobsen, 2006; Reisberg *et al.*, 2009), which define a patterns similar to those observed in 'FUN' inclusion EK-141 (see review by Birck (2004)). In contrast, ordinary chondrites show normal isotopic abundances for these elements, making it more difficult to argue for a nucleosynthetic origin of the negative ^{142}Nd anomalies recorded in these meteorites. It was thus proposed that the ^{142}Nd excess measured in all terrestrial samples compared with ordinary chondrites result from radioactive decay in silicate reservoirs with non-chondritic Sm/Nd ratio, whilst the more negative and more heterogeneous compositions observed in carbonaceous

chondrites also reflect the incorporation of variable amounts of nucleosynthetic material (Andreasen & Sharma, 2006; Carlson *et al.*, 2007). An outstanding question is thus whether the non-chondritic ^{142}Nd composition of terrestrial samples is representative of the bulk Earth composition or only of 'accessible' terrestrial reservoirs (i.e. the uppermost mantle and crust). In the latter scenario, a reservoir 'hidden' in the lower mantle with sub-chondritic ε^{142}Nd composition would be needed to balance the super-chondritic signature of all accessible silicate reservoirs. In contrast, a non-chondritic ^{142}Nd/^{144}Nd signature for the bulk Earth would obviate the need for a hidden reservoir but would contradict the long-held view that the terrestrial planets have perfectly chondritic abundances of refractory lithophile elements.

The hidden reservoir model

The hidden reservoir model was first proposed by Boyet & Carlson (2005). In this model, the authors postulate that the bulk Earth has a perfectly chondritic Sm/Nd and ^{142}Nd/^{144}Nd composition. The more radiogenic ^{142}Nd/^{144}Nd signature of the

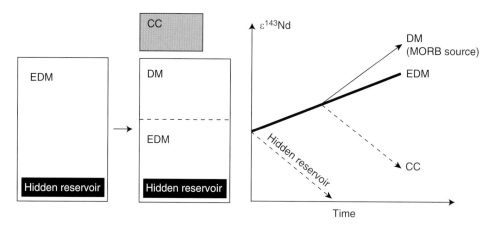

Fig. 1.18 Schematic illustration of the hidden reservoir model and the corresponding ε^{143}Nd evolution of the mantle-crust system. The first stage of differentiation corresponds to the formation of the hidden reservoir and the EDM. This generates heterogeneities in ^{142,143}Nd, with the EDM evolving towards ε^{143}Nd signatures more radiogenic than the chondritic primitive mantle. A second stage of mantle depletion occurs when the continental crust is extracted from the EDM. This further depletes the EDM in incompatible elements, creating a more depleted mantle reservoir (DM = MORB source). As a consequence, the ε^{143}Nd composition of the EDM is necessarily lower than that of the upper mantle.

Earth's 'accessible' mantle and crust must then be balanced by an enriched complementary reservoir with sub-chondritic Sm/Nd and ^{142}Nd/^{144}Nd ratios (Figure 1.18). In order to satisfy ^{146}Sm-^{142}Nd systematics, this hidden reservoir must have formed early (>4.5 Gyr) and remained perfectly isolated from mantle convection since its formation. The most plausible location for storing such a large reservoir is the D'' layer (Tolstikhin et al., 2006) and it was therefore suggested that this hypothetical reservoir could represent a recycled crustal component (Boyet & Carlson, 2005, 2006) or a basal magma ocean (Labrosse et al., 2007), which remained stored at the core-mantle boundary (CMB) owing to a large density contrast with the surrounding mantle. This hypothesis is supported by the presence of positive ^{142}Nd anomaly in early Archean rocks (see previous section), demonstrating the early formation and recycling of a Hadean protocrust (Caro et al., 2003, 2006; Boyet & Carlson, 2005). It was also shown that a dense basaltic crust recycled in the deep mantle would be gravitationally stable, although it is doubtful that such a reservoir would remain

unsampled by mantle convection (Davaille et al., 2002; Bourdon & Caro, 2007).

A major obstacle to the hidden reservoir model lies in the difficulty of establishing a plausible chronology for the formation of this reservoir (Bourdon et al., 2008). This chronology must satisfy coupled ^{146,147}Sm-^{142,143}Nd systematics but should also be consistent with the constraints derived from ^{182}Hf-^{182}W on the timescale of terrestrial accretion and core formation (see the section on 'Hf-W chronology of the accretion and early differentiation of the Earth and Moon'). It is evident that the isolation of a long-lived silicate reservoir at the CMB requires that core formation was complete, which essentially precludes its formation prior to 30 to 35 Myr. It is also unlikely that an early crust could survive the giant impact, and this would delay the formation of long-lived crustal reservoirs on the Earth by >50 Myr after formation of the solar system (Touboul et al., 2007). An additional constraint is that the EDM should have lower Sm/Nd and ^{143}Nd/^{144}Nd ratios than the modern depleted mantle (ε^{143}Nd = +9 ε-units (Salters & Stracke, 2004)), as the latter was further depleted by continental growth (Figure 1.18b). In theory, ^{142}Nd anomalies can be generated until ~4.2 Gyr. However, late formation of the hidden reservoir would require a large degree of Sm/Nd fractionation, which in turn would generate ε^{143}Nd value in the mantle higher than observed in the modern MORB source (Boyet & Carlson, 2005). This constraint puts stringent limits on the minimum age of a hidden reservoir.

A major difficulty in establishing an accurate chronology for the hidden reservoir is to estimate a realistic ε^{143}Nd composition for the EDM. Boyet & Carlson (2005, 2006) estimated a model age of 4.53 ± 0.03 Myr by conservatively assuming that the EDM composition is identical or lower than that of the MORB source. However, Bourdon *et al.* (2008) pointed out that the effect of continental extraction cannot be neglected, as it would further increase the ε^{143}Nd composition of the upper mantle. A more realistic age determination for the hidden reservoir can be obtained in the context of the 4-reservoirs model illustrated in Figure 1.18. The composition of the EDM can be obtained from the following mass balance relationship:

$$\left(\frac{^{143}Nd}{^{144}Nd}\right)_{edm} = \left(\frac{^{143}Nd}{^{144}Nd}\right)_{cc} \varphi_{cc} + \left(\frac{^{143}Nd}{^{144}Nd}\right)_{dm} \varphi_{dm} \quad (23)$$

where:

$$\varphi_{cc} = \frac{[Nd]_{cc}}{[Nd]_{edm}} \frac{M_{cc}}{(M_{cc} + M_{dm})} = 1 - \varphi_{dm} \quad (24)$$

M_{cc} is the mass of the present-day continental crust ($M_{cc} = 2.27 \ 10^{22}$ kg), whilst M_{dm} is the mass of mantle depleted by extraction of the continental crust. $[Nd]_{cc}$ and $[Nd]_{edm}$ are the Nd concentration in the continental crust ($[Nd]_{cc} = 20$ ppm; Rudnick & Fountain, 1995) and the early depleted mantle (EDM). As illustrated in Figure 1.18, the early depleted mantle represents a residual mantle after extraction of the hidden reservoir, whilst the modern depleted mantle (DM) is formed by extraction of the continental crust from the EDM. As a consequence, the DM is more depleted and has higher ^{147}Sm/^{144}Nd and ^{143}Nd/^{144}Nd ratios than

the EDM. The mass of the DM is unconstrained but a conservative model can be established by assuming that the continental crust was extracted from the whole mantle (i.e. $M_{dm} \sim M_{edm} \sim M_{bse}$), which minimizes its impact on the mantle ε^{143}Nd composition. The Nd concentration in the EDM is also unknown but its value is necessarily lower than that of the primitive mantle (i.e. $[Nd]_{edm}$ <1.25 ppm). Using these very conservative parameters, the maximum ^{143}Nd/^{144}Nd composition of the early depleted mantle is estimated to be <0.5130 (i.e. ε^{143}Nd<7 ε-units) for $[Nd]_{edm}$ <1.25 ppm and decreases to <6.5 ε-units for a more realistic Nd concentration in the EDM of 1 ppm (Figure 1.19). The minimum age for the hidden reservoir can then be estimated to 4.57 ± 0.03 Gyr using coupled ^{142,143}Nd chronometry (Figure 1.20). Thus, if we assume that the isotopic shift between terrestrial and meteoritic samples result from radioactive decay, then the Sm/Nd fractionation responsible for this isotopic effect must have taken place within a few million years after formation of the solar system. A corollary is that the hypothetical 'missing' reservoir with subchondritic Sm/Nd and ^{142}Nd/^{144}Nd ratios would need to be older than the core, making it unlikely to result from an internal differentiation process such as magma ocean crystallization and/or crustal formation.

The ^{142}Nd results obtained from meteorites thus open a new perspective, namely that the REE (and by extension, the refractory elements) experienced minor but significant fractionation in the accretion disk, which resulted in the observed non-chondritic ^{142}Nd composition for the bulk Earth. Of course, this would not preclude the preservation of Hadean crust in the deep mantle or in ancient continental roots, but would relax the constraints on the age and size of such residual reservoir, making it easier to account for the homogeneous composition of the modern mantle. Thus, whilst ^{146}Sm-^{142}Nd systematics has been classically interpreted in terms of internal differentiation processes, the possibility that the terrestrial planets accreted from material that was not perfectly chondritic for Sm/Nd must also be considered.

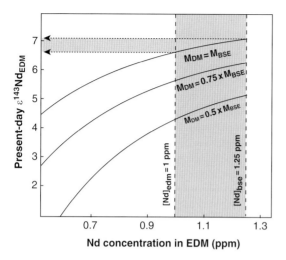

Fig. 1.19 Nd isotope mass balance showing possible $\varepsilon^{143}Nd$ compositions for the early depleted mantle (EDM) complementary to the hidden reservoir. This estimate depends on: i) the Nd concentration in the EDM after extraction of the hidden reservoir; and ii) the mass fraction of depleted mantle (i.e. MORB source) complementary to the continental crust. Both parameters are unconstrained but a maximum value for $\varepsilon^{143}Nd_{EDM}$ can be obtained assuming that the continental crust was extracted from the whole mantle (i.e. $M_{dm} = M_{bse}$) and that the Nd concentration in the EDM is lower than that of the primitive mantle ($Nd_{edm} < 1.25\,ppm$). Using these conservative parameters, a maximum $\varepsilon^{143}Nd_{edm}$ of +7 is obtained.

The non-chondritic Earth model

The non-chondritic Earth hypothesis had initially been discarded based on the generally accepted view that refractory lithophile elements are present in perfectly chondritic abundances in the terrestrial planets. Whilst this is probably a correct approximation within ±10%, the homogeneity of bulk chondrites should not obscure the fact that the REE did experience substantial fractionation in the solar nebula. The constitutive components of chondrites (chondrules, CAIs) show fractionated REE pattern (Figure 1.21) (Ireland et al., 1988; Krestina et al., 1997, 1999; Amelin & Rotenberg, 2004), indicating that the material incorporated in chondrite parent bodies

was not perfectly homogeneous for Sm/Nd. This also demonstrates that the refractory nature of the REE is by no means evidence that these elements did not fractionate in the solar nebula. It should also be noted that the terrestrial planets did not accrete from chondritic material but from differentiated planetesimals (Chambers, 2001). Thus, very early fractionation of refractory lithophile elements could have been produced during the first million years by preferential loss of the planetesimal's crusts during impacts (O'Neill & Palme, 2008) or, alternatively, by explosive volcanism (Warren, 2008), and would thus not necessarily involve volatility-controlled processes.

The magnitude of Sm/Nd fractionation needed to produce an $18 \pm 5\,ppm$ ^{142}Nd excess can be estimated from Equation (11). This yields a $^{147}Sm/^{144}Nd$ value of 0.209 ± 3, which corresponds to a 4 to 7% fractionation compared with the chondritic value. In contrast, the total range of $^{147}Sm/^{144}Nd$ measured in bulk chondrites is ~4% with most values ranging between 0.19 and 0.20 (Jacobsen & Wasserburg, 1984; Patchett et al., 2004; Boyet & Carlson, 2005; Carlson et al., 2007). The Sm/Nd composition required to generate an 18 ppm excess in all terrestrial samples is therefore slightly higher than the average chondritic composition but remain well within the fractionation range observed in chondrite components (Figure 1.21).

A critical test for the non-chondritic Earth model comes from the examination of ^{142}Nd systematics in lunar and Martian samples. It was indeed expected that if the Earth inherited a non-chondritic Sm/Nd composition during its accretion, then the Moon (and possibly Mars) may also be characterized by superchondritic Sm/Nd and $^{142}Nd/^{144}Nd$ ratios. If, on the other hand, these planets formed from chondritic material, then planetary isochrons for Mars and the Moon should have a common intersect corresponding to the chondritic value (i.e. $\varepsilon^{142}Nd = -18\,ppm$, $^{147}Sm/^{144}Nd$ = 0.1966). As illustrated in Figures 1.22a and b, lunar and Martian samples define correlations in an $\varepsilon^{142}Nd$ vs Sm/Nd diagram. If interpreted as an isochron relationship, the array defined by

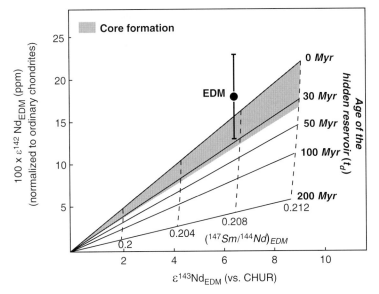

Fig. 1.20 Estimation of the age of the hidden reservoir using coupled [146,147]Sm-[142,143]Nd systematics. The EDM is constrained to have an ε^{142}Nd excess of 18 ± 5 ppm compared with ordinary chondrites. The [143]Nd composition of the EDM is necessarily lower than +7 ε-units (Figure 1.19). This constrains the youngest possible age for the hidden reservoir to 4.57 ± 0.03 Gyr.

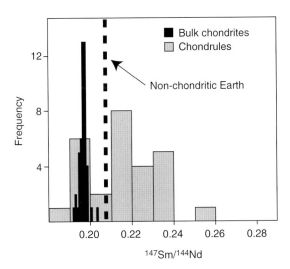

Fig. 1.21 Histogram showing the distribution of Sm/Nd ratios in bulk chondrites and chondrules. Note that chondrule composition is systematically biased towards higher Sm/Nd values. A hypothetical bulk Earth composition consistent with the observed [142]Nd excess of +18 ppm is shown for comparison. (Jacobsen & Wasserburg, 1984; Krestina *et al.*, 1997, 1999; Amelin & Rotenberg, 2004; Patchett *et al.*, 2004; Carlson *et al.*, 2007).

Shergottites yields an age of differentiation for the Martian mantle of 4.52 ± 0.02 Gyr (Figure 1.22a) (Foley *et al.*, 2005a; Debaille *et al.*, 2007; Caro *et al.*, 2008). This is consistent with the presence of small ε^{182}W anomalies in Martian meteorites, as these can only be produced prior to 4.5 Gyr. The [142]Nd systematics of lunar samples and their chronological significance is more ambiguous and is currently a matter of debate. The first attempt to study the [142]Nd composition of lunar basalts showed a well-defined correlation which, if interpreted as an isochron, would yield an age of differentiation of 4.33 ± 0.04 Gyr for the lunar magma ocean (Nyquist *et al.*, 1995; Rankenburg *et al.*, 2006; Boyet & Carlson, 2007) (Figure 1.22b). This general trend has been confirmed using higher precision mass spectrometric techniques (Rankenburg *et al.*, 2006; Boyet & Carlson, 2007; Touboul *et al.*, 2009a). However, the chronological significance of this array has been put into question (Boyet & Carlson, 2007), as the estimated age of differentiation postdates the formation of the oldest lunar anorthosites by >100 Myr (Carlson & Lugmair, 1988). A possible interpretation is that the primordial isochron

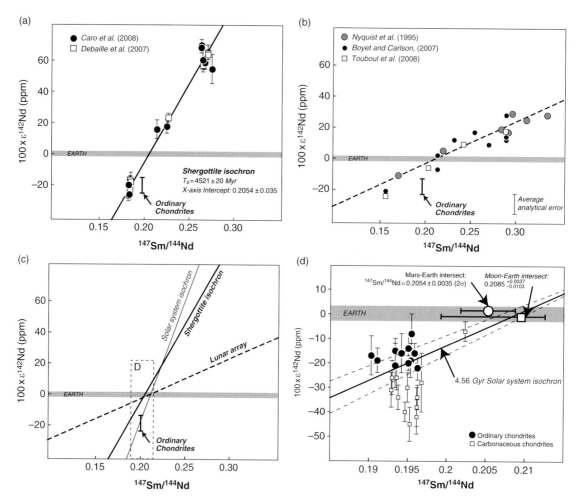

Fig. 1.22 ^{146,147}Sm-^{142,143}Nd arrays for Mars and the Moon compared with terrestrial and chondritic compositions. (a) The isochron relationship defined by Shergottites yields an age of differentiation of 4.52 ± 0.02 Gyr, reflecting crystallization of the Martian magma ocean near the end of accretion. This isochron passes above the chondritic reference value, suggesting that the Martian reservoirs sampled by Shergottites differentiated from a slightly super-chondritic mantle. (b) Lunar array defined by mare basalts (Hi-Ti and Low-Ti) and KREEP basalts. Data with large neutron corrections (>10 ppm) are excluded from the regression. (c) Planetary isochrons obtained from lunar, Martian and terrestrial samples converge towards a common value corresponding to ε^{142}Nd ≈ 0 and ^{147}Sm/^{144}Nd ≈ 0.21. (d) The intersects between the lunar and Martian arrays and the terrestrial composition plot within errors on a solar system isochron passing through the O-chondrite composition. This suggests that Mars and the Earth-Moon system developed a slightly fractionated Sm/Nd composition at a very early stage of planetary accretion.

relationship was partially reset by remelting and mixing of magma ocean cumulates, resulting from gravitational instabilities (Bourdon *et al.*, 2008). Regardless of the chronological signifi-

cance of the lunar array, it is evident that most lunar samples have a superchondritic ^{142}Nd composition, roughly centered on the terrestrial value. This is a crucial observation showing that the super-chondritic ^{142}Nd signature observed in all terrestrial samples is probably not restricted to the uppermost mantle and crust, but more likely represents the isotopic signature of the entire Earth-Moon system.

As illustrated in Figure 1.22c, a more detailed examination of planetary data reveals that lunar and Martian isochrons intersect at an ε^{142}Nd composition of +20 ppm compared with ordinary chondrites, and thus define a triple intersect with the terrestrial composition. In detail, the intersect between the Martian isochron and the terrestrial value corresponds to a slightly super-chondritic ^{147}Sm/^{144}Nd ratio of 0.2054 ± 0.0035, whilst the intersect defined by the lunar isochron corresponds to a similar but less precise value of 0.2085 (Figure 1.22d). Both plot within error on a 4.56 Gyr solar system isochron passing through the chondritic composition, which is precisely what would be expected if all three planets had accreted from material with super-chondritic Sm/Nd ratio. Of course, these compositions remain hypothetical as Mars and the Moon may not necessarily have perfectly 'terrestrial' ^{142}Nd/^{144}Nd signatures (Bourdon *et al.*, 2008). Despite this uncertainty, the seemingly super-chondritic ^{142}Nd compositions of Mars, the Earth and the Moon challenge the conventional view that terrestrial planets have rigorously chondritic abundances of refractory lithophile elements. Future research will thus need to investigate possible mechanisms for refractory elements fractionation during planetary accretion and their consequences in terms of planetary composition. A revision of the Sm-Nd parameters for the bulk Earth would, for example, require a revision of the Lu/Hf, Rb/Sr and possibly Th/U ratios of the bulk silicate Earth, which would modify our perspective on the isotopic systematics in mantle derived rocks. As illustrated in Figure 1.12b, a bulk silicate Earth with Sm/Nd ratio ~6% higher than chondritic would provide a good match to the ^{143}Nd evolution of the mantle prior to 2.5 Gyr, suggesting that the Archean mantle was less depleted than previously thought. The presence of a large continental volume prior to 3 Gyr would thus no longer be required to explain the ^{143}Nd signal in Archean rocks, thereby resolving the long-held paradox of the seemingly depleted signature of the Archean mantle, despite the lack of a large stable continental mass of that age.

CONCLUSIONS

The Earth's mantle exhibits an elevated Hf/W ratio and a small but well resolvable ^{182}W excess relative to chondrites, demonstrating that at least parts of the Earth's core segregated during the lifetime of ^{182}Hf. The ^{182}W excess of the Earth's mantle is about one order of magnitude smaller than excesses found in the silicate part of some of the oldest planetary objects (the parent bodies of differentiated meteorites). This observation is difficult to explain unless formation of the Earth's core was protracted and involved some re-equilibration of newly accreted metal cores within the terrestrial magma ocean. The two-stage model age of the Earth's core of ~30 Myr assumes that the core formed in a single instant at the end of accretion and as such is not appropriate for modeling the continuous segregation of the Earth's core during protracted accretion. The two-stage model age, nevertheless, is useful because it provides the earliest time at which core formation can have been complete.

Determining the rate of the Earth's accretion and timing of core formation based on the ^{182}W excess of the Earth's mantle requires knowledge of a number of parameters, including the degree to which the metal cores of newly accreted bodies equilibrated with the Earth's mantle and the timescales and conditions of core formation of the material that accreted to form the Earth. Our incomplete understanding of these processes makes determining an Hf-W age of the Earth's core uncertain. Additional constraints on the rate of the Earth's accretion are required and are provided by the age of the Moon. The indistinguishable W isotopic compositions of the bulk silicate Moon and Earth in conjunction with their different Hf/W ratios require that the Moon formed late, >50 Myr after solar system formation. The late age of the Moon reveals that the Hf-W systematics of the Earth's mantle can be accounted for in two types of models. Either the Earth's accretion took place by collisions that occurred at a slowly decreasing exponential rate with a mean-life of ~40 Myr, or an initial rapid growth was followed by a long hiatus before the final, late Moon-forming impact. A better understanding of the physics and chemistry involved in metal segregation during large impacts may help distinguishing between these two scenarios.

Ubiquitous ^{142}Nd anomalies in early Archean rocks from West Greenland witness an episode of mantle-crust differentiation 30 to 100 Myr after formation of the solar system. This age constraint is identical within errors to the age of differentiation of the Martian mantle, suggesting that both planets experienced early differentiation following magma ocean crystallization. The preservation of positive ^{142}Nd anomalies in the mantle until 3.6 Gyr requires that the Hadean crust was a long-lived reservoir with a lifetime of ~1 Gyr. The lack of ^{142}Nd heterogeneities in the late Archean mantle, if confirmed, would suggest that this early crust was compositionally more mafic than modern continents and was extensively recycled prior to the major episode of crustal growth 2.5 to 3 Gyr ago.

The radiogenic excess of ~20 ppm measured in all 'accessible' terrestrial reservoirs suggests that part or all of the Earth's mantle evolved with an Sm/Nd ratio 4 to 7% higher than chondritic for at least 4.53 Gyr. This early fractionation thus occurred before the Earth's core had completely segregated and also prior to the giant Moon-forming impact. This is difficult to explain unless the non-chondritic Sm/Nd composition of the Earth was inherited from accreting material (most likely the planetary embryos) rather than generated at a later stage by internal differentiation. As a consequence, the assumption of a chondritic Hf/W ratio of the bulk Earth may not be valid. A non-chondritic Earth would have a Hf/W ratio ~1.5 times higher than chondritic and also the Hf/W ratio of the bulk silicate Earth, which is estimated assuming a chondritic ratios of refractory lithophile elements, would be ~1.5 times higher (Kleine et al., 2009). For a non-chondritic Earth, Hf-W ages become younger compared to a

chondritic Earth (e.g. the two-stage model age is ~40 Myr instead of ~30 Myr) but given the uncertainties inherent in Hf-W models for the Earth's accretion and core formation, this does not affect the main conclusions drawn here.

Martian and lunar samples define linear arrays in an ε^{142}Nd vs Sm/Nd plot, which can be interpreted as planetary isochrons dating the crystallization of magma oceans on these planets. The Martian array yields an age of differentiation of 4.52 Gyr, in agreement with the presence of ^{182}W anomalies in these samples. The lunar array, on the other hand, yields a surprisingly young age of 4.35 Gyr. The chronological significance of this trend is uncertain as it could represent a mixing relationship between magma ocean cumulates rather than a true isochron. An important observation, however, is that the Martian and lunar arrays do not intersect the chondritic reference composition, as would be expected if both planets had chondritic Sm/Nd and ^{142}Nd/^{144}Nd ratios. These arrays define a common intersect with the terrestrial value (at ε^{142}Nd = 0 and ^{147}Sm/^{144}Nd = 0.208), defining a plausible bulk composition for Mars and the Earth-Moon system. This observation supports the view that refractory lithophile elements in all the terrestrial planets may have experienced minor fractionation during accretion, which ultimately resulted in the super-chondritic ^{142}Nd/^{144}Nd signature of the Earth's mantle.

ACKNOWLEDGMENTS

Discussions with B. Bourdon, M. Touboul, B. Marty and many other colleagues are gratefully acknowledged. The paper benefited from constructive reviews by V. Bennett and an anonymous reviewer. Informal review of the manuscript by J. Marin was greatly appreciated.

REFERENCES

Abe Y. 1997. Thermal and chemical evolution of the terrestrial magma ocean. *Physics of the Earth and Planetary Interiors* **100**: 27–39.

Agnor CB, Canup RM, Levison, HF. 1999. On the character and consequences of large impacts in the late stage of terrestrial planet formation. *Icarus* **142**: 219–237.

Albarede F. 1995. *Introduction to Geochemical Modeling*. Cambridge University Press, Cambridge.

Albarede F. 2001. Radiogenic ingrowth in systems with multiple reservoirs: applications to the differentiation of the crust-mantle system. *Earth and Planetary Science Letters* **189**: 59–73.

Albarede F, Rouxel, M. 1987. The Sm/Nd secular evolution of the continental crust and the depleted mantle. *Earth and Planetary Science Letters* **82**: 25–35.

Allegre CJ, Othman DB, Polve M, Richard D. 1979. Nd-Sr isotopic correlation in mantle materials and geodynamic consequences. *Physics of the Earth and Planetary Interiors* **19**: 293–306.

Amelin Y, Rotenberg, E. 2004. Sm-Nd systematics of chondrites. *Earth and Planetary Science Letters* **223**: 267–282.

Amelin Y, Krot AN, Hutcheon ID, Ulyanov AA. 2002. Lead isotopic ages of chondrules and calcium-aluminum-rich inclusions. *Science* **297**: 1679–1683.

Andreasen R, Sharma M. 2006. Solar nebula heterogeneity in p-process samarium and neodymium isotopes. *Science* **314**: 806–809.

Armstrong RL. 1981. The case for crustal recycling on a near-steady-state no-continental-growth Earth. *Philosophical Transactions of the Royal Society* **A301**: 443–472.

Asplund M, Grevesse N, Sauval AJ. 2006. The solar chemical composition. *Nuclear Physics A* **777**: 1–4.

Bennett VC. 2003. Compositional evolution of the mantle. In: *The Mantle and Core*, Carlson RW (ed.), Vol. 2, *Treatise on Geochemistry*, Holland HD, Turekian KK (eds), Elsevier-Pergamon, Oxford.

Bennett VC, Nutman AP, McCulloch MT. 1993. Nd isotopic evidence for transient, highly depleted mantle reservoirs in the early history of the Earth. *Earth and Planetary Science Letters* **119**: 299–317.

Bennett VC, Brandon AD, Nutman AP. 2007. Coupled [142]Nd-[143]Nd isotopic evidence for Hadean mantle dynamics. *Science* **318**: 1907–1910.

Birck J-L. 2004. An overview of isotopic anomalies in extraterrestrial materials and their nucleosynthetic heritage. In: *Geochemistry of Non-traditional Stable Isotopes*. Johnson, C-M, Beard BL, Albarede F. (eds), Mineralogical Society of America, Geochemical Society.

Bourdon B, Caro G. 2007. The early terrestrial crust. *C. R. Geosciences* **339**: 928–936.

Bourdon B, Touboul M, Caro G, Kleine T. 2008. Early differentiation of the Earth and the Moon. *Philosphical Transactions of the Royal Society* **366**: 4105–4128.

Bouvier A, Blichert-Toft J, Moynier F, Vervoort JD, Albarède F. 2007. Pb-Pb dating constraints on the accretion and cooling history of chondrites. *Geochimica et Cosmochimica Acta* **71**: 1583–1604.

Bowring SA, Housh T. 1995. The Earth's early evolution. *Science* **269**: 1535–1540.

Boyet M, Blichert-Toft J, Rosing M, Storey CD, Télouk M, Albarede F. 2004. [142]Nd evidence for early Earth differentiation. *Earth and Planetary Science Letters* **214**: 427–442.

Boyet M, Carlson RW. 2006. A new geochemical model for the Earth's mantle inferred from [146]Sm-[142]Nd systematics. *Earth and Planetary Science Letters* **250**: 254–268.

Boyet M, Carlson RW. 2007. A highly depleted moon or a non-magma ocean origin for the lunar crust? *Earth and Planetary Science Letters* **262**: 505–516.

Boyet M, Carlson RW, 2005. [142]Nd evidence for early (>4.53 Ga) global differentiation of the silicate Earth. *Science* **214**: 427–442.

Boynton WV. 1975. Fractionation in the solar nebula: Condensation of yttrium and the rare earth elements. *Geochimica et Cosmochimica Acta* **39**: 569–584.

Brandon AD, Humayun M, Puchtel IS, Leya I, Zolensky M. 2005. Osmium isotope evidence for an s-process carrier in primitive chondrites. *Nature* **309**: 1233–1236.

Burkhardt C, Kleine T, Palme H, *et al*. 2008. Hf-W mineral isochron for Ca,Al-rich inclusions: Age of the solar system and the timing of core formation in planetesimals. *Geochimica et Cosmochimica Acta* **72**: 6177–6197.

Canup RM. 2004. Dynamics of lunar formation. *Annual Review of Astronomy and Astrophysics* **42**: 441–475.

Canup RM, Asphaug E. 2001. Origin of the Moon in a giant impact near the end of the Earth's formation. *Nature* **412**: 708–712.

Carlson RW, Lugmair GW. 1988. The age of ferroan anorthosite 60025: oldest crust on a young Moon? *Earth and Planetary Science Letters* **90**: 119–130.

Carlson RW, Boyet M, Horan M. 2007. Chondrite barium, neodymium and samarium isotopic heterogeneity and early earth differentiation. *Science* **316**: 1175–1178.

Caro G, Bourdon B, Birck J-L, Moorbath S. 2003. [146]Sm-[142]Nd evidence for early differentiation of the Earth's mantle. *Nature* **423**: 428–432.

Caro G, Bourdon B, Wood BJ, Corgne A. 2005. Trace element fractionation generated by melt segregation from a magma ocean. *Nature* **436**: 246–249.

Caro G, Bourdon B, Birck.-L, Moorbath S. 2006. High-precision ^{142}Nd/^{144}Nd measurements in terrestrial rocks: Constraints on the early differentiation of the Earth's mantle. *Geochimica et Cosmochimica Acta* **70**: 164–191.

Caro G, Bourdon B, Halliday AN, Quitté G. 2008. Superchondritic Sm/Nd ratios in Mars, the Earth and the Moon. *Nature* **452**: 336–339.

Chambers JE. 2001. Making more terrestrial planets. *Icarus* **152**: 205–224.

Chambers JE. 2004. Planetary accretion in the inner solar system. *Earth and Planetary Science Letters* **223**: 241–252.

Chase CG, Patchett PJ. 1988. Stored mafic/ultramafic crust and early Archean mantle depletion. *Earth and Planetary Science Letters* **91**: 66–72.

Christensen UR, Hofmann AW. 1994. Segregation of subducted oceanic crust in the convecting mantle. *Journal of Geophysical Research* **99**: 19,867–19,884.

Condie KC. 2000. Episodic growth models: after thoughts and extensions. *Tectonophysics* **322**: 153–162.

Cottrell E, Walter M.J, Walker D. 2009. Metal-silicate partitioning of tungsten at high pressure and temperature: Implications for equilibrium core formation in Earth. *Earth and Planetary Science Letters* **281**: 275–287.

Dauphas N, Marty B, Reisberg L. 2002. Inference on terrestrial genesis from molybdenum isotope systematics. *Geophysical Research Letters* **29**: 1084.

Davaille A, Girard F, Le Bars M. 2002. How to anchor hotspots in a convective mantle? *Earth and Planetary Science Letters* **203**: 621–634.

Davis AM, Grossman L. 1979. Condensation and fractionation of rare earths in the solar nebula. *Geochimica et Cosmochimica Acta* **43**: 1611–1632.

Debaille V, Brandon AD, Yin QZ, Jacobsen, B. 2007. Coupled ^{142}Nd-^{143}Nd evidence for a protracted magma ocean in Mars. *Nature* **450**: 525–528.

DePaolo DJ. 1980. Crustal growth and mantle evolution: inferences from models of element transport and Sr and Nd isotopes. *Geochimica et Cosmochimica Acta* **44**: 1185–1196.

DePaolo DJ, Wasserburg GJ. 1979. Petrogenetic mixing models and Nd-Sr isotopic patterns. *Geochimica et Cosmochimica Acta* **43**: 615–627.

Foley CN, Wadhwa M, Borg LE, Janney PE, Hines R, Grove TL. 2005a. The early differentiation history of Mars from W-182 Nd-142 isotope systematics in the SNC meteorites. *Geochimica et Cosmochimica Acta* **69**: 4557–4571.

Foley CN, Wadhwa M, Borg LE, Janney PE, Hines R, Grove TL. 2005b. The early differentiation history of Mars from ^{182}W-^{142}Nd isotope systematics in the SNC meteorites. *Geochimica et Cosmochimica Acta* **69**: 4557–4571.

Galer SJG, Goldstein SL. 1991. Early mantle differentiation and its thermal consequences. *Geochimica et Cosmochimica Acta* **55**: 227–239.

Goldstein SL, Galer SJG. 1992. On the trail of early mantle differentiation: ^{142}Nd/^{144}Nd ratios of early Archean rocks. *Eos Transactions AGU* **73**: S323.

Halliday AN. 2004. Mixing, volatile loss and compositional change during impact-driven accretion of the Earth. *Nature* **427**: 505–509.

Halliday AN. 2008. A young Moon-forming giant impact at 70–110 million years acompanied by late-stage mixing, core formation and degassing of the Earth. *Philosophical Transactions of the Royal Society* **366**: 4205–4252.

Halliday A, Rehkämper M, Lee DC, Yi W. 1996. Early evolution of the Earth and Moon: New constraints from Hf-W isotope geochemistry. *Earth and Planetary Science Letters* **142**: 75–89.

Harper CL, Jacobsen SB. 1992. Evidence from coupled ^{147}Sm-^{143}Nd and ^{146}Sm-^{142}Nd systematics for very early (4.5 Gyr) differentiation of the Earth's mantle. *Nature* **360**: 728–732.

Harper CL, Jacobsen SB. 1996. Evidence for ^{182}Hf in the early solar system and constraints on the timescale of terrestrial accretion and core formation. *Geochimica et Cosmochimica Acta* **60**: 1131–1153.

Hofmann AW. 1988. Chemical differentiation of the Earth: the relationship between mantle, continental crust, and oceanic crust. *Earth and Planetary Science Letters* **90**: 297–314.

Horan MF, Smoliar MI, Walker RJ. 1998. ^{182}W and ^{187}Re-^{187}Os systematics of iron meteorites: Chronology for melting, differentiation, and crystallization in asteroids. *Geochimica et Cosmochimica Acta* **62**: 545–554.

Ireland TR, Fahey AJ, Zinner EK, 1988. Trace element abundances in hibonites from the Murchison carbonaceous chondrite: constraints on high-temperature processes in the solar nebula. *Geochimica et Cosmochimica Acta* **52**: 2841–2854.

Jacobsen SB. 1988. Isotopic constraints on crustal growth and recycling. *Earth and Planetary Science Letters* **90**: 315–329.

Jacobsen SB. 2005. The Hf-W isotopic system and the origin of the Earth and Moon. *Annual Review of Earth and Planetary Sciences* 33: 531–570.

Jacobsen B, Wasserburg GJ. 1979. The mean age of mantle and crustal reservoirs. *Journal of Geophysical Research* 84: 7411–7424.

Jacobsen SB, Wasserburg GJ. 1984. Sm-Nd evolution of chondrites and achondrites, II. *Earth and Planetary Science Letters* 67: 137–150.

Kleine T, Münker C, Mezger K, Palme H. 2002. Rapid accretion and early core formation on asteroids and the terrestrial planets from Hf-W chronometry. *Nature* 418: 952–955.

Kleine T, Mezger K, Munker C, Palme H, Bischoff A. 2004a. ^{182}Hf-^{182}W isotope systematics of chondrites, eucrites, and Martian meteorites: Chronology of core formation and early differentiation in Vesta and Mars. *Geochimica et Cosmochimica Acta* 68: 2935–2946.

Kleine T, Mezger, K, Palme H, Scherer E, Münker C. 2004b. The W isotope evolution of the bulk silicate Earth: constraints on the timing and mechanisms of core formation and accretion. *Earth and Planetary Science Letters* 228: 109–123.

Kleine T, Mezger K, Palme H, Scherer E, Münker C. 2005a. Early core formation in asteroids and late accretion of chondrite parent bodies: Evidence from ^{182}Hf-^{182}W in CAIs, metal-rich chondrites and iron meteorites. *Geochimica et Cosmochimica Acta* 69: 5805–5818.

Kleine T, Palme H, Mezger K, Halliday AN. 2005b. Hf-W chronometry of lunar metals and the age and early differentiation of the Moon. *Science* 310: 1671–1674.

Kleine T, Touboul M, Bourdon B, *et al.* 2009. Hf-W chronology of the accretion and early evolution of asteroids and terrestrial planets. *Geochimica et Cosmochimica Acta* 73: 5150–5188.

Krestina N, Jagoutz E, Kurat G. 1997. Sm-Nd system in chondrules from Bjurbole L4 chondrite. *Proceedings of Lunar and Planetary Science XXVIII*, 1659.

Krestina N, Jagoutz E, Kurat G. 1999. The interrelation between core and rim of individual chondrules from the different meteorites in term of Sm-Nd isotopic system. *Proceedings of Lunar and Planetary Science XXX*, 1918.

Labrosse S, Herlund JW, Coltice N. 2007. A crystallizing dense magma ocean at the base of the Earth's mantle. *Nature* 450: 866–869.

Lee DC, Halliday AN. 1995. Hafnium-tungsten chronometry and the timing of terrestrial core formation. *Nature* 378: 771–774.

Lee DC, Halliday AN. 1997. Core formation on Mars and differentiated asteroids. *Nature* 388: 854–857.

Lee DC, Halliday AN, Snyder GA, Taylor LA. 1997. Age and origin of the moon. *Science* 278: 1098–1103.

Lodders K, Fegley B. 1993. Lanthanide and actinide chemistry at high C/O ratios in the solar nebula. *Earth and Planetary Science Letters* 117: 125–145.

Lugmair GW, Marti K. 1977. Sm-Nd-Pu timepieces in the Angra Dos Reis meteorite. *Earth and Planetary Science Letters* 35: 273–284.

Lugmair GW, Scheinin NB, Marti K. 1983. Samarium-146 in the early solar system: Evidence from Neodymium in the Allende meteorite. *Science* 222: 1015–1018.

Lugmair GW, Galer SJG.,1992. Age and isotopic relationships among the angrite Lewis Cliff 86010 and Angra Dos Reis. *Geochimica et Cosmochimioca Acta* 56: 1673–1694.

Markowski A, Quitté G, Halliday AN, Kleine T. 2006. Tungsten isotopic compositions of iron meteorites: chronological constraints vs cosmogenic effects. *Earth and Planetary Science Letters* 242: 1–15.

Markowski A, Quitté G, Kleine T, Halliday A, Bizzarro M, Irving AJ. 2007. Hf-W chronometry of angrites and the earliest evolution of planetary bodies. *Earth and Planetary Science Letters* 262: 214–229.

McCulloch MT, Bennett VC. 1993. Evolution of the early Earth: constraints from ^{143}Nd-^{142}Nd isotopic systematics. *Lithos* 30: 237–255.

McDonough WF, Sun S-S. 1995. The composition of the Earth. *Chemical Geology* 120: 223–253.

Moorbath S, Whitehouse MJ, Kamber BS. 1997. Extreme Nd-isotope heterogeneity in the early Archaean – Fact of fiction? Case hostories from northern Canada and West Greenland. *Chemical Geology* 135: 213–231.

Newsom HE, Sims KWW, Noll P, Jaeger W, Maehr S, Beserra T. 1996. The depletion of W in the bulk silicate Earth: constraints on core formation. *Geochimica et Cosmochimica Acta* 60: 1155–1169.

Nimmo F, Agnor CB. 2006. Isotopic outcomes of N-body accretion simulations: Constraints on equilibration processes during large impacts from Hf/W observations. *Earth and Planetary Science Letters* 243: 26–43.

Nimmo F, Kleine T. 2007. How rapidly did Mars accrete? Uncertainties in the Hf-W timing of core formation. *Icarus* 191: 497–504.

Norman MD, Borg LE, Nyquist LE, Bogard DD. 2003. Chronology, geochemistry, and petrology of a ferroan noritic anorthosite clast from Descartes breccia

67215: Clues to the age, origin, structure, and impact history of the lunar crust. *Meteoritics and Planetary Science* **38**: 645–661.

Nyquist LE, Bansal B, Wiesmann H, Shih C-Y. 1994. Neodymium, Strontium and Chromium isotopic studies of the LEW86010 and Angra Dos Reis meteorites and the chronology of the angrite parent body. *Meteoritics* **29**: 872–885.

Nyquist LE, Wiesmann H, Bansal B, Shih C-Y, Keith JE, Harper CL. 1995. [146]Sm-[142]Nd formation interval for the lunar mantle. *Geochimica et Cosmochimica Acta* **13**: 2817–2837.

O'Neil J, Carlson RW, Francis D, Stevenson RK. 2008. Neodymium-142 evidence for Hadean mafic crust. *Science* **321**: 1821–1831.

O'Neill HSC, Palme H. 2008. Collisional erosion and the non-chondritic composition of the terrestrial planets. *Philosophical Transactions of the Royal Society* **366**: 4205–4238.

Ozima M, Podosek FA. 1999. Formation age of the Earth from [129]I/[127]I and [244]Pu/[238]U systematics and the missing Xe. *Journal of Geophysical Research* **104**: 25,493–25,499.

Pack A, Shelley JMG, Palme H. 2004. Chondrules with peculiar REE patterns: Implications for solar nebular condensation at high C/O. *Science* **303**: 997–1000.

Pahlevan K, Stevenson DJ. 2007. Equilibration in the aftermath of the lunar-forming giant impact. *Earth and Planetary Science Letters* **262**: 438–449.

Patchett PJ, Vervoort JD, Söderlund U, Salters VJM. 2004. Lu-Hf and Sm-Nd isotopic systematics in chondrites and their constraints on the Lu-Hf properties of the Earth. *Earth and Planetary Science Letters* **202**: 345–360.

Prinzhofer DA, Papanastassiou DA, Wasserburg G.J. 1989. The presence of [146]Sm in the early solar system and implications for its nucleosynthesis. *Astrophysics Journal* **344**: L81–L84.

Prinzhofer DA, Papanastassiou DA, Wasserburg GJ. 1992. Samarium-Neodymium evolution of meteorites. *Geochimica et Cosmochimica Acta* **56**: 797–815.

Ranen MC, Jacobsen SB. 2006. Barium isotopes in chondritic meteorites: implications for planetary reservoir models. *Science* **314**: 809–812.

Rankenburg K, Brandon AD, Neal CR. 2006. Neodymium isotope evidence for a chondritic composition of the Moon. *Science* **312**: 1369–1372.

Regelous M, Collerson KD. 1996. [147]Sm-[143]Nd, [146]Sm-[142]Nd systematics of early Archean rocks and implications for crust-mantle evolution. *Geochimica et Cosmochimica Acta* **60**: 3513–3520.

Reisberg L, Dauphas N, Luguet A, Pearson DG, Gallino R, Zimmermann C. 2009. Nucleosynthetic osmium isotope anomalies in acid leachates of the Murchison meteorite. *Earth and Planetary Scence Letters* **277**: 334–344.

Rubie DC, Gessmann CK, Frost DJ. 2004. Partitioning of oxygen during core formation on the Earth and Mars. *Nature* **429**: 58–61.

Rubie DC, Nimmo F, Melosh HJ, Gerald S. 2007. *Formation of Earth's Core, Treatise on Geophysics.* Elsevier, Amsterdam, 51–90.

Rudnick RL, Fountain DM. 1995. Nature and composition of the continental crust: a lower crustal perspective. *Reviews in Geophysics* **33**: 267–309.

Salters VJM, Stracke A. 2004. Composition of the depleted mantle. *Geochemistry Geophysics Geosystems* **5**: doi: 10.1029/2003GC000597.

Schérsten A, Elliott T, Hawkesworth C, Russell S.S, Masarik J. 2006. Hf-W evidence for rapid differentiation of iron meteorite parent bodies. *Earth and Planetary Science Letters* **241**: 530–542.

Schoenberg R, Kamber BS, Collerson KD, Eugster O. 2002. New W-isotope evidence for rapid terrestrial accretion and very early core formation. *Geochimica et Cosmochimica Acta* **66**: 3151–3160.

Sharma M, Papanastassiou DA, Wasserburg GJ, Dymek RF. 1996. The issue of the terrestrial record of [146]Sm. *Geochimica et Cosmochimica Acta* **60**: 2037–2047.

Shirey SB, Hanson GN. 1986. Mantle heterogeneity and crustal recycling in Archean granite-greenstone belts in the Rainy Lake area, superior province, Ontario, Canada. *Geochimica et Cosmochimica Acta* **50**: 2631–2651.

Solomatov VS. 2000. Fluid dynamics of a terrestrial magma ocean. In: *Origin of the Earth and Moon*, RM, Righter K (eds), The University of Arizona Press, Arizona, Tucson 323–338.

Staudacher T, Allegre C.J. 1982. Terrestrial xenology. *Earth and Planetary Science Letters* **60**: 5391–5406.

Stevenson DJ. 1990. Fluid dynamics of core formation. In: *Origin of the Earth*, Newsom HE, Jones JE (eds), Oxford University Press, New York, 231–249.

Taylor SR, McLennan SM. 1985. *The Continental Crust: Its Composition And Evolution.* Blackwell Scientific Publications, Oxford.

Tolstikhin IN, Kramers JD, Hofmann AW. 2006. A chemical Earth model with whole mantle convection: The importance of a core-mantle boundary layer (D″) and its early formation. *Chemical Geology* **226**: 79–99.

Tonks W, Melosh HJ. 1993. Magma ocean formation due to giant impacts. *Journal of Geophysical Research* **98**: 5319–5333.

Touboul M, Kleine T, Bourdon B, Palme H, Wieler R. 2007. Late formation and prolonged differentiation of the Moon inferred from W isotopes in lunar metals. *Nature* **450**: 1206–1209.

Touboul M, Kleine T, Bourdon B. 2008. Hf-W systematics of cumulate eucrites and the chronology of the eucrite parent body. *Lunar and Planetary Science Conference* XXXIX.

Touboul M, Kleine T, Bourdon B, Nyquist LE, Shih C-Y. 2009a. New [142]Nd evidence for a non-chondritic composition of the Moon. *40th Lunar and Planetary Science Conference*, Abstract No, 2269.

Touboul M, Kleine T, Bourdon B, Palme H, Wieler R. 2009b. Tungsten isotopes in ferroan anorthosites: Implications for the age of the Moon and lifetime of its magma ocean. *Icarus* **199**: 245–249.

Upadhyay D, Scherer EE, Mezger K. 2009.[142]Nd evidence for an enriched Hadean reservoir in cratonic roots. *Nature* **459**: 1118–1121.

Wade J, Wood BJ. 2005. Core formation and the oxidation state of the Earth. *Earth and Planetary Science Letters* **236**: 78–95.

Warren PH. 2008. A depleted, not ideally chondritic bulk Earth: The explosive-volcanic basalt loss hypothesis. *Geochimica et Cosmochimioca Acta* **72**: 2217–2235.

Wetherill GW. 1990. Formation of the Earth. *Annual Review of Earth and Planetary Sciences* **18**: 205–256.

Wilde SA, Valley JW, Peck WH, Graham CM. 2001. Evidence from detrital zircons for the existence of continental crust and oceans on the Earth 4.4 Gyr ago. *Nature* **409**: 175–178.

Wood BJ, Wade J, Kilburn MR. 2008. Core formation and the oxidation state of the Earth: Additional constraints from Nb, V and Cr partitioning. *Geochimica et Cosmochimica Acta* **72**: 1415–1426.

Yin QZ, Jacobsen SB, Yamashita K. 2002a. Diverse supernova sources of pre-solar material inferred from molybdenum isotopes in meteorites. *Nature* **415**: 881–883.

Yin QZ, Jacobsen SB, Yamashita K, Blichert-Toft J, Télouk P, Albarède F. 2002b. A short timescale for terrestrial planet formation from Hf-W chronometry of meteorites. *Nature* **418**: 949–952.

Zindler A, Hart S. 1986. Chemical geodynamics. *Annual Reviews in Earth Planetary Science* **14**: 493–571.

2 Diffusion Constraints on Rates of Melt Production in the Mantle

JAMES A. VAN ORMAN[1] AND ALBERTO E. SAAL[2]

[1]Department of Geological Sciences, Case Western Reserve University, Cleveland, OH, USA
[2]Department of Geological Sciences, Brown University, Providence, RI, USA

SUMMARY

Many trace elements diffuse so slowly through mantle minerals that they may not achieve an equilibrium distribution between minerals and melt during partial melting. In principle, the control of diffusion on trace element fractionation can provide information on melting rates. Several models have been developed to address diffusion-controlled fractionation of trace elements and U-series nuclides during mantle melting processes, and these have been applied in a few cases to natural datasets, including abyssal peridotites and basalts. The melting rates inferred using this approach are consistent with physical models of adiabatic decompression melting in a mid-ocean ridge setting, and of melting due to conductive heating by a plume at the base of the lithosphere in an intraplate setting.

INTRODUCTION

Atomic transport by diffusion is a fundamental step in many important geological processes, from mantle convection to the setting and reset-ting of ages in geochronology. In many such processes, diffusion operates in concert with other phenomena; when it is the slowest step in the overall process, it controls the overall rate. For most elements, diffusion in mantle minerals is extremely slow, even at temperatures above the solidus of dry peridotite. For example, the distance over which iron and magnesium interdiffuse in olivine at 1,300°C is only ~10 cm over 1 million years (Chakraborty, 1997). Many trace elements diffuse even more slowly: lanthanides (REE) and actinides (e.g. U and Th) in clinopyroxene, orthopyroxene and garnet may diffuse over distances <1 mm in 1 million years (Van Orman et al., 1998, 2001, 2002b; Tirone et al., 2005; Cherniak & Liang, 2007). Diffusion rates of these trace elements are so slow that equilibrium partitioning between mineral grains and melt may not be established during mantle melting. In this case, trace element fractionation depends on the rate of diffusion relative to the rate of melting; in principle, if the diffusion timescale is known, this provides a basis for inferring the timescale of melt production in the mantle.

Our objectives in this chapter are, first, to illustrate in terms of a simple model how diffusion may influence trace element fractionation during melting, and how information can in principle be extracted on the melting timescale; and second, to review progress that has been made to date in modeling diffusion-controlled fractionation of trace elements, and of U-series radionuclides.

Timescales of Magmatic Processes: From Core to Atmosphere, 1st edition. Edited by Anthony Dosseto, Simon P. Turner and James A. Van Orman.
© 2011 by Blackwell Publishing Ltd.

Whilst the primary focus of this chapter is on conceptual and numerical approaches to the problem, we also discuss specific inferences that have been made regarding melting processes and timescales in the upper mantle.

SIMPLE MODEL

To develop an intuitive understanding about the control diffusion may exert on trace element fractionation during partial melting, it is useful to consider first a highly idealized system – one that is far too simple to represent an actual mantle melting process. Imagine a spherical mineral with isotropic diffusion properties, which undergoes a small degree of partial melting, instantaneously, and is thereafter held in contact with the melt at constant temperature and pressure. Because the melt forms instantaneously, it initially must have exactly the same chemical composition as the mineral, but in practice this is never the equilibrium state of the system. Only in a one-component system would the equilibrium melt composition be identical to the mineral composition. However, one-component systems are not particularly relevant to mantle melting, and even systems that can be approximated in terms of a single component contain trace elements that are partitioned unequally between mineral and melt at equilibrium. Both major elements and trace elements, which are focused on here, must be redistributed between the mineral and melt to achieve an equilibrium distribution.

The processes that lead towards an equilibrium distribution of trace elements begin at the interface between the mineral grain and the melt. Atoms must be attached to and/or detached from the surface of the mineral to bring trace element concentrations towards local equilibrium partitioning. These interface reactions produce concentration gradients on either side of the interface that drive diffusion in each phase. All three processes – interface reactions, diffusion in the mineral, and diffusion in the melt – must operate to bring the system into chemical equilibrium. Modeling the overall chemical exchange process in terms of all three of these processes would be quite involved and computationally intensive, even for this very simple system. Fortunately, it is not necessary to model the kinetics of interface reactions or diffusion in the melt explicitly, because both are so much faster than diffusion in mantle minerals. Diffusion of lanthanides and actinides, for example, is ~10^7 to 10^9 times faster in basaltic liquids (LaTourrette and Wasserburg, 1997) than in pyroxenes (Van Orman *et al.*, 1998, 2001; Cherniak and Liang, 2007) or garnets (Van Orman *et al.*, 2002a; Tirone *et al.*, 2005). Atomic attachment and detachment rates at mineral/melt interfaces have not been as extensively studied as diffusion rates, but are many orders of magnitude faster than diffusion in the melt, with local equilibrium established on the order of seconds or less at magmatic temperatures (Zhang *et al.*, 1989). With diffusion in the melt and interface reactions being so rapid, the rate of the overall trace element equilibration process is governed only by diffusion in the mineral (Dohmen & Chakraborty, 2003). The small amount of liquid can be assumed to re-homogenize continuously as trace elements diffuse out of (or into) the mineral, and the mineral surface can be assumed to adjust its composition continuously so that it always maintains partitioning equilibrium with the melt (Figure 2.1). The ability to make these assumptions allows the system to be modeled in a much simpler way than would be possible if interface reactions or diffusion in the melt occurred at rates similar to solid-state diffusion.

Equations

The behavior of the system described above, and depicted schematically in Figure 2.1, can be expressed mathematically in terms of the following set of equations. The concentration C of a trace element, as a function of the radial position r within the mineral sphere (i.e. distance from the center) and time t, is described by the following differential equation, which is usually

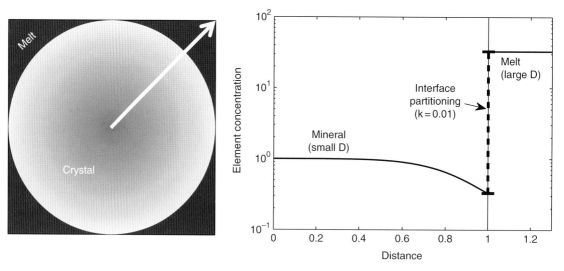

Fig. 2.1 Simple model of incompatible trace element exchange between a spherical mineral and a finite reservoir of melt. Interface reactions are rapid, leading to equilibrium partitioning between the mineral surface and the melt. Diffusion in the melt is also rapid, allowing for continuous re-homogenization of the melt reservoir. Loss of incompatible elements from the minerals is controlled by their diffusion from the mineral interior to its surface.

referred to as Fick's second law of diffusion (Crank, 1975):

$$\frac{\partial C}{\partial t} = D\left(\frac{\partial^2 C}{\partial r^2} + \frac{2}{r}\frac{\partial C}{\partial r}\right)$$

In this equation, D is the diffusion coefficient of the trace element within the mineral, C is its concentration, r is the radial position within the sphere and t is time. Initially, trace elements have equal concentrations in the mineral interior and in the melt, i.e.:

$$C\big|_{t=0,r<R} = C_m\big|_{t=0} = C_0$$

where R is the outer radius of the mineral grain, C_m is the concentration of the trace element in the melt, and C_0 is the concentration of the trace element in the bulk system. The outer surface of the mineral, on the other hand, is constrained to be in partitioning equilibrium with the melt at all times:

$$C\big|_{t=R} = kC_m$$

where k is the equilibrium partition coefficient for the trace element, equal to the mineral/melt concentration ratio at chemical equilibrium. A final constraint on the system is that it is closed to chemical exchange with its surroundings. Mass conservation requires that the change with time in the amount of a trace element in the melt be exactly balanced by the change with time in the amount of the trace element in mineral grain:

$$\frac{d(V_m C_m)}{dt} = -\frac{d}{dt}\int_0^R 4\pi C r^2 dr$$

where V_m is the volume of melt. This set of four equations can be solved in a straightforward way by using a finite difference method, an implicit method (i.e. the Crank-Nicholson method) being preferred for its greater accuracy and unconditional numerical stability. Finite difference methods for the solution of partial differential equations are covered in many textbooks on numerical analysis, and specific applications to diffusion problems are discussed by Crank (1975), Albarède (1995) and Costa *et al.* (2008).

Modeling and discussion

The evolution with time of the concentration of a trace element within a spherical mineral grain that is diffusively exchanging with 1% melt is shown in Figure 2.2a, for a trace element with a mineral/melt partition coefficient $k=0.01$. Diffusion profiles from the center to the rim of the mineral grain are shown for five different times, with the number on each curve referring to the interaction time divided by R^2/D (the square of the grain radius divided by the diffusion coefficient), which is the characteristic diffusion time. As the trace element diffuses out of the mineral, its concentration at the interface with the melt increases, reflecting the increasing concentration of the trace element in the melt. After a time of $0.3R^2/D$, the diffusion profile has become nearly flat and the entire mineral grain is close to partitioning equilibrium with the melt. This time is ~3 million years if the diffusion coefficient is 10^{-20} m²/s and the grain radius is 10^{-3} m, for example.

Although the system considered here is far too simple to approximate a real mantle melting process, Figure 2.2a serves to illustrate how information on the time of mineral-melt interaction can, in principle, be gleaned if diffusion profiles can be recovered from mineral residues of the melting process. In the simple case illustrated in Figure 2.2a, the diffusion profile of a trace element within the mineral records the time during which the mineral and melt interacted, which can be inferred directly if the diffusion coefficient is known and the time is not so long that the mineral approaches equilibrium with the melt. In a more realistic case, where melting is ongoing and melt may be extracted continuously from the system, diffusion profiles in the mineral record information on the *rate* of melting, and depend also on the style of melt extraction (batch, fractional, etc.). In this case, the rate of melting could be inferred by fitting the diffusion profile to an appropriate model of the melting process, again provided that the diffusion parameters in

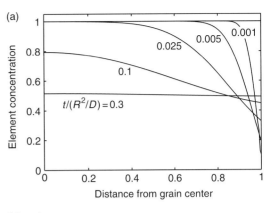

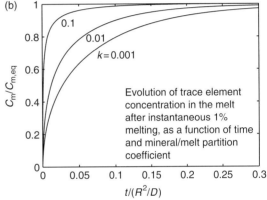

Fig. 2.2 Simulations of trace element evolution in a simple system consisting of a single spherical mineral surrounded by a small amount of melt (1% of the system mass). The mineral and melt begin with the same trace element composition, and evolve towards partitioning equilibrium, with diffusion in the mineral as the rate-limiting step. (a) Diffusion profiles in the mineral for a trace element with partition coefficient $k=0.01$, at 5 different non-dimensional times (timescaled by the characteristic diffusion time). (b) Change with time of the concentration in the melt, for trace elements with 3 different partition coefficients. The concentration in the melt is scaled to the equilibrium concentration. Note that trace elements with small partition coefficients require more time to equilibrate.

the mineral are known, and that the initial grain size can be estimated.

However, few diffusion profiles have actually been measured across mineral grains in peridotites (e.g. abyssal peridotites) that are residues of mantle melting. An exception is orthopyroxene in remarkably fresh peridotites from the Gakkel Ridge, which von der Handt and others have shown to preserve compositional zoning profiles (von der Handt *et al.*, 2007, 2008). In this case, however, the trace element profiles in orthopyroxene clearly reflect diffusive exchange with surrounding minerals under sub-solidus conditions, during a protracted cooling interval associated with slow uplift. Unfortunately, no signature of diffusive exchange during melting appears to remain. Whilst quenched diffusion profiles in the mineral residues of mantle melting would provide the clearest and most robust evidence for diffusion-controlled fractionation, and the strongest diffusion-based constraints on rates of melt production, the appropriate geochemical datasets do not (yet) exist, and such applications remain, for now, hypothetical. Data on the bulk concentrations of trace elements in mineral residues of mantle melting do exist (Johnson *et al.*, 1990), and these have been used to address the effects of disequilibrium during melting, as discussed below.

The melt also may provide information on the time of interaction between minerals and melt, in the simple case considered here, or on the rate of melting in more realistic scenarios. However, because the melt homogenizes continuously on the timescale of interest, much information is lost, and in practice the interpretation of the trace element composition of the melt in terms of disequilibrium, diffusion-controlled exchange must always be at least somewhat ambiguous. In terms of the extremely simple model considered here, if the degree of melting is known, as well as the initial concentration, equilibrium partition coefficient and diffusion coefficient of a trace element, then the concentration of the trace element in the melt is a unique function of the time of interaction. In general, however, the melting degree and the initial concentration of the system are not known, and in more realistic models there is the added complication that the trace element composition depends on the style of melt removal (batch, fractional, etc.). Despite these significant

drawbacks, melts are worthy of consideration because they are far more extensively sampled than the solid residues of mantle melting. As discussed below, diffusion control on the composition of melts becomes particularly important, and powerful as a tool for understanding melting processes, with regard to the U-series isotopes.

Figure 2.2b shows the change in concentration of three incompatible trace elements, with equilibrium partition coefficients $k=0.1$, 0.01 and 0.001, in a melt (which constitutes 1% of the system mass, as in Figure 2.2a) interacting with a spherical mineral. This simple model illustrates an important point: the rate at which the concentration in the melt approaches its ultimate equilibrium value depends on the partition coefficient. Because they are more strongly partitioned into the melt, elements with smaller partition coefficients require a greater diffusive flux and thus take more time to approach equilibrium, for a given diffusion coefficient. Hart (1993) made the same point in reference to an analytical solution for diffusive interaction of a melt cylinder with surrounding minerals. For a more extensive and general treatment of the controls on chemical exchange rates, see Dohmen & Chakraborty (2003).

MORE REALISTIC MODELS

Several models were presented in the early 1990s (Qin, 1992; Iwamori, 1993a, b) that address a more realistic picture of melting than the very simple model described above. These models are based on melting and melt transport models that had been put forward in the 1970s and 1980s (Langmuir *et al.*, 1977; McKenzie, 1985; Richter, 1986), the key difference being that the new disequilibrium models considered trace element exchange explicitly as a diffusion-controlled process rather than assuming continuous equilibration between solid and co-existing melt. As with the simple model described above, the models put forward by Qin (1992) and Iwamori (1993a, b) treat mineral grains as isotropic spheres, with partitioning equilibrium continuously maintained at the interface

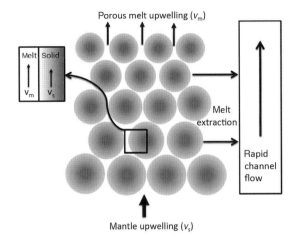

Porous melt upwelling (v_m)

Melt | Solid

v_m | v_s

Melt extraction

Rapid channel flow

Mantle upwelling (v_s)

Fig. 2.3 A general picture of disequilibrium decompression melting models, wherein minerals interact with porous melt via diffusion. In the Iwamori (1993a) model, there is a differential flow between the porous melt and solid mantle (with $v_m > v_s$). The model of Iwamori (1993b) includes both porous flow and melt extraction into channels.

between minerals and melt. The solid is implicitly treated as mono-mineralic, with the behavior of a trace element governed by a single diffusion coefficient and partition coefficient that remain constant throughout the melting and diffusive exchange process. Where these models differ from the simple model described above is in their treatment of melting and melt extraction (Figure 2.3). Instead of considering melting and diffusive exchange separately, as a two-step process with the melting step being instantaneous, melting is considered to occur continuously, at a constant rate. The number of mineral grains remains constant during melting, and grains maintain their spherical symmetry as they dissolve; this requires that the boundary between mineral and melt shift towards the grain center as melting progresses. The assumption that grains retain their spherical geometry implies that grains remain undeformed during melting, or at least that deformation has no significant influence on the diffusion profiles that develop within the grains. This is an important assumption, but one that is difficult to test.

Significant deformation of grains during melting might promote more rapid trace element exchange between minerals and melting.

Where the models of Qin (1992) and Iwamori (1993a, b) differ is in their treatment of melt transport. The model of Qin (1992) is a 'critical melting' or 'dynamic melting' model, in which melt remains in the system until a threshold melt fraction is exceeded. Thereafter, melt is extracted at the rate that leaves the fraction of melt in the residue constant. Extracted melt is considered to pool in an isolated reservoir, which experiences no further chemical exchange with the solid. The Qin (1992) model can describe a range of melt extraction styles between two end-member scenarios: when the threshold melt fraction is set higher than the total degree of melting, the model is equivalent to 'batch' melting, where melt always remains in chemical contact with the solid; when the threshold melt fraction is set to zero, melting is 'fractional', i.e. melt is extracted from the system as soon as it is produced. The model of Iwamori (1993a) considers simultaneous melting and one-dimensional porous flow. Melt is considered to percolate continuously upwards, at a rate greater than the rate of solid mantle flow, interacting with minerals as it ascends via diffusion-controlled exchange. The model of Iwamori (1993b) adds to this porous flow model the possibility of melt extraction into chemically isolated channels. Like the Qin (1992) model, this model can simulate the full range of conditions between the two end-member physical melting models of fractional melting (perfectly efficient extraction of melt into channels) and batch melting (no extraction of melt into channels).

Several important points arise from the Qin and Iwamori models. One is that the degree of chemical disequilibrium during melting, where chemical exchange is controlled by diffusion in the solid, depends upon the ratio of the characteristic diffusion time, R^2/D, and a characteristic melting time, i.e. the time required to produce a certain degree of melting. This result can be appreciated in terms of the simpler model shown in Figure 2.2, where the time of mineral/melt interaction is scaled to the characteristic diffusive exchange time. Regardless of differences in the style of melting, the equilibration process is essentially governed by the relative timescales for diffusion and mineral/melt interaction. Another important point is that the greater the degree of disequilibrium during melting, the higher the concentration of an incompatible element in the residual solid. This result can also be appreciated in terms of Figure 2.2a – the integrated concentration of an incompatible trace element in the solid grain steadily decreases as diffusive exchange proceeds, reaching a minimum at equilibrium. What this implies is that, all else being equal, the concentration of an incompatible element in the residual solid will increase with the rate of melting. In principle, the concentrations of incompatible trace elements in the residual solid may thus record information on the melting rate. However, the style of melting also has a strong influence on the degree to which incompatible elements are stripped from the solid – highly incompatible elements are removed more efficiently by equilibrium fractional melting than by equilibrium batch melting. In terms of the bulk concentrations of trace elements in the residual solid, the effect of disequilibrium is mimicked by the effect of inefficient melt extraction, and it is not easy to determine which of these effects may dominate in a particular case.

Iwamori (1993b) compared the results of two different models to trace element data from abyssal peridotite clinopyroxene grains (Johnson et al., 1990; Johnson & Dick, 1992). In one model, equilibrium is maintained between the solid and residual melt during melting, and what is varied is the efficiency of melt extraction; in the other, melt extraction into channels is perfectly efficient (i.e. melting is fractional) and what is varied is the degree of chemical disequilibrium (considered to be the same for each element considered, since relevant diffusion data were not available at that time). Iwamori (1993b) showed that reasonable fits to the abyssal peridotite data could be obtained in either case: with equilibrium assumed, the data were best modeled with ~80% of the melt being extracted from the residue by channelized flow;

assuming fractional melting, the data were best modeled with a significant degree of chemical disequilibrium, with the diffusion time R^2/D being approximately equal to the total melting time, represented by Iwamori (1993a, b) as H/V_s, where H is the height of the mantle column undergoing adiabatic decompression melting, and V_s is the mantle upwelling velocity. The ambiguity in interpretation of the data prevented any definite conclusions from being made on the degree of disequilibrium. Even if such a determination had been possible, the lack of appropriate data on trace element diffusivities in mantle minerals would have made an inference of the melting rate and/or grain size in the mantle highly speculative.

Further developments

The availability of new experimental data on trace element diffusion in clinopyroxene (Cherniak, 1998, 2001; Van Orman et al., 1998, 2001) and garnet (Van Orman et al., 2002a; Tironi et al., 2005) and of new thermodynamic analyses of the melting process (Asimow et al., 1997) led to the development of disequilibrium melting models that address more specifically the conditions that prevail during melting beneath mid-ocean ridges. Van Orman et al. (2002b) presented a model that considers both clinopyroxene and garnet, the two primary hosts of rare earth elements in the upper mantle. During melting, each mineral exchanges trace elements with the melt, and each of which may enter the melt at a different rate. The greater availability of experimental data also made it appropriate to consider variations in the diffusivity of each element as a function of temperature and pressure during decompression melting, i.e.:

$$D_i^j = D_{0,i}^j \exp\left(\frac{-(E_i^j + PV_i^j)}{RT}\right)$$

where D_i^j is the diffusion coefficient of element i in mineral j, D_0, E and V are the pre-exponential factor, activation energy and activation volume for diffusion, respectively, P is the pressure, T is the absolute temperature, and R is the gas (or

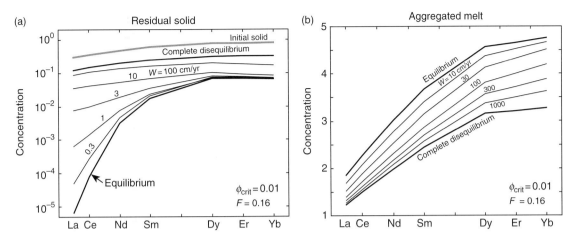

Fig. 2.4 Influence of mantle upwelling rate on the REE composition of (a) the residual solid and (b) the aggregated extracted melt after 16% disequilibrium critical melting in the spinel stability field (Van Orman *et al.*, 2002b). The initial grain radius in the simulations is 2 mm, and the threshold melt fraction is 0.01. The numbers on the curves indicate the upwelling rate in cm/yr. Reprinted from *"Earth and Planetary Science Letters"*, **198**, Van Orman J.A., Grove T.L. and Shimizu N. Diffusive fractionation of trace elements during production and transport of melt in Earth's upper mantle. 93–112, 2002 with permission from Elsevier.

molar Boltzmann) constant. See the chapter by Costa and Morgan for further discussion about variations in, and fundamental controls on, D. The diffusion coefficients for REE in clinopyroxene decrease by one to two orders of magnitude with decreasing temperature along an isentropic decompression mantle melting path; this implies that the degree of disequilibrium and the effective partitioning behavior of the REE vary as melting proceeds. Van Orman *et al.* (2002b) also considered variations in the rate of melt production along an isentropic decompression melting path, following the thermodynamic analysis of Asimow *et al.* (1997). In particular, the melting rate is inferred to increase strongly at lower pressures, where REE diffusion coefficients are slow. The combination of rapid melting rates and slow diffusion rates strongly favors disequilibrium partitioning at shallow depths; conversely, relatively slow melting rates and faster diffusion coefficients promote equilibrium near the base of the melting column.

Van Orman *et al.* (2002b) incorporated these refinements into a model that is similar in other respects to the dynamic melting model developed by Qin (1992), and used it to simulate REE frac-

tionation during adiabatic decompression melting in the mantle. These authors found the REE composition of the solid residue melting to be a strong function of the mantle upwelling rate, when melting is nearly fractional (Figure 2.4a). When upwelling is very slow, ~0.1 cm/yr or less, for grain radii of 2 mm, equilibrium between minerals and melt is maintained during melting for all of the REE. This results in a strong depletion of the residual solid in light rare earth elements, which are highly incompatible. As the upwelling rate increases beyond ~0.3 cm/yr, light rare earth elements are unable to maintain equilibrium concentrations in the melt, due to their slow diffusivities and small partition coefficients (cf. Figure 2.2b), whilst faster diffusing, less incompatible heavy rare earth elements continue equilibrating at upwelling rates an order of magnitude faster. As a result, light rare earth element abundances in the residual solid shift upwards, becoming comparable to heavy rare earth element abundances at upwelling rates of 10 cm/yr or more.

In contrast, the aggregated melt (Figure 2.4b) is fairly insensitive to disequilibrium partitioning of the REE. At very rapid upwelling rates the

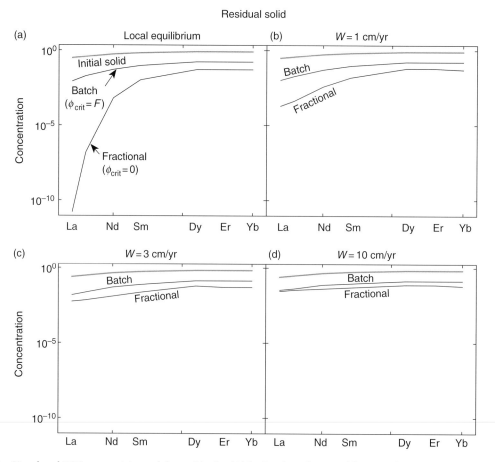

Fig. 2.5 Simulated REE compositions of the residual solid for batch melting and fractional melting, after 16% melting in the spinal peridotite field (Van Orman *et al.*, 2002b). With increasing deviation from partitioning equilibrium (a–d) the solid residue of fractional melting becomes similar to that from batch melting. The difference in La concentration after batch melting vs fractional melting is ~10^8 at equilibrium, but <2 at an upwelling rate of 10 cm/yr. The initial grain radius in the simulation is 2 mm. Reprinted from *"Earth and Planetary Science Letters"*, **198**, Van Orman J.A., Grove T.L. and Shimizu N. Diffusive fractionation of trace elements during production and transport of melt in Earth's upper mantle. 93–112, 2002 with permission from Elsevier.

abundances of REE decrease somewhat in the aggregated melt compared to their abundances at equilibrium, but there is very little change in the shape of the REE pattern. The reasons for this are discussed by Van Orman *et al.* (2002b); the practical implication is that rare earth elements in basaltic melts are unlikely to divulge much information on mantle melting rates.

When melting is a batch process, with melt remaining in contact with the minerals until it is removed *en masse*, rather than a near-fractional process, the REE composition of the solid residue is not very sensitive to variations in the degree of disequilibrium. As the upwelling rate increases, REE concentrations in the solid residue of batch melting increase only slightly, whilst for fractional melting the REE (and particularly the LREE) concentrations increase strongly with upwelling rate (Figure 2.5). The result is that the solid residue of fractional melting becomes very similar to the solid residue of batch melting, as the degree of disequilibrium increases; for upwelling rates of

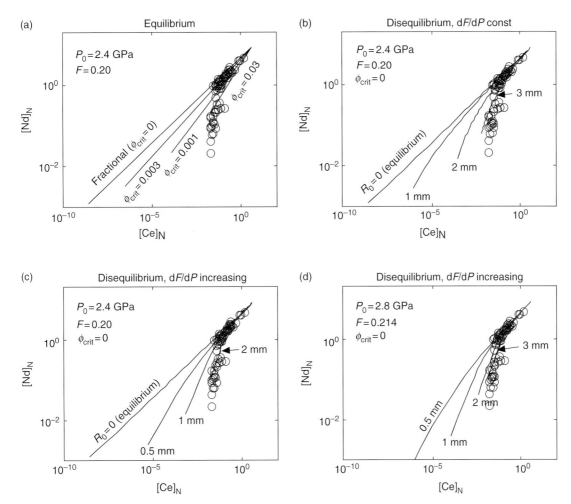

Fig. 2.6 Comparison of model predictions (Van Orman *et al.*, 2002b) of the clinopyroxene composition in the residual solid with observed chondrite-normalized Ce and Nd concentrations in cpx grains from abyssal peridotites (open circles, data from Johnson *et al.*, 1990; Johnson & Dick, 1992). All calculations assume an upwelling rate of 0.9 cm/ yr, consistent with the plate spreading rate of the sampling locations. In (a), the curves show model residual cpx concentrations for 0 to 20% melting beginning at 2.4 GPa, in the spinel stability field, with equilibrium maintained during melting and threshold porosities between 0 and 3%. (b) Disequilibrium fractional melting with constant melt productivity. Melting begins at 2.4 GPa, and the curves are for different grain sizes between 0 and 3 mm. (c) Same as in (b) but with increasing melt productivity at pressures below 1.0 GPa. (d) Same as in (c) but with melting beginning at 2.8 GPa, in the garnet stability field. In general, the abyssal peridotite data are consistent with disequilibrium fractional melting, with initial mantle grain sizes of 2 to 3 mm and either constant melting rate, or variable melting rate constrained by thermodynamic considerations (Asimow *et al.*, 1997). Reprinted from *"Earth and Planetary Science Letters"*, **198**, Van Orman J.A., Grove T.L. and Shimizu N. Diffusive fractionation of trace elements during production and transport of melt in Earth's upper mantle. 93–112, 2002 with permission from Elsevier.

3 cm/yr or more, the REE patterns in the solid residue have nearly identical shapes for fractional and batch melting (with grain radii of 2 mm, and a total degree of melting of 16%).

Following Iwamori (1993b), Van Orman *et al.* (2002b) re-examined trace element data for clinopyroxene from abyssal peridotites, focusing on the light rare earth elements Ce and Nd. Because these elements have different diffusion parameters (and partition coefficients), they respond somewhat differently to variations in melting rate or grain size. The result is that Ce vs Nd trends in solid residues of variable degrees of melting are curved for disequilibrium, whereas they are straight for equilibrium melting (Figure 2.6). Van Orman *et al.* (2002b) found a somewhat better match to the abyssal peridotite data for disequilibrium models than for equilibrium models, with best fits obtained for grain radii of 2 to 3 mm, at an upwelling rate of 0.9 cm/yr (inferred from the plate spreading rate in the vicinity of the sampling site).

Most disequilibrium melting models have focused on decompression melting, which is relevant to magma production at mid-ocean ridge and intraplate settings. However, Liang (2003) investigated the influence of disequilibrium on eclogite melting within a subducting slab. The REE compositions of adakites were found to be reasonably consistent with slab melting under either equilibrium or disequilibrium conditions. However, under disequilibrium conditions a significantly larger degree of slab melting was necessary to explain the data, ~25 to 30% rather than ~10% in the case of equilibrium melting.

DIFFUSIVE FRACTIONATION OF U-SERIES ISOTOPES

The U-series isotopes provide a unique opportunity to combine kinetic constraints from both diffusion and radioactive decay. The chapters by Bourdon and Elliott, Dosseto and Turner, and Turner and Bourdon introduce basis aspects of U-series isotopes and their applications to magmatic processes. In this section we briefly address the diffusive fractionation of U-series isotopes during melting; a more detailed discussion of this topic, and presentation of the equations involved, is given by Van Orman *et al.* (2006).

As discussed above, chemical disequilibrium of a stable trace element during melting depends on the timescale of diffusion in the minerals relative to the timescale of melting. For U-series nuclides, a third timescale is involved as well – the timescale of radioactive decay. For the intermediate daughters in the U-series, it is necessary to consider radioactive production and decay occurring simultaneously with diffusive exchange. This makes their behavior during melting qualitatively different than the behavior of stable trace elements.

A distinct advantage of the U-series system is that the relative concentrations of the radioactive nuclides in the series, in the mantle source prior to the onset of melting, is well constrained. In a system that has remained closed for sufficient time, the activity of each nuclide in the series (which is equal of the product of its decay constant and its concentration) is the same. This steady state condition, referred to as secular equilibrium, applies to mantle that has remained undisturbed for a few hundred thousand years prior to the onset of partial melting. During partial melting, nuclides in the decay series may be fractionated, leading to relative activities in the melt that deviate from those expected at secular equilibrium. As with stable trace elements, the fractionation of U-series nuclides depends on their equilibrium partition coefficients when local equilibrium is maintained during melting, but depend also on their diffusion coefficients under disequilibrium conditions.

Qin (1992) presented, in an appendix, a model for diffusion-controlled transfer of ^{230}Th and ^{226}Ra during melting. The model is essentially the same as the model for stable trace elements presented in the same paper (and described above) but also considers the simultaneous radioactive production and decay of ^{226}Ra and the decay of ^{230}Th (for simplicity, radioactive production of ^{230}Th is ignored). Qin (1992) suggested that Ra (which is divalent) was likely to diffuse more rapidly in mantle minerals than tetravalent Th, and presented numerical calculations in which the diffusion coefficient for Ra was assumed to be 10 times higher than for Th. Under these assumptions, the calculations show that Ra is fractionated more efficiently from Th when melting occurs under disequilibrium conditions, producing larger ^{226}Ra excesses in the melt than when partitioning equilibrium between solid and melt is maintained. It is important to add the caveat that this is the case only when the melting rate is not too high – if melting is so rapid that no diffusive transfer between minerals and melt is possible, there is no fractionation of Ra from Th during the melting process. As discussed below, however, the melt in this case could still inherit a Ra excess or deficit (where excess and deficit denote greater and lesser activities than ^{230}Th, respectively), if ^{226}Ra and/or ^{230}Th is distributed unequally among the minerals when melting begins, and the minerals enter the melt in unequal proportions.

Iwamori (1994) also considered U-series radioactive disequilibria produced during melting, using a model that is based on the one presented by Iwamori (1993b) where melt is transported by porous and channel flow, but is modified to consider radioactive production and decay of U-series isotopes. It was shown that ^{230}Th excess in the melt during fractional melting increased compared to the equilibrium case if Th diffused significantly faster than U, but decreased if Th and U diffused at the same rate, or if U diffused faster than Th. Similarly, it was shown that Ra could be strongly fractionated from Th if it diffused more rapidly in the solid, resulting in large ^{226}Ra excesses in the melt, but fractionation was shown to diminish with increasing chemical disequilibrium if Ra and Th diffused at similar rates, or Th diffused faster than Ra. Based on sparse diffusion data that were available at the time, Iwamori (1994) suggested that chemical disequilibrium effects were unlikely to be very important in the fractionation of U-series isotopes. Later experimental work (Van Orman *et al.*, 1998) showed diffusion of U and Th in clinopyroxene to be several orders of magnitude slower than assumed by Iwamori (1994), re-opening the possibility that diffusive fractionation of U-series isotopes during mantle melting could be important.

Saal & Van Orman (2004) re-considered the effects of diffusion on ^{226}Ra/^{230}Th radioactive disequilibrium, focusing on shallow interactions between melts and cumulates in the crust-mantle transition zone. Based on observed relationships between the diffusion coefficient of an element and its ionic radius, it was inferred that Ra likely diffused significantly faster than either Th or U in both clinopyroxene and plagioclase. In this case, diffusive interaction of melt with either clinopyroxene or plagioclase cumulates was shown to produce large ^{226}Ra excesses in the melt, whilst having only a minor influence on the ^{230}Th/^{238}U activity ratio. These authors therefore suggested a plausible scenario in which ^{230}Th/^{238}U disequilibria in mid-ocean ridge and ocean island basalts are produced in the deep mantle, whilst ^{226}Ra excesses are produced mainly by shallow melt-cumulate interactions. In this case, ^{226}Ra/^{230}Th radioactive disequilibria in basalts may yield little or no information on the timing of melting production in the mantle, or on the timing of melt transport from the deep upper mantle. On the other hand, radioactive disequilibria involving ^{226}Ra and its shorter-lived daughters might yield information on the residence time of melts at shallow levels, and their transport through the crust.

The influence of multiple minerals

All of the models discussed above deal with interaction between only two phases, a solid and a melt. More complicated behavior arises in disequilibrium melting models if more than one mineral is a significant host of U-series isotopes during melting. In the equilibrium case, it is not necessary to consider the distribution of U-series nuclides among the minerals in the solid – all of the minerals are in equilibrium with the melt, and fractionation can be modeled simply in terms of the bulk partitioning of elements between the solid and the melt. When equilibrium partitioning is not attained, due to slow diffusion in one or more of the minerals, it is necessary to consider in detail how nuclides are distributed among (and within) the minerals, both prior to and during partial melting. Prior to melting, the solid mantle as a whole can be assumed to be in secular equilibrium. However, a wide range of possibilities exists for the distribution of U-series nuclides between the minerals in the mantle. At one extreme, daughter nuclides may be distributed among the minerals according to their equilibrium partition coefficients. This situation arises if diffusion in the minerals is very rapid compared to radioactive decay rates. At the other extreme, when diffusion rates are very slow, each mineral is internally in secular equilibrium. In this case, there is no partitioning of the intermediate daughter nuclides among the minerals; instead each mineral is effectively isolated. In general, when diffusion is neither very fast not very slow compared to radioactive decay, the distribution of daughter nuclides among the minerals reaches a steady state between these two extremes, in which diffuse fluxes are balanced by radioactive production and decay.

Feineman and DePaolo (2003) considered the distribution of U-series nuclides among clinopyroxene and hydrous minerals (phlogopite and amphibole) prior to melting, and discussed the possible influence of the initial distribution on ^{226}Ra in subduction zone lavas. They presented an analytical model for the steady-state distribution of ^{226}Ra and ^{230}Th between clinopyroxene and phlogopite or amphibole, which assumes that diffusion in the hydrous phase is rapid enough to remain homogeneous. For realistic partition and diffusion coefficients, mineral proportions and grain sizes, the hydrous mineral was found to develop an activity ratio $(^{226}Ra)/(^{230}Th) \gg 1$, complemented by $(^{226}Ra)/(^{230}Th) \approx 1$ in the clinopyroxene (the bulk system being in secular equilibrium, with $(^{226}Ra)/(^{230}Th) = 1$). During melting, the melt may inherit excess ^{226}Ra from the hydrous phase, provided that its modal abundance in the melt is sufficiently high and the melt is separated (and formed) rapidly enough. Feineman & DePaolo (2003) suggested that large ^{226}Ra excesses in island arc lavas may be generated by such a process in the hydrated mantle wedge above the subducting slab. These authors also noted that $^{226}Ra/^{230}Th$ fractionation in the opposite sense, with the melt acquiring $(^{226}Ra)/(^{230}Th) < 1$, could result if melt extraction were insufficiently rapid, and a hydrous mineral remained in the residue. The transition from ^{226}Ra excesses at high melting rates to ^{226}Ra deficits at slower melting rates, during melting of hydrated mantle, was later investigated by Bourdon & Van Orman (2009), as discussed below (Figure 2.7).

Van Orman *et al.* (2006) presented a numerical model that extended the work of Feineman & DePaolo (2003) to address both the steady-state distribution of U-series nuclides among minerals below the solidus, and their diffusion-controlled redistribution during melting. The model is an extension of the model of Van Orman *et al.* (2002b) for diffusion-controlled fractionation of stable trace elements, modified to account for simultaneous radioactive production and decay. The set of equations that govern the distribution of U-series nuclides among all phases, both prior to and during melting, are given in Van Orman *et al.* (2006). Bourdon and Van Orman (2009) applied this model to investigate $^{226}Ra/^{230}Th$ fractionation in lavas from an intraplate setting, the Pitcairn seamounts, which are characterized by ^{226}Ra deficits relative to ^{230}Th. Based on water contents and trace element systematics of the lavas, it was inferred that phlogopite was present in the source region of the Pitcairn seamount lavas, and was presumed to be within the lithosphere because phlogopite is unstable at asthenospheric temperatures. The initial sub-solidus steady-state distributions of $^{226}Ra/^{230}Th$ and $^{230}Th/^{238}U$ between phlogopite and garnet were first calculated, followed by simulation of the diffusion-controlled redistribution of these nuclides during partial melting. Melting at the base of the lithosphere was treated as an isobaric process under conditions of increasing temperature, with melt extraction being continuous above a threshold porosity. It was found that simultaneous ^{226}Ra deficits and relatively large ^{230}Th excess in the Pitcairn lavas could only be explained if the rate of melting was $<\sim 10^{-7}$/yr (Bourdon & Van Orman, 2009). This is consistent with the melting rate predicted from a simple model of conductive heating at the base of the lithosphere by an underlying thermal plume.

^{210}Pb

Diffusive fractionation of relatively long-lived intermediate daughters, such as ^{230}Th (with a half-life of 75,000 years) and ^{226}Ra (with a half-life

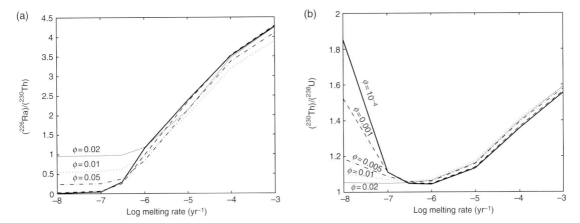

Fig. 2.7 Results of numerical simulations of disequilibrium ^{226}Ra-^{230}Th-^{238}U fractionation during partial melting of hydrated peridotite at the base of the lithosphere (Bourdon & Van Orman, 2009). (a) (^{226}Ra/^{230}Th) and (b) (^{230}Th/^{238}U) as a function of melting rate, in the aggregated extracted melt after 3% melting of a hydrated peridotite containing 10% garnet and 1.25% phlogopite. The numbers on the curves show the threshold porosity used in the simulation, i.e. the melt fraction beyond which melt is continuously extracted. The mineral grain radii used in the simulations are 1 mm. Similar results were obtained for 1 and 7% melting of the same hydrated peridotite. Pitcairn lavas with the largest radioactive disequilibria have (^{226}Ra/^{230}Th) activity ratio of ~0.9 with (^{230}Th/^{238}U) of ~1.37 (Bourdon & Van Orman, 2009). These data can only be explained by melting rates of <10^{-7}/yr (i.e. 10^{-5}%/yr). Reprinted from "*Earth and Planetary Science Letters*", **277**, Bourdon B., Van Orman J.A. Melting of enriched mantle beneath Pitcairn seamounts: Unusual U-Th-Ra systematics provide insights into melt extraction processes. 474–481, 2009, with permission from Elsevier.

of 1,600 years) depends strongly on their diffusivities in the minerals, and on any chemical disequilibrium that exists prior to, or develops during, melting or mineral-melt interaction. The behavior of the very short-lived radionuclide ^{210}Pb, which has a half-life of only 22 years, is quite different. For ^{210}Pb, decay is so rapid that it usually provides the primary driving force for diffusive exchange, overriding the effects of chemical disequilibrium. Van Orman & Saal (2009) investigated the fractionation between ^{210}Pb and/^{226}Ra during the interaction of melts with crustal cumulates, using the finite difference model presented by Van Orman *et al.* (2006). Interaction of basaltic melt with plagioclase, clinopyroxene or gabbroic cumulates was found to be capable of inducing a strong ^{210}Pb deficit in the melt, with the deficit developing quickly, on a timescale comparable to or shorter than the half-life of ^{210}Pb. Whilst ^{226}Ra may develop either

an excess or a deficit in the melt after interacting with clinopyroxene or plagioclase, depending on the sense of the initial trace element disequilibrium between the mineral and melt (Saal & Van Orman, 2004; Van Orman *et al.*, 2006), ^{210}Pb deficits develop under a broad range of conditions, regardless of the direction of the initial disequilibrium. Also, unlike ^{226}Ra, the behavior of ^{210}Pb during mineral/melt interaction turns out not to be sensitive to its diffusivity relative to its parents in the decay chain; ^{210}Pb deficits develop in the melt whether the diffusivity of Pb is greater than or less than the diffusivity of Ra. The rapid decay of ^{210}Pb has such a strong control on the diffusive fluxes that it effectively buffers the system against variations in the diffusivity of Pb and Ra.

An important practical implication of the modeling results (Van Orman & Saal, 2009) is that a ^{210}Pb deficit in the melt can arise quite readily during diffusive interactions in the crust. This suggests caution in interpreting ^{210}Pb disequilibria in basalts as a mantle signature. The short timescale, <100 years, indicated by the presence of ^{210}Pb deficits in basalts (Sigmarsson, 1996; Rubin *et al.*, 2005), may reflect crustal rather than mantle processes.

CONCLUDING REMARKS

Recent experimental and theoretical developments on the kinetics of chemical exchange between minerals and melt have provided a basis for constraining the rates of mantle melting processes. However, few determinations of mantle melting rates have yet been made using this approach. Rare earth element data from abyssal peridotites at mid-ocean ridges (Johnson *et al.*, 1990; Johnson & Dick, 1992) can be modeled successfully as residues of melting with diffusion-controlled chemical exchange (Van Orman *et al.*, 2002b). The inferred rates of melting are consistent with the rates predicted from plate spreading rates, along with thermodynamic constraints on the productivity of peridotite. In an intraplate setting, the melting rate was constrained to be <10^{-5}%/yr based on a comparison of model results with ^{226}Ra-^{230}Th-^{238}U systematics in Pitcairn seamount lavas (Bourdon & Van Orman, 2009). This result is consistent with partial melting due to conductive heating from a mantle plume at the base of the lithosphere.

The primary obstacle in applying diffusion constraints to mantle melting problems is the paucity of natural data that can be modeled in terms of diffusion-controlled fractionation during melting. Diffusion constraints are much more readily applied to infer rates of magma chamber processes (see the chapter by Costa and Morgan), because such processes are often recorded with high fidelity as well-preserved zoning profiles in phenocrysts. Unfortunately, this is not the case for solid residues of mantle melting; abyssal peridotites typically have undergone significant sub-solidus chemical processing that

overprints the melting signature, and diffusion profiles representative of melt-solid chemical exchange have not yet been recovered. Basalts, the liquid products of mantle melting, although not as easy to interpret uniquely in terms of the diffusion-controlled chemical exchange processes involved in their production, nonetheless provide a promising avenue for future developments, particularly with regard to their U-series systematics.

REFERENCES

Albarède F. 1995. *Introduction to Geochemical Modeling.* Cambridge University Press, Cambridge 543 pp.

Asimow PD, Hirschmann MM, Stolper EM. 1997. An analysis of variations in isentropic melt productivity. *Philosophical Transactions of the Royal Society of London* **A355**: 244–281.

Bourdon B, Van Orman JA. 2009. Melting of enriched mantle beneath Pitcairn seamounts: Unusual U-Th-Ra systematics provide insights into melt extraction processes. *Earth and Planetary Science Letters* **277**: 474–481.

Chakraborty S. 1997. Rates and mechanisms of Fe-Mg interdiffusion in olivine at 980°–1,300°C. *Journal of Geophysical Research* **102**: 12,317–12,331.

Cherniak DJ. 1998. Pb diffusion in clinopyroxene. *Chemical Geology* **150**: 105–117.

Cherniak DJ. 2001. Pb diffusion in Cr diopside, augite, and enstatite, and consideration of the dependence of cation diffusion in pyroxene on oxygen fugacity. *Chemical Geology* **177**: 381–397.

Cherniak DJ, Liang Y. 2007. Rare earth element diffusion in natural enstatite. *Geochimica et Cosmochimica Acta* **71**: 1324–1340.

Costa F, Dohmen R, Chakraborty S. 2008. Timescales of magmatic processes from modeling the zoning patterns of crystals. *Reviews in Mineralogy and Geochemistry* **69**: 545–594.

Crank J. 1975. *The Mathematics of Diffusion*, 2nd edn, Oxford University Press, London 414 pp.

Dohmen R, Chakraborty S. 2003. Mechanism and kinetics of element and isotopic exchange mediated by a fluid phase. *American Mineralogist* **88**: 1251–1270.

Feineman MD, DePaolo DJ. 2003. Steady-state ^{226}Ra/^{230}Th disequilibrium in mantle minerals: Implications for melt transport rates in island arcs. *Earth and Planetary Science Letters* **215**: 339–355.

Hart SR. 1993. Equilibration during mantle melting: A fractal tree model. *Proceedings of the National Academy of Sciences* **90**: 11,914–11,918.

Iwamori H. 1993a. Dynamic disequilibrium melting model with porous flow and diffusion controlled chemical equilibration. *Earth and Planetary Science Letters* **114**: 301–313.

Iwamori H. 1993b. A model for disequilibrium mantle melting incorporating melt transport by porous and channel flows. *Nature* **366**: 734–737.

Iwamori H. 1994. ^{238}U-^{230}Th-^{226}Ra and ^{235}U-^{231}Pa disequilibria produced by mantle melting with porous and channel flows. *Earth and Planetary Science Letters* **125**: 1–16.

Johnson KTM, Dick HJB. 1992. Open system melting and temporal and spatial variation of peridotite and basalt at the Atlantis-II fracture-zone. *Journal of Geophysical Research* **97**: 9219–9241.

Johnson KTM, Dick HJB, Shimizu N. 1990. Melting in the oceanic upper mantle: An ion microprobe study of diopsides in abyssal peridotites. *Journal of Geophysical Research* **95**: 2661–2678.

Langmuir CH, Bender JF, Bener AE, Hanson GN, Taylor SR. 1977. Petrogenesis of basalts from the FAMOUS area: Mid-Atlantic Ridge. *Earth and Planetary Science Letters* **36**: 133–156.

LaTourrette T, Wasserburg GJ. 1997. Self diffusion of europium, neodymium, thorium, and uranium in haplobasaltic melt: The effect of oxygen fugacity and the relationship to melt structure. *Geochimica et Cosmochimica Acta* **61**: 755–764.

Liang Y. 2003. On the thermo-kinetic consequences of slab melting. *Geophysical Research Letters* **30**: Art. No. 2270.

McKenzie D. 1985. ^{230}Th-^{238}U disequilibrium and the melting processes beneath ridge axes. *Earth and Planetary Science Letters* **72**: 149–157.

Qin Z. 1992. Disequilibrium partial melting model and its implications for trace element fractionations during partial melting. *Earth and Planetary Science Letters* **112**: 75–90.

Richter FM. 1986. Simple models for trace element fractionation during melt segregation. *Earth and Planetary Science Letters* **77**: 333–344.

Rubin KH, van der Zander I, Smith MC, Bermmanis EC. 2005. Minimum speed limit for ocean ridge magmatism from ^{210}Pb-^{226}Ra-^{230}Th diseqilibria. *Nature* **437**: 534–538.

Saal AE, Van Orman JA. 2004. The ^{226}Ra enrichment in oceanic basalts: Evidence for diffusive interaction processes within the crust-mantle transition zone. *Geochemistry Geophysics Geosystems* **5**: Art. No. Q02008.

Sigmarsson O. 1996. Short magma chamber residence time at an Icelandic volcano inferred from U-series disequilibria. *Nature* **382**: 440–442.

Tirone M, Ganguly J, Dohmen R, Langenhorst F, Hervig R, Becker H-W. 2005. Rare earth diffusion kinetics in garnet: Experimental studies and applications. *Geochimica et Cosmochimica Acta* **69**: 2385–2398.

Van Orman JA, Grove TL, Shimizu N. 1998. Uranium and thorium diffusion in diopside. *Earth and Planetary Science Letters* **160**: 505–519.

Van Orman JA, Grove TL, Shimizu N. 2001. Rare earth element diffusion in diopside: Influence of temperature, pressure and ionic radius, and an elastic model for diffusion in silicates. *Contributions to Mineralogy and Petrology* **141**: 687–703.

Van Orman JA, Grove TL, Shimizu N, Layne GD. 2002a. Rare earth element diffusion in a natural pyrope single crystal at 2.8 GPa. *Contributions to Mineralogy and Petrology* **142**: 416–424.

Van Orman JA, Grove TL, Shimizu N. 2002b. Diffusive fractionation of trace elements during production and transport of melt in Earth's upper mantle. *Earth and Planetary Science Letters* **198**: 93–112.

Van Orman JA, Saal AE. 2009. Influence of crustal cumulates on ^{210}Pb disequilibria in basalts. *Earth and Planetary Science Letters* **284**: 284–291.

Van Orman JA, Saal AE, Bourdon B, Hauri EH. 2006. Diffusive fractionation of U-series radionuclides during mantle melting and shallow level melt-cumulate interaction. *Geochimica et Cosmochimica Acta* **70**: 4797–4812.

von der Handt A. 2008. Deciphering petrological signatures of reactive melt stagnation and cooling in the oceanic mantle underneath ultraslow-spreading ridges. Ph.D. Dissertation, Johannes-Gutenberg-Universität, Mainz.

von der Handt A, Hellebrand E, Snow JE. 2007. Cooling rates of mantle peridotites estimated from lithophile trace element diffusion in orthopyroxene. Eos Trans. AGU 88(52), Fall Meet. Suppl., Abstract T53B-1316.

Zhang Y, Walker D, Lesher CE. 1989. Diffusive crystal dissolution. *Contributions to Mineralogy and Petrology* **102**: 492–513.

3 Melt Production in the Mantle: Constraints from U-series

BERNARD BOURDON[1] AND TIM ELLIOTT[2]

[1]Institute of Geochemistry and Petrology, ETH Zurich, Switzerland
[2]Bristol Isotope Group, University of Bristol, Bristol, UK

SUMMARY

U-series decay chains provide novel insights into mantle melting processes, notably on melting rates and on the porosity at which magma can separate from melting solid and the depth of melting. We review how such physical information can be derived from these geochemical measurements. U-series analyses are most clearly interpreted in the products of decompression melting, the most common mechanism of melting but also provide useful constraints on other modes of mantle melting such as isobaric heating and flux melting. There is tantalizing potential for U-series to constrain key aspects of mantle dynamics. The fraction of melt retained in the mantle informs on and influences its physical properties and U-series analyses clearly show that the residual melt porosity is ubiquitously low, only a few parts per thousand. Determination of melting rates from U-series measurements provides information about mantle upwelling velocities and mantle lithologies. This approach is of particular appeal in trying to understanding mantle plumes, although such scenarios are still under-constrained.

Timescales of Magmatic Processes: From Core to Atmosphere, 1st edition. Edited by Anthony Dosseto, Simon P. Turner and James A. Van Orman.
© 2011 by Blackwell Publishing Ltd.

INTRODUCTION

The production of melt in the Earth's mantle is one of the most fundamental processes operating in the solid Earth. First, it is the process whereby the oceanic and continental crusts are built. Second, it is a direct sample of the convective mantle at the surface of the Earth and thereby reveals aspects of the composition and working of the planet's interior. Third, it contributes greatly to the heat release from the mantle by rapidly carrying hot material to the surface. Providing a deeper understanding of melt production in the mantle is thus essential for understanding how our planet functions.

U-series nuclides are unique geochemical probes of melting processes since, unlike other geochemical tracers, they strongly depend on the rates and timescales of melting. Specifically, the half-lives of ^{230}Th and ^{231}Pa, the daughters of ^{238}U and ^{235}U, respectively, are of the same order of magnitude as the timescale of melting. Prior to melting, we can safely assume that the solid mantle is in secular equilibrium (i.e. has been undisturbed for >350 ka) and so all measured disequilibria can usefully be attributed to the current process of melting. The degree of ^{230}Th-^{238}U and ^{231}Pa-^{235}U fractionation in erupted magmas will depend on the rate of melt production and segregation. Thus, U-series measurements are likely to shed a different light on the process of mantle melting than provided by traditional major and trace element approaches. However, understanding the dynamic information provided by U-series measurements requires an introduction to time-dependent formulations of melting and so we initially review different melting models.

SIMPLE VS. COMPLEX MODELS FOR MELT PRODUCTION IN THE MANTLE

Three main modes of melt production

The main processes for the production of melt in the mantle can be distinguished as:
1 decompression melting;
2 flux melting; and
3 isobaric heating.

In decompression melting, upwelling mantle moves along a trajectory in pressure and temperature space (typically following an adiabat) that intersects its solidus (that has a shallower slope than the adiabat) and melts as it rises above this point (Figure 3.1).

Solidii are composition-dependent and the composition of the mantle can thus have an important effect on the depth of onset of melting. The driving force for upwelling, that ultimately leads to melting, is either the response to overlying plate motions (passive upwelling) or inherent buoyancy (active upwelling), for example due to higher temperature or compositional differences.

The second mode of melting is very different: hydrous fluids or melts released from subducting slabs move upwards and react with the hotter subsolidus mantle domain. As the solidus temperature is a function of fluid content, the solidus temperature is thereby lowered to an extent that depends on the amount of fluid added (Figure 3.1). However, in situations in which flux melting is implicated, adiabatic upwelling may also occur and it is difficult to clearly distinguish the relative roles of these two processes (Conder *et al.* 2002). Moreover, the flux of other elements carried by the fluids can make signatures of melting difficult to disentangle from those of subduction contributions.

In the third melting mode, hot upwelling mantle heats cold, overlying lithosphere with a lower

solidus, due to the presence of metasomatic, hydrous phases, for example. Thus the melting mantle is static and this scenario is a case of isobaric heating (Figure 3.1).

In principle, these three mechanisms have different controls on their melting rates and thus U-series disequilibria. In the first case, mantle upwelling plays a dominant role in the melting rate for a given mantle composition. In the second case, the melting rate should be controlled by the rate of fluid addition. If fluid addition is a function of dehydration reactions in the slab, the melting rate could potentially depend on the subduction rate, all other parameters being equal. In the third case, the melting rate will be a function of the rate of lithospheric heating, which will depend on plate velocity and the heat flux from the asthenosphere. Below, we examine how the expectations we have from U-series to inform melting processes are realized in natural scenarios.

The description of melt production

There are several parameters that are used to describe the production of melt in the literature and we shall here briefly summarize these definitions.

An often used parameter is the degree of melting (F). F represents the mass fraction of liquid (M_L) relative to the initial mass of mantle involved in the melting processes (M_0):

$$F = \frac{M_L}{M_0}$$

The degree of melting is an average value of mixed, erupted melts that have sampled a wide range of sources that have undergone different degrees of melting. More generally, we can define the degree of melting as:

$$F = \frac{1}{M_0} \iiint_D \frac{dM_L}{dM_{mantle}} dM_{mantle}$$

where D represents the mantle domain undergoing melting with mass M_0. Note that F is

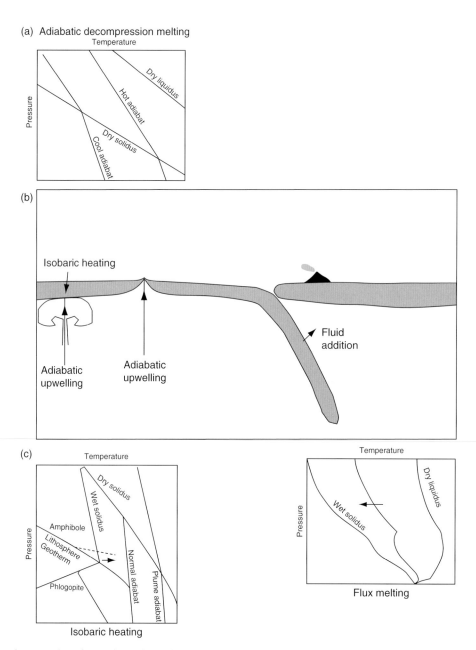

Fig. 3.1 Three modes of mantle melting: (a) adiabatic decompression melting, rising mantle (shown as adiabat curves) intersects the dry solidus at a pressure depending on mantle temperature (b) flux melting due to fluid addition in the mantle wedge; the addition of fluid pushes the solidus to lower temperature and (c) isobaric heating due to heating of the base of a static lithosphere. In this case, the geotherm is disturbed by heating due to a plume and intersect the dry or wet solidus. The three modes of melting are also illustrated in P-T diagrams.

non-dimensional. *F* should not be confused with the matrix porosity (or instantaneous volume fraction of melt), a parameter that is defined locally as:

$$\Phi = \frac{V_{melt}}{V_{solid} + V_{melt}}$$

where V_{melt} and V_{solid} represent the volume of melt and solid matrix, respectively. Melt production (in km³/yr) is also an integral quantity. It is typically defined at a larger length scale and is often defined for a given volcano, or for a given arc (or ridge) system (in km³/km of arc/year) and is usually estimated by calculating erupted volumes. As shown below, this parameter should in some way be related to the degree of melting and to the melting rate.

The melting rate (Γ or sometimes denoted $\dot{M}$) is a parameter that has been defined specifically in the context of U-series studies (McKenzie, 1985), because it is usually not accessible via other observational methods. It is simply defined as the mass of melt produced per unit of volume per unit of time (kg m⁻³ yr⁻¹). It can be related to *F* by the following equation:

$$\Gamma = \frac{\rho_s F}{\Delta t}$$

where ρ_s is the solid density. This last equation can be rewritten in the case of decompression melting as a function of the mantle upwelling rate:

$$\Gamma = \frac{\rho_s F W}{\Delta z} = \rho_s W r_m$$

where *W* is the mantle upwelling velocity and Δz the depth variation during decompression melting. r_m is defined as the melt productivity in weight % of melt per km of adiabatic decompression (this parameter can also be given as % melting per kbar of pressure release). Given experimental constraints on the melting behavior of suitable mantle compositions, the melting rate is a function of the mode of melting, defined above, and can be quantitatively linked to the melting process.

Controls on melting rate

The degree of melting and the melting rate are not constant between different tectonic settings and can be influenced by a number of parameters that we shall now describe.

In the case of mantle melting, the main lithology is supposed to be lherzolite and the main minerals hosting U-series nuclides are clinopyroxene, garnet and accessory phases if they are present. When mantle has been metasomatized, hydrous accessory phases such as amphibole or phlogopite can also be present and they can host a significant amount of U, Th and more importantly Ra (Blundy & Wood, 2003). The melting rates of typical lherzolites have been extensively studied experimentally (Kinzler & Grove, 1992; Hirschmann *et al.* 1999), while there are fewer modern studies of peridotites with hydrous phases (Thibault *et al.* 1992).

A common assumption, based on experimental petrology, has been that melt productivities of 1 to 1.2%/kbar are appropriate for adiabatic decompression melting of dry lherzolite (Langmuir *et al.* 1992). However, Asimow *et al.* (1995) showed both theoretically and based on the MELTS software that this parameter is not constant and that phase transitions for example, should induce large variations in the rate of melting. Furthermore, melt productivity is much smaller for incipient melting (Asimow *et al.*, 1997). These considerations are particularly relevant here because, as highly incompatible elements, U-series nuclides are especially sensitive to the first few percent of melting.

Even in the context of mid-ocean ridge and hotspot melting, water can have a strong influence on melting. A small amount of water (~100 ppm) can greatly increase the depth of the onset of melting and decrease the melting rate and melt productivity (Asimow & Langmuir, 2003). Since there are relatively few measurements of the abundance of water in oceanic magmas, these effects have often been ignored in the U-series literature (Kokfelt *et al.*, 2003: Bourdon *et al.*, 2005). Similarly, the effect of CO_2 on lowering the solidus also needs to be taken into account (Dasgupta & Hirschmann, 2007). A distinction can be made between the case of arc or back-arc melting and intra-plate oceanic islands (e.g. so-called 'wetspots', i.e. the Azores) or mid-ocean ridges. The effect of water at the latter setting is to decrease the rate of melting (Asimow & Langmuir, 2003), while in convergent settings, the addition of water overall increases the degree of melting and melting rate. This difference in the effect of water on melting can be rationalized by the very different mantle trajectories and temperatures in the two scenarios (Hirschmann *et al.*, 1999; Kelley *et al.*, 2006). In the case of plumes or ridges, any intrinsic water in these upwelling sources initiates melting at greater depth than an anhydrous equivalent, but the solidus is not greatly overstepped. If melting is fractional and water supply is limited in the case of ridges, the contrast with arc settings is further enhanced. However, in convergent settings, the parcel of hydrated mantle is initially heated along its trajectory, which greatly enhances melting (Kelley *et al.*, 2006). In this case, the degree of melting is directly related to the amount of water that has been added.

It has been argued also that mafic lithologies are present in the mantle and greatly influence melt production because these lithologies are more fertile. As shown by Hirschmann & Stopler (1996) or Peterman & Hirschmann (2004), mafic lithologies have lower solidii and considerably higher melting rates than peridotites, although the magnitude of these effects depends on the exact compositions. Fertile lithologies such as eclogite may have solidus temperatures 200°C lower than peridotites, whilst wehrlites can start to melt at similar temperatures (Hirschmann & Stopler, 1996). Melts from mafic assemblages should also have distinctive trace element fractionations compared to peridotitic melts, as recently discussed by Stracke & Bourdon (2009). The picture might be more complex if melts from mafic assemblages react with the surrounding peridotite (Yaxley & Green, 1998). Prytulak & Elliott (2009) invoked such a scenario to account for the apparent absence of high melt productivities in the U-series signatures of some ocean island basalts. This in turn contrasts with the work of Sobolev *et al.* (2005, 2007) that argues for a large contribution (15–80%) of melts from mafic lithologies in erupted intra-plate lavas. Such divergent opinions require reconciliation.

U-series melting models

It was recognized about 25 years ago that U-series fractionations observed in mantle-derived rocks could provide important clues about melting rates (McKenzie, 1985). Since this pioneering study, there has been several models proposed for describing time-dependent melting processes based on U-series observations (Qin, 1993, Spiegelman & Elliott, 1993; Iwamori, 1994; Lundstrom, 2000; Jull *et al.*, 2002; Zou & Zindler, 2000; Feineman & DePaolo, 2003; Dosseto *et al.*, 2003; Thomas *et al.*, 2002) and several reviews (Williams & Gill, 1989; Elliott, 1997; Zou & Zindler, 2000; Bourdon *et al.*, 2003; Lundstrom *et al.*, 2003; Elliott & Spiegelman, 2003). Questions we might ponder are:
• Whether the wealth of models that have been proposed are all required; and
• Whether we could cut down the complexity of models to identify key parameters and processes. The review of Lundstrom (2003) presents a rather complete picture of existing models. In this chapter, we focus on melt production and we have attempted to extract the basic features of the models.

Two important end-member models that have been proposed are dynamic melting (McKenzie, 1985) and equilibrium transport melting (Spiegelman & Elliott, 1993). In the first case, melt is extracted from the solid matrix once it reaches a given threshold and thereafter the matrix porosity stays constant. Extracted melt is instantaneously transported to the surface. In equilibrium transport melting, by contrast, the melt is in constant equilibrium with the matrix and the matrix permeability controls the rate of melt migration (Spiegelman & Elliott, 1993).

Melting rates based on U-series are determined in comparison to the rate of radioactive decay. The timescale of melting is of the order of:

$$t_m = \frac{\rho_s F}{\Gamma}$$

The ratio of the melting and radioactive decay timescales must be >1 to expect any significant 'in-grown' disequilibrium (i.e. disequilibrium above normal trace element fractionation effects):

$$\frac{\lambda \rho_s F}{\Gamma} > 1$$

This constraint should be independent of the chosen model and we examine here both the dynamic melting model and equilibrium transport model. If we assume that the partition coefficients are small relative to the degree of melting, it can be shown that the melting rates deduced from dynamic melting or equilibrium transport melting are equivalent, provided some approximations are made. The simplified analytical solution obtained by Spiegelman & Elliott (1993) for equilibrium transport melting (assuming rapid melt transport) is:

$$\left(\frac{^{230}Th}{^{238}U}\right) \approx \left(\frac{^{230}Th}{^{238}U}\right)_{BM} \left(1 + \lambda_{Th}\left(t_U - t_{Th}\right)\right)$$

where $\left(^{230}Th/^{238}U\right)_{BM}$ is the activity ratio obtained for batch melting. We can write a simplified expression for t_i, the residence time of element i in the melting column:

$$t_i = \frac{1}{\Gamma} \frac{\rho_f \phi + \rho_s (1-\phi) D_i}{F + (1-\phi) D_i}$$

By simplifying the equations given in Bourdon & Sims (2003) for dynamic melting, we also obtain the following expressions:

$$\left(\frac{^{230}Th}{^{238}U}\right) \approx \frac{F(1-D_U)}{F(1-D_{Th})} \frac{1 + \lambda_{Th} t_U \dfrac{F + (1-\phi)D_U}{F(1-D_U)}}{1 + \lambda_{Th} t_{Th} \dfrac{F + (1-\phi)D_{Th}}{F(1-D_{Th})}}$$

$$\left(\frac{^{230}Th}{^{238}U}\right) \approx \frac{F(1-D_U)}{F(1-D_{Th})} \left(1 + \lambda_{Th}\left(t_U - t_{Th}\right)\right)$$

These equations confirm the basic intuition that excess daughter nuclide is simply a function of radioactive decay, element residence times in the melting column and melting timescale for both models. It should be noted that the approximations made here are rather crude and the two models do not yield identical results (Bourdon & Sims, 2003). The objective is to illustrate the similarities in concept. This is further illustrated in Figure 3.2, where the curves for dynamic and equilibrium transport melting are compared. In both cases, what controls the magnitude of the Th-U fractionation is the difference in residence times between U and Th in the melting column. In the case of dynamic melting, final disequilibrium is generally smaller because the differences in residence times are only generated near the base of the melting column, while they are generated along the whole melting column in equilibrium transport melting.

In this context, it is also interesting to consider double-porosity models, such as proposed by Lundstrom (2000) and Iwamori (1994) or Jull *et al.* (2002). These models include a component of equilibrium transport melting and a component of dynamic melting and as such represent hybrids between these two extremes. If we assume that the fraction of melt transport in channels is close to one, the model is a dynamic melting, while if this same fraction is 0, the model is an equilibrium transport model. Since the two end-member models are conceptually similar (at least for ^{230}Th and ^{231}Pa disequilibrium), the value of these double-porosity models is to redistribute the locus of production of U-series excesses.

A somewhat different class of models has recently been proposed that relies on the slow diffusion of U-series isotopes relative to melt extraction rates (Van Orman *et al.*, 1998). In such

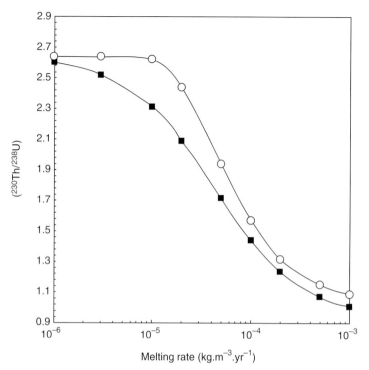

Fig. 3.2 Comparison of equilibrium transport melting (open circles) and dynamic melting (solid squares) as a function of melting rate. For equilibrium transport melting, the simplified equation of Spiegelman & Elliott (1993) was used, while for dynamic melting the equation of Williams & Gill (1989) was used. Matrix porosity was 1‰, bulk partition coefficients for U and Th were 5.3×10^{-3} and 2.8×10^{-3}, respectively. Maximum degree of melting was 0.1.

situations, we need to consider non-equilibrium melting models that include the effects of diffusion. Several generations of such models have been developed (Qin, 1993; Iwamori, 1994; Feineman & DePaolo, 2003; Saal & Van Orman, 2004; Van Orman et al., 2006; Bourdon & Van Orman, 2009). If diffusion is fast relative to the melting rate, these models behave like equilibrium melting models. Interesting behavior arises when diffusion becomes a limiting factor, and especially if the diffusivities of parent and daughter nuclides are different. In such a case, we can consider that the effective partition coefficients of U and Th, for example, depend on the ratio of diffusion coefficients and that U/Th fractionation is enhanced if this ratio differs greatly from 1. In practical terms, this is the case for ^{226}Ra relative to ^{230}Th, but not ^{238}U and ^{230}Th. The diffusion of Th^{4+} is relatively slow while the diffusion of the alkali earth Ra^{2+} should be faster, as gauged from the chemically analogous Ba. The consequence is that ^{226}Ra excesses can be generated by rapid dif-

fusion of ^{226}Ra out of minerals that host the bulk of its parent ^{230}Th (e.g. garnet or clinopyroxene) into adjacent minerals (e.g. plagioclase) that have an affinity for Ra (Saal & Van Orman, 2004; Van Orman et al., 2006). This can lead to large ^{226}Ra excesses during disequilibrium melting of such assemblages. More applications to the study of MORB or OIB melting are given below and in the chapter by Van Orman and Saal (this volume).

HOW DO WE DETERMINE THE MELTING RATE?

Melting rates based on major element geochemistry

Following the approach of Langmuir et al. (1992), it is possible to determine a degree of melting for mid-ocean-ridge basalts using the simplifying assumption that Na_2O concentration in the mantle is relatively constant (~0.3%wt). If the mineral melt bulk partition coefficient for Na during melting is known, we can determine a degree of melting with the following equation for batch melting:

$$F = \frac{1}{1 - D_{Na}}\left[\frac{Na_0}{Na} - D_{Na}\right]$$

where F is the degree of melting, Na content in melt has been corrected for the effect of fractional crystallization, Na_0 is the Na content of the initial mantle source, D_{Na} is the bulk partition coefficient of Na. In the case of mid-ocean ridges, if we assume a corner flow (in a melting regime with an upper boundary at $45°$ to the horizontal), the melting rate Γ can then be determined. If W is the mantle upwelling velocity, the melting rate is equal to:

$$\Gamma = \frac{\rho_s W F}{z}$$

where z is the depth of melting, and ρ_s is the solid matrix density. For the mid-Atlantic ridge, assuming an upwelling rate equal to a typical half-spreading rate of $1\,km/yr$, an onset of melting at $80\,km$ and a total of 10% melting, the melting rate is equal to $\sim 4 \times 10^{-5}\,kg/m^3/yr$. As we shall see below, this estimate is close to other estimates based on U-series disequilibria. In the case of hotspot melting, such an estimate cannot be calculated because there is no simple estimate of W, the mantle upwelling velocity or the degree of melting F. The same situation applies to melting at convergent margins where no geophysical observable can be safely assumed to scale with the melting rate.

Melting rate determination based on U-series

U-series disequilibria measured in basalts can potentially be used to determine melting rates in a wider variety of settings. The general principle is as follows: prior to melting, the U-series decay chain should be in secular equilibrium. During melting the various elements found in the chain (U, Th, Ra and Pa) will fractionate, one relative to another,

due to differences in their mineral-melt partition coefficients. If the degree of melting is large, there is little *net* fractionation between ^{232}Th and ^{238}U as the mineral-melt partition coefficients are small. In this case, degrees of disequilibria are primarily a function of melting rate and of the porosity at which melt segregates. In theory, the system can thus be constrained using only two U-series nuclide pairs, for example ($^{230}Th/^{238}U$) and ($^{231}Pa/^{235}U$) pairs. As outlined above, there are several possible melting model formulations and so the melting rate determinations will be model-dependent.

Key input parameters for the determination of melting rates are the U-Th-Ra-Pa partition coefficients. Since Ra and Pa are found to be extremely incompatible in typical mantle assemblages, it turns out that in most cases, the partition coefficients for U and Th are the most relevant. There is now abundant literature on this topic, as summarized in Blundy & Wood (2003). As shown in Figure 3.3, U-Th partition coefficients for the same mineral structure vary by over an order of magnitude. Clearly, accurate bulk partition coefficients require a good knowledge of the mineral compositions involved. Assessing appropriate mineral modes to calculate bulk partition coefficients is another issue to address. The abundance of U-Th bearing phases, i.e. garnet, clinopyroxene, phlogopite or amphibole (or possibly accessory phases) is not known *a priori* but is frequently estimated from a generic mantle peridotite composition. Thus, there remains large uncertainties in U-series partitioning at the onset of melting.

Potentially more problematic in quantifying the melting process from U-series measurements is that many believe the mantle source to be lithologically heterogeneous. As discussed above, mafic lithologies melt at a faster rate than host peridotite during adiabatic decompression (Petermann & Hirschmann, 2003, Kogiso & Hirschmann, 2006). This should result in low U-series disequilibria in melts from mafic components. However, mafic lithologies also have larger garnet and clinopyroxene modes, which affect the fractionation of some nuclide pairs (e.g. ^{230}Th-^{238}U) in an opposite sense to the effect of faster melting rates. Figure 3.4 illustrates the

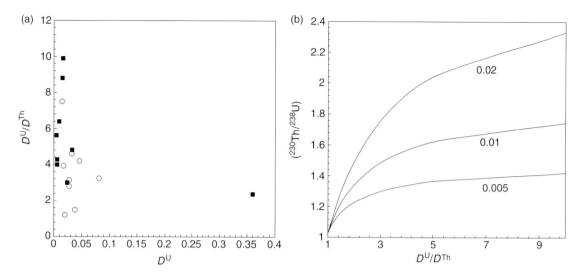

Fig. 3.3 (a) Compilation of U-Th partition coefficients determined experimentally for garnet/melt based on the compilation by Stracke *et al.* (2006). (b) (^{230}Th/^{238}U) in a melt produced by dynamic melting as a function of the partition coefficients. The curves are labeled for various values of D^U. Solid squares are relevant for mafic compositions while open circles are relevant for ultramafic compositions.

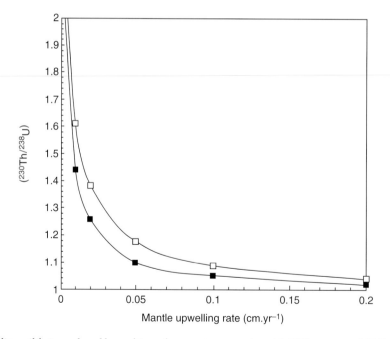

Fig. 3.4 ^{230}Th-^{238}U disequilibria produced by melting of a garnet pyroxenite with 20% garnet and 80% clinopyroxene (open squares). The melting rate varies during melting using the parameterization of Peterman & Hirschmann (2003) from 16%/GPa to 70%/GPa. If we increase the initial melting rate (e.g. see results for another pyroxenite in Kogiso and Hirschmann 2006), the ^{230}Th-^{238}U disequilibria is even smaller (solid squares).

^{230}Th-^{238}U fractionation in a melt from a mafic source as a function of mantle upwelling rate. Overall, the influence of pyroxenite melts on (^{230}Th/^{238}U) activity ratios of averaged erupted melts is relatively limited and does not necessarily lead to large ^{230}Th or ^{231}Pa excesses in the melt, as argued by Hirschmann & Stopler (1996). Nevertheless, it introduces considerably more variables (proportion of mafic component, degree of melting, melting rate and porosity of melt segregation for the mafic phase) in modeling erupted U-series disequilibria, which makes extracting quantitative information about the melting process considerably more difficult. On the other hand, if the pyroxenite melt reacts with the peridotite, as argued by Prytulak & Elliott (2009), the U-series signature of pyroxenite melting may no longer be detectable in the final erupted melt.

As raised earlier, the water content of the mantle source is another important consideration. Thanks to the detailed parameterization of Asimow & Langmuir (2003), the role of water on melting rate can be assessed more precisely. Water in the mantle source lowers the solidus temperature, which means that the initial depth of melting in a rising mantle column will increase. However, in this context, once the mantle starts melting, the melt productivity (and the corresponding melting rate) are extremely low. The net effect of water is thus to increase the U-series fractionation (i.e. greater ^{230}Th/^{238}U and ^{231}Pa/^{235}U). An additional complexity arises from the type of melt extraction. In the case of fractional melting, the removal of water from the system will lead to an even slower melting rate because the melt-matrix assemblage will become drier. This will not be the case for batch or equilibrium transport melting because the removal of water is less efficient.

Based on this discussion, it is apparent that derivation of melting rate based on U-series can be rather under-constrained, although under favorable circumstances important information can be gleaned. Moreover, assessing relative variations in melting rates over a limited area, rather than absolute values, is a potentially valuable exercise. This approach is further illustrated below. Furthermore, source heterogeneities

can usually be identified with radiogenic isotopes such as Sr, Nd and Pb and in this case, the effect of these heterogeneities on U-series systematics can be assessed at least qualitatively (Bourdon *et al.*, 1996, Sims *et al.*, 1999). In the end, a combination of various observations (U-series, major and trace elements) will provide the most robust constraints on melting rates (Sims *et al.*, 2002).

MELTING RATES AND GEODYNAMIC PROCESSES

In this section, we describe how U-series disequilibria measured in recent lavas have been used to determine melting rates in various geodynamic settings based on the melting models describe above.

Linking melting rate with upwelling rates

Early studies of U-series in mid-ocean ridge and ocean island basalts have realized that U-series disequilibria were sensitive to mantle upwelling rates and thereby melting rates (Cohen *et al.*, 1993; Bourdon *et al.*, 1996; Lundstrom *et al.*, 1998; Bourdon *et al.*, 1998; Sims *et al.*, 1999). For example, the observations of a relationship between ^{230}Th excess in OIB and buoyancy fluxes was interpreted as due to variations in the mantle upwelling velocity (Chabaux & Allègre, 1994; Cohen *et al.*, 1993; Bourdon *et al.*, 1998). Similarly, the increasing ^{230}Th and ^{231}Pa excesses as a function of distance to plume center, found by Sims *et al.* (1999), and also by Kokfelt *et al.* (2003), were interpreted as reflecting a radial decreases in upwelling velocities. Since there is no simple method to determine the upwelling velocity of mantle at intra-plate settings based on geophysical observables (as the relationship between buoyancy flux and upwelling velocity is also a function of excess temperature which is also unknown), the U-series observations at hotspots provide a potentially important tool to look at the thermal and velocity structure in mantle plumes.

Bourdon *et al.* (2006) showed that both ^{231}Pa and ^{230}Th excesses correlate negatively with buoyancy fluxes and that this trend could be explained by variations in mantle upwelling velocities as buoyancy fluxes are proportional to W^2, although this negative array is pinned by the large Hawaiian plume at one end and ridge-centered plumes at the other. In addition, U-series data indicate that mantle plumes with large buoyancy fluxes seem to have higher excess temperatures. As shown by Bourdon et al. (2006), these observations are consistent with the idea that smaller plumes experience more cooling during their rise through the mantle, as inferred by Albers & Christensen (1996).

Such gross trends should not hide the complexity of mantle melting and, in some cases (i.e. Azores), the presence of water or of more fertile components can obscure the simple U-series systematics (Bourdon et al., 2005). For example, the presence of water will slow down melting in the first increments of melting only near the center of the plume and the net result will be to lower the contrast in ^{230}Th excess as a function of radial distance. While several plumes do show clear U-series systematics as a function of radial distance (Sims et al., 1999; Kokfelt et al., 2003), others such as Pitcairn or the Canaries do not (Thomas et al., 1999; Bourdon & Van Orman, 2009; Lundstrom et al., 2003). In both of these cases, melting of the lithosphere has been invoked and this will complicate the U-series systematics. A more extensive discussion of lithospheric melting is given below.

For mid-ocean ridge basalts, Bourdon et al. (1996) observed that ^{230}Th excess were correlated negatively with the axial ridge depth and interpreted this observation as a result of deeper melts, erupted at shallower ridges having a stronger garnet signature (i.e. larger ^{230}Th excess). A depth dependence of U-Th fractionation can also be accounted for by the increasing D^U/D^{Th} of clinopyroxene with pressure (Wood et al., 1999, Landwehr et al., 2001). Interestingly there was no obvious spreading rate dependence on the global dataset, which seemed consistent with theoretical arguments made by Richardson and McKenzie for a two-dimensional melting regime. A more recent compilation of U-series disequilibria in MORB by

Elliott & Spiegelman (2003) shows little correlation of (^{230}Th/^{238}U), and more importantly of (^{231}Pa/^{235}U), with spreading rates. Samples derived from actively upwelling mantle (i.e. moving faster than dictated by plate spreading rates), however, should plot below the ^{230}Th excess vs axial depth trend, as could be the case for Iceland (Bourdon et al., 1996).

In contrast, Lundstrom et al. (1998) emphasized aspects of the ^{230}Th-^{238}U disequilibrium in MORB (namely the slope defined by individual localities on an 'equiline' plot) that seemed to be controlled by ridge spreading rates. In detail, Lundstrom et al. (1998) modeled this effect using an equilibrium porous flow model and mixing melts from two sources (of mafic and peridotitic compositions respectively) with different melting properties as discussed above. More complex, double porosity models have also been used to account for negative trend between ^{226}Ra and ^{230}Th excess found in the East Pacific Rise. These models involve a combination of porous flow at grain boundaries and channel flow (Lundstrom, 2000; Jull et al., 2002; Iwamori, 1993). While these models can explain the data, Stracke et al. (2006) illustrated that a simple dynamic melting involving explicit consideration of melt transport times could also account for these observations. Thus some of these inferences are model-dependent.

Melting beneath arc volcanoes

The melt production beneath arc volcanoes is controlled by a radically different set of parameters compared to hotspot and ridge melting. The release of water from the subducted slab lowers the solidus temperature of the mantle wedge, thereby causing its melting (Figure 3.1). Beyond this relative simple picture, there are still many outstanding questions regarding the exact process of melting in the mantle wedge. We initially outline the main hypotheses that have been formulated in the literature. Given that the depth to the surface of the slab beneath arc was thought approximately constant, melting was inferred to be controlled by a pressure-dependent reaction (Gill, 1981). In this case it has been suggested that the breakdown of amphibole was the main trigger for mantle melting (Tatsumi, 1986). However, later work by Schmidt & Poli (1998) emphasized that the loss of water occurred all along the subduction of the slab in the mantle. A more recent study by England & Wilkins (2004) (see also Abers *et al.*, 2006; Grove *et al.*, 2009) argues for a temperature control of the position of the arc. Another hypothesis has been that melting should be viewed as a flux melting process whereby the dehydration leads to continuous melting of mantle overriding the slab (Thomas *et al.*, 2002). In this case, the narrow zone of volcanoes along the arc would be due to an efficient focusing of melt to the active arc region. More recently, seismic observations beneath Japan (Tamura *et al.*, 2002) and the modeling work of Gerya & Yuen (2003) has suggested rising diapirs due to the buoyancy of the subducted sediment in the mantle wedge.

An exact delimitation of the melting zone inside the mantle wedge has also proven difficult. As argued by Grove *et al.* (2002), there could be two zones of melting, the hydrated mantle above the wedge and the ascending hot asthenosphere between arc and back arc region, as has been shown in numerical models with temperature dependent viscosity. This brief overview clearly illustrates our lack of understanding of even the fundamental style of melting in the mantle wedge.

From the U-series viewpoint, the melting beneath arc volcanoes is still a very active area of research and several models that fit more or less rigorously the various possibilities outlined above have been proposed. First, it has been clearly shown that both ^{238}U-^{230}Th and ^{226}Ra-^{230}Th disequilibria are dominated by a slab 'fluid' signature (see review in Turner *et al.* 2003 and chapter by Turner and Bourdon this volume for detailed arguments). Although the experimental evidence is not so clear-cut, it is generally thought that U (especially if it is in an oxidized form) will partition preferentially in a fluid phase compared with Th. As shown by Kessel *et al.* (2005), these simple systematics break down at higher pressure and temperature when fluids become supercritical. More recently the importance of accessory

minerals such as allanite in controlling the U-Th 'fluid' fractionations, rather than the nature of the 'fluid' itself, has been stressed (Klimm *et al.*, 2008; Hermann & Rubatto, 2009). In principle, a 'fluid' signature derived from an allanite bearing slab (Hermann, 2002) should be characterized by large ^{238}U and ^{226}Ra excesses.

In contrast with U-Th disequilibrium, the U-Pa systematics in arcs reveal a clear melting signature as they generally show a large Pa enrichment, despite being affected in parallel by a slab fluid signature. It can be argued that the melting signature in this case overwhelms any previous ^{235}U enrichment relative to ^{231}Pa. Thus, U-Pa systematics in arcs have been used to investigate melting processes at subduction zones (Bourdon *et al.*, 1999, 2003; Thomas *et al.* 2002; Dosseto *et al.* 2003; Turner *et al*, 2006; Huang & Lundstrom, 2007). In principle, U-Pa systematics can be used to determine melting rate, although the difficulties described above in constraining melting rate even in the OIB case, where the melting process is better understood and all measured disequilibria can be related to melting, should be borne in mind.

A key observation is that the melting rates cannot be too slow if the slab-derived fluid signature witnessed by ^{226}Ra excesses is to be preserved (unless the different disequilibria are generated by the independent components that contribute to arc volcanism). This consideration enables some bounds to be placed on melting rates in arcs (Bourdon *et al.*, 2003), as melting rates also need to be of the order of 10^{-4} kg/m^{-3}/yr or lower to produce significant ^{231}Pa excesses. The exact parameter controlling the melting rate is still a matter of debate. Turner *et al.* (2006) have argued that the melt production is directly linked to the convergence rate of the subduction zone, as ^{231}Pa excesses in a compilation of arcs showed a very broad negative array with convergence rates. However, the slope of this array is pinned by the rather anomalous Kick 'em Jenny volcano (a highly unusual alkalic arc volcano in the Lesser Antilles) and nearly the full range of ^{231}Pa excesses are evident in individual arc suites at locations with *constant* subduction rate (Huang &

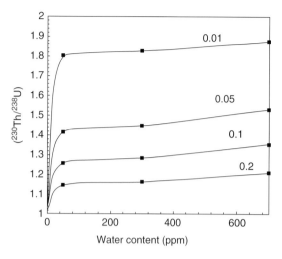

Fig. 3.5 Effect of water on ^{238}U-^{230}Th melting systematics as a function of water content in the source in ppm, based on the parameterization of Asimow & Langmuir (2003) and the modeling approach of Bourdon *et al.* (2005) The modeling is based on dynamic melting with variable melting rates. The matrix porosity is 1‰ and the partition coefficients for wet and dry melting are those used in Bourdon *et al.* (2005). The curves are labeled with the upwelling rate in cm/yr.

Lundstrom, 2007). The basic concept underlying these studies is that the melt production simply follows the progressive heating of hydrated mantle as it is being dragged downward by the slab. This implicitly assumes that the coupling between wedge and slab is identical everywhere. Alternatively, we could consider that the rate of melting reflects the rate at which the slab gets heated and loses fluids by continuous dehydration. This would then be consistent with the model of flux melting proposed by Thomas *et al.* (2002), whereby the rate of fluid addition controls the rate of melting. However, the model proposed by Thomas *et al.* (2002) was based on a significant mobility of Pa in fluids, which is not supported by evidence from the low-T Pa chemistry, although high P,T fluids need not have the same properties (Kessel *et al.*, 2005).

Another source of information about melting rates could possibly come from ^{230}Th-^{238}U systematics (Figure 3.5). If U-Pa systematics are affected

by melting, so too should ^{230}Th-^{238}U fractionations. The relative partitioning of U-Th and U-Pa and the half-lives of Th and Pa are not identical but we can predict that melting should modify the slope of the U-Th array in the equiline diagram (Turner *et al.*, 2003). A slow melting rate will lead to a steep slope in the U-Th isochron diagram, while a large melting rate will only increase the slope of array slightly. If the melting is relatively slow, then the slope of the array records the in-growth of ^{230}Th following fluid addition during the melting process. An alternative interpretation could be that the slope of the array in the U-Th isochron diagram is mostly controlled by the time since fluid dehydration (Figure 3.6) and this would provide little constraint on melting rate.

This discussion illustrates some of the current dilemma in the interpretation of U-series in arc volcanoes. There is still ambiguity due to an incomplete knowledge of the underlying melting processes, partitioning behavior in the subduction zone environment and the relative roles of slab additions relative to melting effects. The diversity of geodynamical models precludes straightforward interpretations.

Melting rate due to lithospheric heating

In some intraplate settings, it has been argued that melting of the base of the lithosphere can play a significant role in magma genesis (Class & Goldstein 1997; Class *et al.*, 1998). Such lithospheric melting is rather distinct from adiabatic decompression melting. A classical example would be a mantle plume that heats up the base of the lithosphere and therefore causes isobaric melting (Gallagher & Hawkesworth, 1992). In this case, the rate of melting depends on the rate of vertical heat transfer as well as the velocity of plate motion that controls the timescale over which this external heat source is applied to the base of the lithosphere. If the base of the lithosphere has been previously metasomatized, then melting of a hydrated lithosphere will be facilitated by a lower solidus temperature.

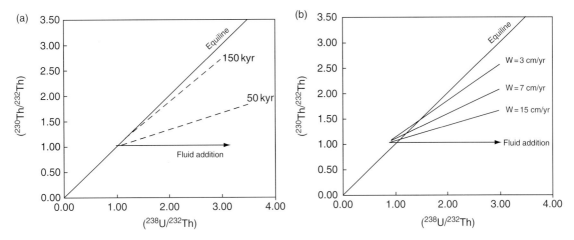

Fig. 3.6 U-Th fractionation in subduction zone setting. (a) Classical interpretation of ^{238}U-^{230}Th fractionation in convergent margins. The enrichment in ^{238}U is due to fluid addition and the slope of the data array in the U-Th equiline diagram is the age since fluid addition. (b) The fluid addition produces a horizontal vector that is then modified by melting processes. The slope of the array is a function of melting rate and down-welling rate (here given in cm/yr).

The study of U-series disequilibria can provide additional insight into lithospheric melting as:
• The extent of U-series nuclides fractionation is again sensitive to the melting rate;
• The direction of ^{226}Ra-^{230}Th fractionation may be sensitive to the presence of hydrous phases such as phlogopite or amphibole (Claude-Ivanaj *et al.*, 1998).
Following the earlier work of Class & Goldstein (1997), Claude Ivanaj *et al.* (1998) investigated U-series systematics of the Grande Comore island and found a clear difference in the U-Th-Ra data for the Karthala and La Grille volcanoes. Based on isotope and trace element analyses, Class & Goldstein (1997) had concluded that the La Grille lavas showed clear evidence for lithospheric melting and further argued for the presence of amphibole in the source of La Grille. Interestingly, recent la Grille lavas also exhibited smaller ^{226}Ra excess and this was attributed to the influence of residual amphibole, which should preferentially retain Ra relative to Th.

Studies of continental basalts in the southwest United States by Asmerom *et al.* (2000) have suggested that lithospheric melting leads to large ^{231}Pa-^{235}U but small ^{230}Th-^{238}U fractionations. The

large ^{231}Pa excesses were attributed either to a very small degree of melting (0.1% range) or to chromatographic exchange during melt percolation through the lithosphere. If this occurs within the spinel stability field, only modest ^{230}Th-^{238}U fractionations would be expected. Fittingly, melts from the same area, inferred to be derived from deeper, garnet-bearing asthenosphere, have much higher ^{230}Th-^{238}U fractionations.

More recently, a study of Pitcairn seamounts by Bourdon & Van Orman (2009) has shown that the Pitcairn lavas can be explained by a mixture between lithospheric and plume-derived melts. Strikingly, the lithospheric melts are characterized by ^{226}Ra deficits that have rarely been observed in other settings. The presence of a substantial amount of water in the lavas, coupled with an unusual Nb/La fractionation was used to argue that the ^{226}Ra deficits are due to the presence of phlogopite in the source, as Ra is more compatible than Th in phlogopite. Furthermore, the presence of ^{230}Th excesses indicates a garnet-bearing source and in this case, most of the ^{226}Ra budget must have been initially hosted in the garnet that holds most of the ^{238}U in the rock. Prior to melting, diffusion of Ra from garnet to phlogopite will lead to a large enrichment in Ra in the phlogopite. Modeling by Bourdon & Van Orman (2009) demonstrated that, depending on the melting rate, the aggregated melt can have either ^{226}Ra excesses or deficits. In this special case, the melting rate of the lithosphere can be further constrained by ^{230}Th-^{238}U disequilibria and shown to be consistent with conductive heating of the lithosphere by the underlying mantle plume. It is possible that a similar scenario of lithosphere heating and relatively slow melting could also explain the observations of Asmerom et al. (2000) in the southwest United States. Interestingly, the resulting melting rate is not greatly different to that produced by adiabatic decompression melting.

POROSITY OF THE MELTING REGION

Observations of U-series disequilibria in mid-ocean ridge and ocean islands basalts also provide key constraints on the porosity of the melting region. As U and Th are both highly incompatible in mantle minerals, the existence of excess ^{230}Th, ^{231}Pa and ^{226}Ra in erupted lavas puts some rather tight bounds on the porosity of the melting region (McKenzie, 1985; Williams & Gill, 1989; Spiegelman & Elliott, 1993). A simplified treatment of dynamic melting derived by McKenzie (2000) shows that the activity ratio r in the melt can be approximated with the following equation:

$$r = \frac{1 + D^{parent} / \Phi + (\tau_{1/2}^{daughter} / 0.693\tau_{melt})}{1 + D^{daughter} / \Phi + (\tau_{1/2}^{daughter} / 0.693\tau_{melt})}$$

where $\tau_{1/2}^{daughter}$ and τ_{melt} are the mean life of the daughter nuclide and the residence time of the melt, respectively. This equation shows that an excess of the daughter in the melt can only be significant if $D^{daughter}$ (the solid/melt partition coefficient for the daughter nuclide) is distinct from D^{parent} and if D^{parent} is sufficiently relative to the matrix porosity Φ. In particular, in order to explain the relatively large ^{231}Pa and

^{226}Ra excesses in MORB (Rubin & MacDougall, 1988; Goldstein et al., 1993) and OIB (Oversby & Gast, 1968; Pickett & Murrell, 1997), the matrix porosity in the melting region has to be of the order of a few ‰. Clearly, a porosity of a few % would be too large. In the context of dynamic melting, as melt has no further interaction with the solid once it is extracted, the matrix porosity would have to be low only at the base of the melting region where a significant fraction of uranium is still residing in the solid matrix. Above that region, the matrix is so depleted in uranium that it should have little influence on the U-series disequilibria in the melt.

The inferred matrix porosity for the melting region has major implications for our understanding of melt extraction processes (see also chapter by Turner and Bourdon, this volume and O'Neill and Spiegelman, this volume).

CONCLUSIONS

Conceptually U-series disequilibria offer an attractive means of investigating melt production in various geodynamic settings, as they place constraints on aspects of melting not probed by more traditional tracers. As illustrated in this chapter, some of the expectations have already been fulfilled but large areas still remain obscure. Unfortunately, the information we can extract from U-series disequilibria is model-dependent and the parameters required to provide clear-cut solutions to the melting scenarios are rarely available.

An outstanding constraint derived from U-series studies is that the matrix porosity during mantle melting is of the order of a few permil, which is significantly smaller than previously thought. The studies of hotspots also illustrates significant variation in melting rates that are probably relative to variations in mantle upwelling. These melting rates increase with the size of hotspots and generally, the melting rate at the center of hotspots is faster than at ridges.

Our understanding of melting at convergent margins is still imperfect and critically depends on a better knowledge of partitioning behavior of U-series nuclides between fluid, melt and solids. As an area of active research, more progress is likely to arise in the next five years.

REFERENCES

Abers GA, van Keken PE. Kneller EA, Ferris A, Stachnik JC. 2006. The thermal structure of subduction zones constrained by seismic imaging: Implications for slab dehydration and wedge flow. *Earth and Planetary Science Letters* **241**: 387–397.

Albers M, Christensen U.-R. 1996. The excess temperature of plumes rising from the core-mantle boundary. *Geophysical Research Letters* **23** & **24**: 3567–3570.

Asimow P, Hirschmann MM, Ghiorso MS, O'Hara MJ, Stolper EM. 1995. The effect of pressure-induced solid-solid phase-transitions on decompression melting of the melting. *Geochimica et Cosmochimica Acta* **59**: 4489–4506.

Asimow PD, Hirschmann MM, Stolper EM. 1997. An analysis of variations in isentropic melt productivity. *Philosophical Transactions of the Royal Society A* **355**: 255–281.

Asimow PD, Langmuir CH. 2003. The importance of water to oceanic mantle melting regimes. *Nature* **421**: 815–820.

Asmerom Y, Cheng H, Thomas R, Hirschmann M, Edwards RL. 2000. Melting of the Earth's lithospheric mantle inferred from protactinium-thorium-uranium isotopic data. *Nature* **406**: 293–296.

Blundy JD, Wood BJ. 2003. Mineral-melt partitioning of Uranium, Thorium and their daughters. In: *U-series Geochemistry 52*, Bourdon B, Lundstrom C, Henderson G. Turner SP. (eds), Mineralogical Society of America, pp. 59–118.

Bourdon B, Langmuir C, Zindler A. 1996. Ridge-hotspot interaction along the Mid-Atlantic ridge between 37°30 and 40°30 N: the U-Th disequilibrium evidence. *Earth and Planetary Science Letters* **142**: 175–196.

Bourdon B, Zindler A, Elliott T, Langmuir CH. 1996. Constraints on mantle melting at mid-ocean ridges from global ^{238}U-^{230}Th disequilibrium data. *Nature* **384**: 231–235.

Bourdon B, Joron J-L, Claude-Ivanaj C, Allègre CJ. 1998. U-Th-Pa-Ra systematics for the Grande Comore vol-

canics: melting processes in an upwelling plume. *Earth and Planetary Science Letters* **164**: 119–133.

Bourdon B, Turner S, Allègre CJ. 1999. Melting dynamics beneath the Tonga-Kermadec island arc inferred from ^{231}Pa-^{235}U systematics. *Science* **286**: 2491–2493.

Bourdon B, Sims KWW. 2003. U-series constraints on intraplate basaltic magmatism, Uranium-series geochemistry. *Reviews in Mineralogy and Geochemistry* **52**: 215–254.

Bourdon B., Turner, S, Dosseto, A. 2003. Dehydration and partial melting in subduction zone: constraints from U-series disequilibria. *Journal of Geophysical Research* **108**: ECV3-1–19.

Bourdon B, Turner SP, Ribe NM. 2005. Partial melting and upwelling rates beneath the Azores from a U-series isotope perspective. *Earth and Planetary Science Letters* **239**: 42–56.

Bourdon B, Ribe NM, Stracke A, Saal AE, Turner SP. 2006. Dynamics of mantle plumes: insights from U-series geochemistry. *Nature* **444**: 713–717.

Bourdon B, Van Orman JA. 2009. Melting of enriched mantle beneath Pitcairn seamounts: Unusual U-Th-Ra systematics provide insights into melt extraction processes. *Earth and Planetary Science Letters* **277**: 474–481.

Chabaux F, Allègre CJ. 1994. ^{238}U-^{230}Th-^{226}Ra disequilibria in volcanics- a new insight into melting conditions. *Earth and Planetary Science Letters* **126**: 61–74.

Class C, Goldstein SL. 1997. Plume-lithosphere interactions in the ocean basins: constraints from the source mineralogy. *Earth and Planetary Science Letters* **150**: 245–260.

Class C, Goldstein SL, Altherr R, Bachelery P. 1998. The process of plume-lithosphere interactions in the ocean basins – The case of Grande Comore. *Journal of Petrology* **39**: 881–903.

Claude-Ivanaj C, Bourdon B, Allègre CJ. 1998. Ra-Th-Sr isotope systematics in Grande Comore Island: a case study of plume-lithosphere interaction. *Earth and Planetary Science Letters* **164**: 99–117.

Cohen AS, O'Nions RK. 1993. Melting rates beneath Hawaii: Evidence from uranium series isotopes in recent lavas. *Earth and Planetary Science Letters* **120**: 169–175.

Conder JA, Wiens DA, Morris J. 2002. On the decompression melting structure at volcanic arcs and back-arc spreading centers. *Geophysical Research Letters* **29**: 1727, doi:10.1029/2002GL015390.

Dasgupta R, Hirschmann MM. 2007. Effect of variable carbonate concentration on the solidus of mantle peridotite. *American Mineralogist* **92**: 370–379.

Dosseto A, Bourdon B, Joron JL, Dupré B. 2003. U-Th-Pa-Ra study of the Kamchatka arc: new constraints on genesis of arc basalts *Geochimica et Cosmochimica Acta* **67**: 2857–2877.

Elliott T. 1997. Fractionation of U and Th during mantle melting: a reprise. *Chemical Geology* **139**: 165–183.

Elliott T, Spiegelman M. 2003. Melt Migration in Oceanic Crustal Production: A U-series Perspective *Treatise on Geochemistry* **3**: 465–510, doi:10.1016/B0-08-043751-6/03031-0.

England P, Wilkins C. 2004. A simple analytical approximation to the temperature structure in subduction zones. *Geophysics Journal International* **159**: 1138–1154.

Feineman MD, DePaolo DJ. 2003. Steady-state $^{226}Ra/^{230}Th$ disequilibrium in mantle minerals: Implications for melt transport rates in island arcs. *Earth and Planetary Science Letters* **215**: 339–355.

Gallagher K, Hawkesworth C. 1992. Dehydration melting and generation of continental flood basalts. *Nature* **358**: 57–59.

Gerya TV, Yuen DA. 2003. Rayleigh-Taylor instabilities from hydration and melting propel 'cold plumes' at subduction zones. *Earth and Planetary Science Letters* **212**: 47–62.

Gill J. 1981. *Orogenic Andesites and Plate Tectonics, Minerals and Rocks, 16.* Springer-Verlag, Berlin, 390 pp.

Goldstein SJ, Murrell MT, Williams RW. 1993. ^{231}Pa and ^{230}Th Chronology of Mid-ocean Ridge Basalts. *Earth and Planetary Science Letters* **115**: 151–159.

Grove TL, Parman SW, Bowring SA, Price RC, Baker MB. 2002. The role of an H_2O-rich fluid component in the generation of primitive basaltic andesites and andesites from the Mt. Shasta region, N California. *Contributions to Mineral Petrology* **142**: 375–396.

Grove TL, Till CB, Lev E, Chatterjee N, Médard E. 2009. Kinematic variables and water transport control the formation and location of arc volcanoes. *Nature* **459**: 694–697.

Hermann J. 2002. Allanite: thorium and light rare earth element carrier in subducted crust. *Chemical Geology* **192**: 289–306.

Hermann J, Rubatto D. 2009. Accessory phase control on the trace element signature of sediment melts in subduction zones. *Chemical Geology* **265**: 512–526.

Hirschmann MM, Stolper EM. 1996. A possible role for pyroxenite in the origin of the garnet signature in MORB. *Contributions to Mineral Petrology* **124**: 185–208.

Hirschmann MM, Asimow PD, Ghiorso MS, Stolper EM. 1999. Calculation of peridotite partial melting form thermodynamic model of minerals and melts III. Controls on isobaric melt production and the effect of water on melt production. *Journal of Petrology* **40**: 831–851.

Huang F, Lundstrom CC. 2007. ^{231}Pa excesses in arc volcanic rocks: Constraint on melting rates at convergent margins. *Geology* **35**: 1007–1010.

Iwamori H. 1993. Dynamic disequilibrium melting model with porous flow and diffusion controlled-chemical equilibration. *Earth and Planetary Science Letters* **114**: 301–313.

Iwamori H. 1994. ^{238}U-^{230}Th-^{226}Ra and ^{235}U-^{231}Pa disequilibria produced by mantle melting with porous and channel flows. *Earth and Planetary Science Letters* **125**: 1–16.

Jull M, Kelemen P, Sims KWW. 2002. Consequences of diffuse and channeled porous melt migration on uranium series disequilibria. *Geochimica et Cosmochimica Acta* **66**: 4133–4148.

Kelley KA, Plank T, Grove TL, Stolper EM, Newman S, Hauri E. 2006. Mantle melting as a function of water content beneath back-arc basins. *Journal of Geophysical Research* **111**: B09208.

Kessel R, Schmidt MW, Ulmer P, Pettke T. 2005. Trace element signature of subduction-zone fluids, melts and supercritical liquids at 120–180 km depth. *Nature* **437**: 724–727.

Kinzler RJ, Grove TL. 1992. Primary magmas of mid-ocean ridge basalts. 1. Experiments and methods. *Journal of Geophysical Research* **97**: 6885–6906.

Klimm K, Blundy JD, Green TH. 2008. Trace element partitioning and accessory phase saturation during H_2O-saturated melting of basalt with implications for subduction zone chemical fluxes. *Journal of Petrology* **49**: 523–553.

Kogiso T, Hirschmann MM. 2006. Partial melting experiments of bimineralic eclogite and the role of recycled mafic oceanic crust in the genesis of ocean island basalts. *Earth and Planetary Science Letters* **249**: 188–199.

Kokfelt TF, Hoernle KA, Hauff F. 2003. Upwelling and melting of the Iceland plume from radial variation of ^{238}U-^{230}Th disequilibria in postglacial volcanic rocks. *Earth and Planetary Science Letters* **214**: 167–186.

Landwehr D, Blundy J, Chamorro-Perez E, Hill E, Wood B. 2001. U-series disequilibria generated by partial melting of spinel lherzolite. *Earth and Planetary Science Letters* **188**: 329–348.

Langmuir CH, Klein EM, Plank T. 1992. Petrological systematics of mid-ocean ridge basalts – constraints on melt generation beneath ocean ridges. In: *Mantle flow and melt generation at mid-ocean ridges*, Morgan JP, Blackman DK, Sinton JM. (eds), Book Series: Geophysical Monograph Series **71**: 183–280.

Lundstrom C. 2000. Models of U-series disequilibria generation in MORB: the effects of two scales of melt porosity. *Physics of the Earth and Planetary Interiors* **121**: 189–204.

Lundstrom CC. 2003. Uranium-series disequilibria in mid-ocean ridge basalts: Observations and models of basalt genesis, in Uranium-series geochemistry. *Reviews in Mineralogy and Geochemistry* **52**: 175–214.

Lundstrom CC, Williams Q, Gill JB. 1998. Investigating solid mantle upwelling rates beneath mid-ocean ridges using U-series disequilibria, 1: a global approach. *Earth and Planetary Science Letters* **157**: 151–165.

Lundstrom CC, Hoernle K, Gill JB. 2003. U-series disequilibria in volcanic rocks from the Canary Islands: Plume versus lithospheric melting. *Geochimica et Cosmochimica Acta* **67**: 4153–4177.

McKenzie D. 1985. ^{230}Th-^{238}U disequilibrium and the melting processes beneath ridge axes. *Earth and Planetary Science Letters* **72**: 149–157.

McKenzie D. 2000. Constraints on melt generation and transport from U-series activity ratios. *Chemical Geology* **162**: 81–94.

Oversby VM, Gast PW. 1968. Lead isotope composition and uranium decay series disequilibrium in recent volcanic rocks. *Earth and Planetary Science Letters* **5**: 199–206.

Peterman M, Hirschmann MM. 2003b. Partial melting experiments on a MORB-like pyroxenite between 2 and 3 GPa: Constraints on the presence of pyroxenite in basalt source regions from solidus location and melting rate. *Journal of Geophysical Research* **108**: doi:10.1029/2000JB000118.

Peterman M, Hirschmann MM, Hametner K, Günther D, Schmidt MW. 2004. Experimental determination of trace element partitioning between garnet and silica-rich liquid during anhydrous partial melting of MORB-like eclogite. *Geochemistry, Geophysics, Geosystems* **5**: doi:10.1029/2003GC000638.

Pickett DA, and Murrell MT. 1997. Observations of ^{231}Pa/^{235}U disequilibrium in volcanic rocks. *Earth and Planetary Science Letters* **148**: 259–271.

Prytulak J, Elliott T. 2009. The melt productivity of ocean island basalt sources from 238U-230Th and 235U-231Pa disequilibria. *Geochimica et Cosmochimica Acta* **73**: 2103–2122.

Qin ZW. 1993. Dynamics of melt generation beneath midocean ridge axes-theoretical analysis based on ^{238}U-^{230}Th-^{226}Ra and ^{235}U-^{231}Pa disequilibria *Geochimica et Cosmochimica Acta* **57**: 1629–1634.

Rubin KH, Macdougall JD. 1988. ^{226}Ra excesses in mid-ocean-ridge basalts and mantle melting. *Nature* **335**: 158–161.

Saal, AE, Van Orman JA. 2004. The ^{226}Ra enrichment in oceanic basalts: evidence for diffusive interaction within the crust-mantle transition zone. *Geochemistry, Geophysics, Geosystems* **5**: Q02008, doi: 02010.01029/02003GC000620, pp. 000617.

Schmidt M.W, Poli S. 1998. Experimentally based water budgets for dehydrating slabs and consequences for arc magma generation. *Earth and Planetary Science Letters* **163**: 361–379.

Sims KWW, DePaolo DJ, Murrell MT, *et al.* 1999. Porosity of the melting zone and variations in the solid mantle upwelling rate beneath Hawaii: inferences from ^{238}U-^{230}Th-^{226}Ra and ^{235}U-^{231}Pa disequilibria. *Geochimica et Cosmochimica Acta* **63**: 4119–4138.

Sims KWW, Goldstein SJ, Blichert-Toft J, *et al.* 2002. Chemical and isotopic constraints on the generation and transport of magma beneath the East Pacific Rise. *Geochimica et Cosmochimica Acta* **66**: 3481–3504.

Sobolev AV, Hofmann AW, Sobolev SV, Nikogosian IK. 2005. An olivine-free mantle source of Hawaiian shield basalts. *Nature* **434**: 590–597.

Sobolev A.V, Hofmann AW, Kuzmin DV, *et al.* 2007. The amount of recycled crust in sources of mantle-derived melts. *Science* **316**: 412–417.

Spiegelman M, Elliott T. 1993. Consequences of melt transport for uranium series disequilibrium in young lavas. *Earth and Planetary Science Letters* **118**: 1–20.

Stracke A, Bourdon B, McKenzie D. 2006. Constraints on melt extraction mechanisms in the Earth's mantle from U-Th-Pa-Ra studies in oceanic basalts. *Earth and Planetary Science Letters* **244**: 97–112.

Stracke A. Bourdon B. 2009. The importance of melt extraction for tracing mantle heterogeneity. *Geochimica et Cosmochimica Acta* **73**(1): 218–238.

Tamura Y, Tatsumi Y, Zhao DP, Kido Y, Shukuno H. 2002. Hot fingers in the mantle wedge: New insights into magma genesis in subduction zones. *Earth and Planetary Science Letters* **197**: 105–116.

Tatsumi Y. 1986. Formation of the volcanic front in subduction zones. *Geophysical Research Letters* **13**: 717–720.

Thibault Y, Edgar AD, Lloyd FE. 1992. Experimental investigation of melts from a carbonated phlogopite lherzolite – Implications for metasomatism in the continental lithospheric mantle. *American Mineralogist* **77**: 784–794.

Thomas RB, Hirschmann MM, Cheng H, Reagan MK, Edwards RL. 2002. (^{231}Pa/^{235}U)-(^{230}Th/^{238}U) of young mafic volcanic rocks from Nicaragua and Costa Rica and the influence of flux melting on U-series systematics of arc lavas. *Geochimica et Cosmochimica Acta* **66**: 4287–4309.

Thomas LE, Hawkesworth CJ, Van Calsteren P, Turner SP, and Rogers NW. 1999. Melt generation beneath ocean islands: a U-Th-Ra isotope study from Lanzarote in the Canary Islands. *Geochimica et Cosmochimica Acta* **63**: 4081–4099.

Turner S, Bourdon B, Gill J. 2003. Insights into magma genesis at convergent margins from U-series isotopes, in Uranium-series geochemistry. *Reviews in Mineralogy and Geochemistry* **52**: 255–315.

Turner S, Regelous M, Hawkesworth C, Rostami K. 2006. Partial melting processes above subducting plates: Constraints from ^{231}Pa–^{235}U disequilibria. *Geochimica et Cosmochimica Acta* **70**: 480–503.

Van Orman JA, Grove TL, Shimizu N. 1998. Uranium and thorium diffusion in diopside *Earth and Planetary Science Letters* **160**: 505–519.

Van Orman JA, Saal AE, Bourdon B, Hauri EH. 2006. Diffusive fractionation of U-series radionuclides during mantle melting and shallow-level melt-cumulate interaction. *Geochimica et Cosmochimica Acta* **70**: 4797–4812.

Williams RW, Gill JB. 1989 Effects of partial melting on the uranium decay series. *Geochimica et Cosmochimica Acta* **53**: 1607–1619.

Wood BJ, Blundy JD, Robinson JAC. 1999. The role of clinopyroxene in generating U-series disequilibrium during mantle melting. *Geochimica et Cosmochimica Acta* **63** (**10**): 1613–1620.

Yaxley GM, Green DH. 1998. Reactions between eclogite and peridotite: mantle refertilisation by subduction of oceanic crust Schweiz. Mineral. Petrogr. Mitt. **78**: 243–255.

Zou HB, Zindler A. 2000. Theoretical studies of ^{238}U-^{230}Th-^{226}Ra and ^{235}U-^{231}Pa disequilibria in young lavas produced by mantle melting. *Geochimica et Cosmochimica Acta* **64**: 1809–1817.

4 Formulations for Simulating the Multiscale Physics of Magma Ascent

CRAIG O'NEILL[1] AND MARC SPIEGELMAN[2]

[1]GEMOC ARC National Key Centre, Department of Earth and Planetary Sciences, Macquarie University, Sydney, NSW, Australia
[2]Lamont-Doherty Earth Observatory of Columbia University, Palisades, NY, USA

SUMMARY

Melt migration from a mantle source to eventual eruption may involve a variety of processes operating over a range of scales. For instance, melt ascent begins with grain-scale interactions, which, for larger fluxes may be described by permeable flow. Field evidence demonstrates that melt eventually coalesces to forms vein or dyke networks, which may rapidly ascend and be erupted, or be emplaced at depth, where they may feed into crustal magma chambers. A variety of mathematical formulations exist to describe these processes at different scales, and this contribution draws these diverse approaches together to form a framework for quantifying magma ascent from source to surface.

INTRODUCTION

The migration of melt from its source zone is a critical process for surface volcanism, crustal formation and differentiation. However, the detailed processes involved are in many ways still poorly understood. One reason for this is the large range of scales that melt migration encompasses, from the grain scale to the emplacement of plutons. Whilst the field evidence may be equivocal for the relative importance and interactions of these processes, many formulations exist describing the different aspects of the melt ascent system. The key to developing a quantitative understanding of melt migration processes is to develop numerical models of the highly non-linear coupled equation sets describing the dynamics of this fluid-solid system. Fundamental to this, though, is having a mathematical formulation of the problem wee want to model. The purpose of this contribution is to bring together a number of these formulations for modeling magma migration, and discuss their implications for the ascent of melt from its source to its final emplacement.

Melting initiates with supersolidus melting of a heterogenous mineral assemblage, either in the mantle or crust. Whilst the details of this melting process are not the primary concern here, U-series disequilibria have shown that this incipient melt migrates from its source at very low melt fractions, and so the discussion of melt ascent begins with physics of grain-scale melt migration and interaction. At the macroscale, this interstitial flow leads naturally to a permeable flow formulation for melt migration. Such formulations have

Timescales of Magmatic Processes: From Core to Atmosphere, 1st edition. Edited by Anthony Dosseto, Simon P. Turner and James A. Van Orman.

been widely adopted to study melt migration in the mantle (McKenzie, 1984), with particular focus on spreading ridge systems (Scott & Stevenson, 1989; Spiegelman, 1996). We will present and discuss one of the better understood formulations, known as the 'McKenzie Equations' (McKenzie, 1984). However, field evidence illustrates that permeable flow is not a complete description of melt migration, as partial melt generically coalesces to form networks of veins (Sleep, 1988), and at a critical scale, dykes, which are capable of rapid ascent via fracture propagation (Rubin, 1995). Larger dykes may erupt or be emplaced at depth, depending on their buoyancy, flow rate and temperature. Alternatively, they may feed into crustal magma systems, such as mid/upper crustal magma chambers. Such magma chambers may produce a more mixed magma supply for a volcanic plumbing system and eventually erupt or, depending again on their temperature, composition and overpressure, they may in emplaced at depth (Burov et al., 2003). Another alternative is that partially molten plutons may be able to migrate to shallower crustal levels. The mechanism by which this very large-scale ascent occurs is controversial, and proposed mechanisms include diapiric ascent, ballooning by continual magma supply, dyking/stoping or tectonically-related migration (Weinberg & Podladchikov, 1994).

Here we review the most commonly implemented formulations of these physical processes, with a specific view towards their numerical implementation. Where numerous implementations exist, we have generally opted for the more widespread, generic or easily understood. We also review the implications of the models for the dominant physical processes, major bottlenecks for migration, and ascent rates.

GRAIN-SCALE PROCESSES

Melting initiates at grain boundaries, forming discrete microscopic pockets. As the melt fraction increases, eventually these discrete pods of melt will interconnect, forming a network and thus allowing the melt to migrate through the system. The point at which this occurs depends not only on the melt fraction, but also on the way in which melt is distributed along the grain boundaries. At this scale, the distribution of melt along the grains depends strongly on surface tension and capillary effects; the melt may either minimize its volume due to its own surface tension and form small discrete pods on the grain boundaries, or it may do the opposite and spread out like a film (Cheadle et al., 2004; Laporte & Watson, 1995). Historically, this effect has been encapsulated in a parameter called the dihedral angle (Figure 4.1; Laporte & Watson, 1995). For low dihedral angles ($\theta < 60°$), the melt spreads out and interconnects, whilst for large dihedral angles, the melt tends to pod. In the latter case, melt cannot form an interconnected network and migrate from the source until a large melt

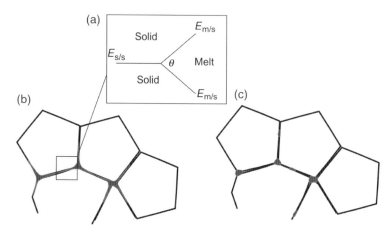

Fig. 4.1 The effect of the dihedral angle on the distribution of melt on grain boundaries. (a) The dihedral angle set by the surface energies between the melt and the grains $(E_{m/s})$, and the grains themselves $(E_{s/s})$. It is defined as $\cos(\theta/2) = E_{s/s}/2E_{m/s}$. (b) Distribution of melt if the dihedral angle is <60°. Melt forms an interconnected network at low melt fractions, whereas (c) shows the distribution of melt for dihedral angle >60°. The dihedral angle for melt coexisting with olivine is <50°. See Plate 4.1 for a colour version of these images.

fraction is reached. For an olivine-melt system, experimental evidence suggests that the dihedral angle is <~50°, allowing melt to migrate at very low melt fractions (Daines & Kohlstedt, 1993).

Hier-Majumder *et al.* (2006), following Bercovici *et al.* (2001) and Ricard *et al.* (2001), explore the effect of grain boundary tension of the distribution and migration of melt in a two-phase composite. Generally, they find that strong tension on the grain-melt interface tends to result in a small dihedral angle. This implies a well-connected distribution of melt and high permeability. However, Hier-Majumder *et al.* (2006) suggest that in this case melt may be retained in tubules along grain junctions due to capillary action, leading to somewhat inefficient melt extraction and the decay of melt pulses during ascent. In contrast, they suggest that aggregates with large dihedral angles may in fact give rise to self-separation of the melt from the matrix due to strong grain-melt tension effects. They also found the behavior of melt was strongly dependent on the 'disaggregated melt fraction' – the melt fraction at which the grain boundaries were completely wetted. In addition, since the dihedral angle depends on the minimization of surface energies, they also show strong temperature and pressure dependence (Holness, 1993).

For mantle melt migration, the physics of the grain-scale processes reduces to the situation where very small incipient melt fractions can form an interconnected network due to their low dihedral angle, consistent with U-series constraints (see Turner and Bourdon, and Bourdon and Elliot, this volume). At this point they will migrate by porous flow from their source region, and the mathematical basis for this is developed next.

POROUS FLOW IN A DEFORMABLE MEDIA

Formulation

At a macroscopic scale, the complexities of grain-scale migration reduce to a description of porous flow of a liquid through a permeable but deformable solid matrix. The problem has a long history, and one of the fundamental, and widely adopted,

equation sets was produced by McKenzie (1984). In its basic form, it considers the macroscopic conservation of mass, momentum and energy for coexisting liquid and solid-matrix continua. As derived by McKenzie (1984), the equations reduce to a coupled system for viscous mantle deformation by Stoke's flow, and porous flow as per Darcy's law for the fluid. The mass and momentum conservation equations can be written as:

$$\frac{\partial \rho_f \phi}{\partial t} + \nabla \cdot [\rho_f \phi v_f] = \Gamma$$

$$\frac{\partial \rho_s (1 - \phi)}{\partial t} + \nabla \cdot [\rho_s (1 - \phi) v_s] = -\Gamma$$

$$\phi(v_f - v_s) = -\frac{K}{\mu} [\nabla P - \rho_f g]$$

$$\nabla P = \nabla \cdot (\eta [\nabla v_s + \nabla v_s^T]) + \nabla [(\zeta - \frac{2}{3} \mu) \nabla \cdot v_s] + \bar{\rho} g$$

Here ρ_f, ρ_s are the solid and fluid densities, ϕ the porosity, v_f and v_s the fluid and solid velocities, Γ the melting (or crystallization) rate, K the permeability, μ the melt viscosity, P the fluid pressure, g the gravitational acceleration, η and ζ the shear and bulk viscosities of the solid, and $\bar{\rho} = \rho_f \phi + \rho_s (1 - \phi)$ the mean density of the solid and fluid phases (Spiegelman & McKenzie, 1987).

The first two equations describe the conservation of mass for the fluid and solid, respectively. In contrast to the simple incompressibility constrain generally applied in mantle convection simulations (Moresi & Solomatov, 1998), here the solid mass can diminish in size as the porosity (and hence melt) increases, and the rate at which mass is converted from the solid to fluid phase is governed by the melting rate Γ. The third equation is a modified form of Darcy's law, describing the separation of the melt from the solid. The rate at which this separation proceeds depends on the permeability K, and the pressure gradient available to drive fluid flow. The equation above is the momentum equation for the viscous matrix, assuming it is inertia free (i.e. Stoke's flow) and in this case compressible.

Constitutive relationships

The system of the four equations above possess inherent non-linearities giving rise to complex dynamical behavior. This non-linear complexity arises primarily due to constitutive relations of the permeability, and the shear and bulk viscosities η and ζ. The permeability is generally assumed to be some function of the porosity ϕ, and in most cases is modeled by a simple power law relationship, i.e. $K = a\phi^n$, where $n = 2\sim5$, and a is a proportionality constant.

The bulk viscosity describes how a material resists volume changes, and so should trend to infinity for an incompressible material (as assumed for the solid matrix). However, in the case of porous flow, the bulk viscosity is a property of the solid-fluid aggregate, and is generally assumed to be of a similar magnitude to the shear viscosity. The shear viscosity of the solid, η, is not only a function of temperature and strain-rate, but also of porosity in a two-phase situation. Detailed constitutive relationships for permeability, shear and bulk viscosities are still somewhat lacking, and these parameters are often assumed constant or to vary in a simple well-defined manner for most cases (Spiegelman, 1993a).

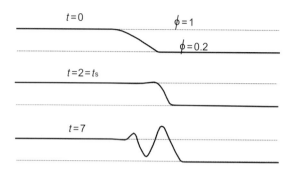

Fig. 4.2 Schematic of the evolution of a non-dimensional solitary porosity wave, for an induced porosity contrast of 1.0 to 0.2. The porosity distribution evolves into a shock front at t_s (in reality, the viscosity of the matrix prevents a true shock wave from forming). This viscous resistance causes a dispersion of the shock front, resulting in a series of porosity waves.

ple, the dimensionless pressure-based version of the four equations above (Spiegelman 1993b) can be written as:

$$\frac{D\phi}{Dt} = P$$

$$-\nabla \cdot \phi^n \nabla P + P = \nabla \cdot \phi^n \hat{g}$$

Porous flow examples

Magmatic solitary waves

One of the most interesting physical implications of this melt transport formulation, is that they predict magma solitons – non-linear porosity waves, which propagate faster than the melt (Scott & Stevenson, 1984). Though these do not transport the melt themselves, they may impact on fluid pathways, and so are important in any discussion of porous compaction driven flow.

The equations above allow for the matrix to either compact as a result of melt flux out of an element, or dilate to allow melt flux into a local area. Since the relationship between porosity and permeability is a power law, this introduces a non-linearity into the migration of porosity zones, and thus the equations support non-linear porosity waves, or magma solitons. As an exam-

These assume the limit of small porosity ($\phi \ll 1$), no melt production ($\Gamma = 0$), constant η and $\zeta = 1$, and again assuming the permeability $K = \phi^n$. These coupled equations give rise to non-linear solitary waves in one, two- or three-dimensional simulations. The equations are valid for a near-static solid, except for a small compressibility on the order of the porosity. The waves propagate over a low uniform porosity background, with a constant speed (i.e. constant phase velocity) and fixed form. An example is shown in Figure 4.2.

The speed of porosity waves depends on the relationship of porosity with permeability, and the viscous resistance of the matrix (Spiegelman, 1993b). For most commonly used power-law permeability relationships, $\partial K/\partial \phi$ is positive, and so melt flux variations will steepen into non-linear waves, or even porosity shocks in the absence of viscous effects. However, as noted, the porosity

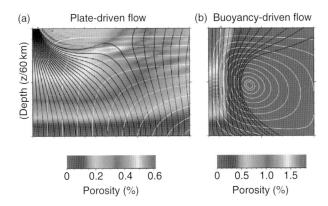

(a) Plate-driven flow (b) Buoyancy-driven flow

Depth (z/60 km)

0   0.2   0.4   0.6
Porosity (%)

0   0.5   1.0   1.5
Porosity (%)

Fig. 4.3 Porosity and flow-lines for two mid-ocean ridge models. (a) Assumes purely plate driven flow. White lines represent mantle flow lines, black lines represent melt flowlines. (b) For lower shear viscosity, buoyancy of the melt can dominate the flow-field, resulting in a focused higher viscosity beneath the ridge crest. The buoyancy number R, which denotes the relative contribution of buoyancy and plate-driven flow, is 0.26 is the plate-driven flow case, and 169 in the buoyancy driven flow case (modified from Spiegelman, 1996). See Plate 4.3 for a colour version of these images.

waves travel faster than the melt, and so do not transport melt themselves. This complex porosity behavior, however, impinges on fluid flow paths and, potentially, the rates at which fluids flow through the background porosity field.

A variety of numerical implementations for this phenomenon exist, and further details are provided in Spiegelman (1993c), Barcilon & Richter (1986) and Wiggins & Spiegelman (1995).

Mid-ocean ridge melt focussing

Spiegelman (1996) and Scott & Stevenson (1986, 1989) discuss the physics of melt extraction beneath a mid-ocean ridge. One of the most important questions for MOR melt transport, is how melt, which is produced in a wide melting zone beneath the ridge swell, is focused beneath the ridge so that it erupts within ~2 km of the ridge crest (Spiegelman, 1996; Buck & Su, 1989; Sparks & Parmentier, 1991).

Two mechanisms contribute to melt focusing beneath a ridge, shear induced by the overriding plate motions, and shear due to the buoyancy of the melt. In the case of negligible melt buoyancy, flow is essentially a corner-driven flow, as shown in Figure 4.3. Large shear is induced in the viscous matrix by plate motions, which creates pressure gradients that focus melt flow. However, if the plate motions (or viscosities) are lower, then variations in melt content can result in buoyancy driven flow beneath the ridge, focusing melt beneath the ridge crest. Here the low density of

the melt, and the melt residuum, relative to the asthenosphere, result in large buoyancy forces beneath the ridge crest where most melting occurs, driving a much faster upwelling directly beneath the ridge, focusing the melt column (Scott & Stevenson, 1989). Spiegelman (1996) argues on the basis on stable trace element predictions that buoyancy-driven flow must have limited effect in real-world MOR systems.

The rate of melt extraction beneath a MOR is limited by the total melt fraction reached, and the melt productivity – which is a factor of mantle temperature and upwelling velocity. McKenzie (1984) estimated the extraction rate for simple matrix compaction to be ~1 cm/yr for porosities of around ~3%. Seismic studies of MOR systems constrain the regional melt fractions to be <~3% (Forsyth *et al.*, 1998), so this is a reasonable estimate for the rate at which porous flow in the matrix feeds into higher permeability channels. Obviously, flow within the channels themselves, which have a much higher local melt fraction, will be significantly higher.

Development of shear bands

Katz *et al.* (2006) describe the formation of melt shear bands in a matrix with variable shear viscosity. In this case melt production and buoyancy are set to zero ($\Gamma = 0$, g = 0). The viscosity is dependent on both the strain rate and the porosity. An example from Katz *et al.* (2006) is shown in Figure 4.4.

Porosity (simulation), vol. frac. gamma = 2.79

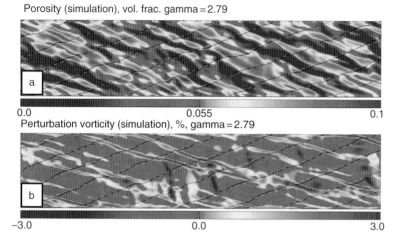

Fig. 4.4 Porosity and perturbation vorticity (the vorticity minus the simple shear contribution) fields demonstrating the development of low-angle shear bands in a numerical experiment based on a laboratory core deformation experiment. In this experiment $n = 6$, $\alpha = -27$, and the shear strain is 2.79. Black lines are tracers which were initially vertical at commencement of shear, white dotted lines demonstrate the expected position due to only simple shear (modified from Katz *et al.*, 2006). See Plate 4.4 for a colour version of this image.

When the shear viscosity depends solely on porosity weakening, existing melt is localized in pure and simple shear bands, in orientations determined by the fastest opening direction (Stevenson, 1989; Spiegelman, 2003). Katz *et al.* (2006) present a relationship between the shear viscosity and porous weakening plus strain rate. They assumed a relationship of the form:

$$\eta(\phi, \dot{\varepsilon}) = \eta_0 \, e^{\alpha(\phi - \phi_0)} \, \dot{\varepsilon}_{II}^{\frac{1-n}{n}}$$

where η_0 is the reference shear viscosity, at the reference porosity. The porosity weakening coefficient $\alpha = 28$, is experimentally derived (Hirth & Kohlstedt, 1995; Mei *et al.*, 2002). $\dot{\varepsilon}_{II}$ is the second invariant of the strain-rate tensor (i.e. the effective strain rate). The strain weakening exponent n was found to fit experiments best when equal to 6 (Katz *et al.*, 2006).

The localization of melt into shear bands is an important effect in tectonically active regions, and results in drastically faster melt migration velocities than for pure percolative flow. In the experiments where strain-rate weakening dominated porosity weakening, low-angle high porosity bands where found to evolve stably at angles 15 to 25° to the plane of shear (Katz *et al.*, 2006). For the MOR melt focusing problem, pure Newtonian rheologies (i.e. no strain weakening) would result in melt bands aligned away from the ridge axis, and so would not focus melt towards the ridge. However, the rotation of the melt bands observed for highly non-Newtonian rheologies (n = 6, with strain-rate weakening), would result in band orientations that would, if interconnected, focus melt directly towards the ridge axis. There is a clear length scale dependence here, which will be expanded on in the following sections.

The development of shear bands allows an interconnectivity of melt at much lower melt fractions than is ordinarily the case, rapidly sourcing surrounding porous melt, and enabling more rapid and focused melt extraction than would be the case without deformation of the matrix.

Reactive flow

Extensive dunite veins in the Oman ophiolite have been interpreted to be reactive melt transport zones (see Spiegelman *et al.* (2001) and Kelemen *et al.* (1997) for summaries). This is consistent with the observations:
• That MOR basalt is not in equilibrium with shallow abyssal peridotite opx – it must have been sourced deeper and not reacted, which is the case if it moved through the dunite zones (Kelemen *et al.*, 1997);
• MOR basalt is in equilibrium with residual cpx in the dunite zones, suggesting they were conduits; and

• Field evidence suggests the dunite zones were reaction/melt infiltration fronts (Kelemen *et al.*, 1997).

Spiegelman *et al.* (2001) summarize the mechanism known as reactive flow, where migrating melt preferentially reacts with (and melts) one component (here pyroxene, as pyroxene solubility increases as melt ascends). Thus regions of larger melt flux dissolve more pyroxene, increasing permeability, which again enhances the local flux – and a positive feedback ensues, resulting in channelization.

In order to model this, Spiegelman *et al.* (2001) assumes that the two above equations can be cast in terms of the concentration of the soluble phase of the solid, c^s (i.e. pyroxene), and the melt, c^f – rather than as a simple one component solid-fluid mix as before. To do so, we need a relationship for the conservation of composition, or how solid pyroxene might melt. The linear dimensionless dissolution rate may be written as:

$$\Gamma = DaA'(c_{eq}^f - c^f)$$

where

$$Da = \frac{R\delta}{\rho_s \phi_0 w_0} \qquad A' = \frac{c^s(1 - \phi_0 \phi)}{c_0^s (1 - \phi_0)}$$

where c_{eq}^f is the concentration of the fluid at equilibrium (i.e. reaction stops). Da is known as the Damköhler number, which describes the amount of reaction that occurs as the fluid moves one compaction length (Spiegelman *et al.*, 2001). If Da is large, a lot of reaction happens, if small, little reaction occurs over that time. R is the reaction rate constant, and w_0 the velocity at the reference porosity ϕ_0. For the full non-dimensional equations and implementation, see Spiegelman *et al.* (2001).

The models of Speigelman *et al.* (2001) show channel localization of 90% of the melt flux into <20% or the area. Melt reactions create permeability pathways, and solid compaction of the matrix outside the channels forces the flow into these structures. They suggest such channel systems

Fig. 4.5 Schematic of a bifurcating vein network. A large number of smaller veins upstream tap the background porous flow, and coalesce to form one large transporting dyke. See Plate 4.5 for a colour version of this image.

could form on scales of ~100,000 years on Earth, and would have a spacing of around 1 to 200 m.

VEIN NETWORKS

The ability of the extended McKenzie equations to generate channelized flow highlights the close relationship between descriptions of flow through networks of veins, and continuum descriptions of two-phase flow, which do, in fact, parameterize the behavior of a microscopic vein network in their permeability description. Here we consider a more stylized division between the two mechanisms, and look at the ability of background porous flow to fill a growing vein (Figure 4.5).

Veins are characterized as low pressure regions compared to the surrounding rock, which tend to cause background melt to flow towards them (Ribe, 1986; Sleep, 1988; Stevenson, 1989). This pressure gradient surrounding a dyke extends to a distance comparable to the compaction length δ_c (McKenzie, 1984), and this is given by:

$$\delta_c^2 = K \frac{(\lambda_2 + 2\eta_2)}{\eta_m}$$

where K is the permeability, λ_2 is the Lame coefficient (an elastic coefficient) and η_2 the shear viscosity of the solid grains, and η_m is the melt viscosity. Following from Weinberg (1999) and Sleep (1988), we adopt a permeability description $K = Ca^2f^3$, where $C = 10^{-3}$ is a constant, a is the grain size, and f the melt fraction. For $a = 1\,\mathrm{mm}$, permeability will be $10^{-15}\,\mathrm{m}^2$ for 1% melt, or $\sim 10^{-13}\,\mathrm{m}^2$ for $\sim$6% melt. Assuming the latter, and letting $\lambda_2 + 2\eta_2 = 10^{18}\,\mathrm{Pa/s}$ (Sleep, 1988), and adopting a low basaltic melt viscosity of $1\,\mathrm{Pa/s}$ (Rubin, 1995), then we get a compaction length of over $300\,\mathrm{m}$ – sufficient to drain a large volumes of partially molten source.

Another constraint on the growth of veins or dykes is the porous infilling rate. This infilling rate can be estimated from the pressure gradient driving flow towards the vein, ∇P:

$$v = \frac{\nabla P.k}{\eta_m}$$

Using again a permeability of $10^{-13}\,\mathrm{m}^2$, a viscosity of $1\,\mathrm{Pa/s}$, and a pressure gradient of $10\,\mathrm{MPa/m}$ (Wickham, 1987; Etheridge et al., 1984), the infilling velocity is $\sim$31 m/yr. Again, this is fast enough to ensure fairly rapid accumulation of melts in a distributed vein network on the order of a few years.

Flow through a vein or tributary dyke network can be considered as a contribution from a finite number of individual dykes. The structure of such a network has been discussed in terms of a bifurcating system, similar to tributary river networks. A large number of smaller veins exist upstream, tapping the porous flow source. These coalesce to form larger veins/dykelets, which eventually merge to form one large transporting dyke. The average velocity in an individual dyke (for laminar flow, see Petford et al. (1993) and Weinberg (1999)) can be written as:

$$\bar{v} = \frac{\Delta\rho g w^2}{12\eta_m}$$

Here $\Delta\rho g$ is the density dependent pressure gradient which drives melting, w is the dyke width, and η_m is the melt viscosity. The flow rate through the dyke is then given by:

$$Q = \bar{v}.w$$

For a mantle-derived basaltic melt, we can use the parameters $\Delta\rho = 300\,\mathrm{kg/m}^3$, $w \sim 0.1\,\mathrm{m}$, and $\eta_m \sim 10\,\mathrm{Pa/s}$, to obtain an average flow velocity of $v = 24\,\mathrm{cm/s}$, and $Q = 0.024\,\mathrm{m}^2/\mathrm{s}$. For basaltic style melts, flow rates through macroscopic dykes and veins are extremely fast, and able to drain a melt source efficiently over short timescales. It should be noted that this rate is extremely sensitive to melt viscosity and vein diameter, and can potentially be faster or slower than the infilling rate.

Finally, what structure might a feeding vein network have? This is a difficult question for mantle derived melts, or even for felsic migmatitic sources (Weinberg, 1999). For the case of river systems, the tributary network can be classified in terms of a self-similar system, and if characteristics such as tributary dimensions, spacing, diameter exponent (determining how the flow rates add), maximum tributary order and bifurcation ratio of the system are known, then the system structure can be constrained (Hart, 1993). Existing field evidence suggests this may be an appropriate analogy for feeding vein networks (Kelemen et al., 2000). As veins/dykelets coalesce, their flow rates simply add together, and thus in steady state, the flow rate through the 'highest order' transporting vein is equal to the sum of the flow rates through the smallest, lowest order veins (Weinberg, 1999).

DYKING

The term dyking here is used to refer specifically to the ability small volume magma filled fractures to propagate by elastic cracking of the host rock – differentiating dykes from vein networks and porous shear bands. The transition from diffuse melt (or, indeed, from a large melt body) into a critical volume dyke, which may then propagate vertically, is a difficult problem, and not fully understood. The mechanics of dyking has a long vintage, and a thorough review of this literature is provided by Rubin (1995). In this particular work, we again focus on the problem of mantle melt extraction, and formulations of dyke propagation focusing on extraction of mantle-derived melts.

The problem of dyke formation is really an extension of melt localization in a vein network, with the end result an elongated body of melt that is able generate sufficient stress to fracture the surrounding rock. There are two aspects to the problem, the ability to generate a mature feeder network for the dyke, as discussed in the last section, and the ability of the nascent dyke to efficiently drain this network and propagate (Weinberg, 1999). This is the starting point for the following discussion.

Dyke propagation formulisms

The fundamentals of dyke propagation are outlined in Rubin (1995), and a specific application of dyke propagation to the problem of mantle melt extraction was explored by O'Neill *et al.* (2007), which we follow here.

Whether or not a dyke will be able to self-propagate depends on whether the elastic stress generated by the dyke exceeds the fracture toughness of a rock, K_c. This condition is encompassed by the crack tip intensity factor, K, where:

$$K = 1.12\, \Delta P_c (l)^{1/2}$$

where ΔP_c is the critical overpressure, l is the dyke length, and $K > K_c$ for fracture propagation.

For crustal dykes, the overpressure may depend on the excess pressure from a feeder magma chamber; however, as noted by Rubin (1995), for most mantle dykes, the overpressure comes primarily from the dyke's inherent buoyancy with respect to the mantle. To borrow the example of O'Neill *et al.* (2007), a typical critical K_c of ~5 MPa/m$^{1/2}$, this condition is reached for fairly minor melt volumes of around ~10 m^3. If we adopt a nominal overpressure of ~1 MPa, then this fracture criterion is met for dyke lengths of a few meters, and thus should be met in most mantle melting events.

However, Rubin (1995) notes the limiting criterion to dyke propagation is not elastic fracture, but that the flow rate through the dyke is sufficient so that the dyke widens faster than it narrows to freezing of the margins. If not, as a dyke propagates into cold host rock, it will cool, become more viscous, slow, and eventually freeze. This freezing condition is encapsulated in the dimensionless parameter β:

$$\beta = 2\left(\frac{3\kappa\eta_m}{\pi\,\Delta P_c}\right)^{\frac{1}{2}} \frac{c\,|\,dT_{wr}\,/\,dz\,|}{L}\left(\frac{\Delta P_c}{M_{elas}}\right)^{-2}$$

where κ is the thermal diffusivity, c is the specific heat, η_m is the magma viscosity, L is the latent heat of melting, dT_{wr}/dz is the temperature gradient in the wall rock in the dyke path, and M_{elas} is the elastic stiffness. If the calculated value of β is <0.15, then the dyke will not freeze out. O'Neill *et al.* (2007) calculates the critical volume (via overpressure) for dyke propagation under mantle conditions to be ~21 m^3. Again, this is a fairly low melt volume.

The critical β, however, does not describe how far a dyke will ascend. The melt production rate in the mantle is tied to mantle flow rates, which are significantly slower than dyke ascent rates, and so for the mantle problem it is most relevant to consider the case of dyke propagation from a finite volume source (Lister, 1994). Here, neglecting elasticity, and assuming the freezing of the dyke walls is the fundamental control on ascent, the maximum ascent height can be calculated using:

$$h_f = 0.48\lambda^{\frac{2}{5}}\left(\frac{3}{4}A\right)^{\frac{4}{5}}\left(\frac{\Delta\rho g}{\eta_m\kappa}\right)^{\frac{1}{5}}$$

where A is the cross-sectional area of the dyke, $\Delta\rho$ is the density contrast between the melt and host rock, g is gravity, η_m is the magma viscosity, and κ is the thermal diffusivity. λ comes from the Stefan problem for dyke solidification (Turcotte & Schubert, 1982), and is defined by the transcendental equation:

$$\frac{e^{-\lambda^2}}{\lambda(1+erf\lambda)}=\frac{L\sqrt{\pi}}{c(T_m-T_{wr}(z))}$$

where erf refers to the error function, T_m is the magma temperature, and $T_{wr}(z)$ is the temperature of the wall rock at depth z. This function can be either solved numerically or can be linearized (for small λ) to:

$$\lambda=\frac{c(T_m-T_{wr}(z))}{\pi^{1/2}L}$$

For small magma volumes (~0.001 km³), dykes may be able to propagate ~24 km before freezing. If h_f exceeds the depth of the source, the dyke may erupt, and the volume erupted can be estimated by multiplying the source volume of $(h_f-d)/h_f$.

Although many small volume dykes are unable to propagate significant distances, near to melt sources dykes may be found in multitude, and an important consideration is how dyke swarms interact. If enough dykes coalesce, then their combined volume of melt allows them to migrate significantly further than is possible individually. Ito & Martel (2002) considered this problem by exploring how dykes interact with the local stress field. The combination of distortions in the stress field of two nearby dykes may be either repulsive or attractive, depending on the horizontal and vertical displacement between the dykes. The behavior is summarized in Figure 4.6. If the dykes are too widely spaced, no intersection of the dyke paths will occur. If the two dykes are in close proximity and at similar depths, the stress fields interact such that the path of least resistance for the propagating

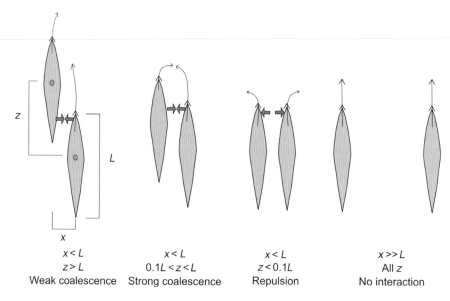

$x<L$	$x<L$	$x<L$	$x\gg L$
$z>L$	$0.1L<z<L$	$z<0.1L$	All z
Weak coalescence	Strong coalescence	Repulsion	No interaction

Fig. 4.6 Rising dykes interact with the local stress field, and can thus affect nearby dykes in a number of ways, depending on their spacing. Here L is the length of the dyke, z their vertical spacing, and x their horizontal spacing. For all $x<L$, dykes will interact. If they are at similar depths ($z<0.1\,L$), they will tend to repulse each other. If their depths are different, they will tend to attract, and the strength of attraction depends on z. See Plate 4.6 for a colour version of this image.

fracture is away from the other dyke, and the interaction will be repulsive. If they are at different depths, then this interaction of stress fields reverses, the dykes will coalesce, and the strength of the interaction will depend on the spacing between them.

The ascent rates of dykes is one of the most rapid geological processes, and has been constrained by field observations on the settling of xenoliths, or by mineralogical reactions such as the reversion of diamond to graphite in kimberlites (see O'Reilly & Griffin, this volume). These calculated velocities are generally in the range of 0.01 to 10 m/s for mantle derived dykes (Rubin, 1995).

DIAPIRISM

At the largest scale, molten, partially molten or even solidified magma chambers can potentially rise through the mid- to upper crust due to their buoyancy, given appropriate host rock properties. As summarized by Burov *et al.* (2003), there are two mechanisms at work: viscous diapiring in the lower/mid crust, and by brittle ascent in the upper crust, at depths less than the brittle-ductile transition (BDT) (Figure 4.7).

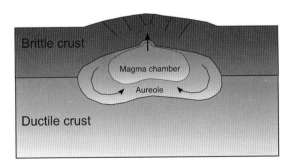

Fig. 4.7 Schematic of the emplacement of viscoelastic-plastic magmatic pluton. A partially molten diapir may rise ductilely through the lower crust. It is more impeded by the cool, brittle upper crust but may induce uplift and fracturing of the brittle crust as it migrates upwards at rates of millimeters per year (based on the work of Burov *et al.*, 2003). See Plate 4.7 for a colour version of this image.

Newtonian viscous flow

At its simplest, diapiric ascent can be considered as a viscous, buoyant sphere rising through a viscous medium. This allows us to use Stoke's equation to determine the velocity of ascent for a Newtonian fluid, given as:

$$W = \frac{C \Delta \rho g r^2}{3 \mu_{wr}}$$

where μ_{wr} is the wall rock viscosity, $\Delta \rho$ the density contrast, r the radius. C is a constant and depends on the internal and wall rock viscosities:

$$C = \frac{\mu_r + \mu_d}{\mu_r + (3/2)\mu_d}$$

where μ_d is the diapir's viscosity. As argued by Burov *et al.* (2003), this velocity is only weakly dependent on the viscosity contrast, but rather sensitive to the absolute wall rock viscosity – and hence temperatures, and thus the amount of heat transferred from the diapir to the country rock. To use the example of Burov *et al.* (2003), if we assume a thermal diffusion time for the diapir of $t_c \sim r^2/\kappa$, over this time the diapir will have migrated a vertical distance equivalent to $W t_c$. For a 1-km radius diapir ($\kappa = 10^{-6}$ m^2/s), with a low-end wall rock viscosity of 10^{18} Pa/s, and extreme viscosity contrast of 400 kg/m^3, the ascent distance is around 4 km. Given the extreme values used here, it is clear that only very large diapirs (>1 km) will ascend through the crust.

Non-Newtonian diapiric ascent

Weinberg & Podladchikov (1994) explored the ascent of diapirs in a non-Newtonian fluid more appropriate to upper crustal rocks. In this case the strain rate is given by:

$$\dot{\varepsilon} \equiv \frac{\partial \varepsilon}{\partial t} = A \sigma^n e^{\frac{-Q}{RT}}$$

where σ is the deviatoric stress, A, R, n and Q are constants for a given material, and T is temperature. Assuming the magnitude of the deviatoric stress due to the buoyant diapir is $\sim\Delta\rho gr$, then the effective viscosity can be written as:

$$\mu_{eff} = \frac{6^{n-1}\exp[Q/RT]}{3^{(n-1)/2}A[\Delta\rho gr]^{n-1}}$$

Then the ascent velocity for the non-Newtonian case would be:

$$V = \frac{\Delta\rho gr^2}{3\mu_{eff}}$$

This not only implies a stronger dependence on the diapir's radius r, but also a sensitivity to the temperature T. An important implication of this, noted by Weinberg & Podladchikov (1994), is that for temperatures $<\sim200°C$, the ascent velocity of the diapir drops to zero.

Brittle ascent mechanisms

Burov et al. (2003) extended models for diapir emplacement by considering brittle fracture of the uppermost crust. If a diapir is assumed to represent a line load P_0 at the base of the elastic lithosphere, then:

$$P_0 = 2\pi\Delta\rho g r^2$$

where r is the radius of the half cylinder representing the flattened diapir. The elasic layer itself is defined by its flexural rigidity D and the flexural parameter α:

$$D = \frac{Eh^3}{12(1-v^2)}$$

$$\alpha = \left(\frac{4D}{\Delta\rho_c g}\right)^{1/4}$$

where E is Young's modulus, v is Poisson's ratio, h the elastic lithospheric thickness, and $\Delta\rho_c$ the density contrast between the material above and below the elastic layer. Turcotte & Schubert (1982) present the solution to this problem. The maximum stresses are reached at the top and bottom of the flexing layer, and are given by:

$$\sigma_{max} = \frac{3P_0\alpha}{2h^2}$$

Whether or not the diapir will be able to fracture the elastic layer depends on the yield strength of the crust. If we assume a Byerlee style yield law, where yield strength depends on confining pressure P, then:

$$\sigma_{yield} = C_0 + \zeta P$$

where ζ is the coefficient of friction ($\zeta \sim0.6$–0.85) and C_0 is the cohesion. The brittle layer will fail and the diapir will ascend when $\sigma_{yield} > \sigma_m$. As shown by Burov et al. (2003), this condition is satisfied for shallow crustal diapirs on the scale of kilometres.

Full diapiric ascent formulation

The full formulation of diapiric ascent involves solving the coupled equations of mass conservation and momentum with an energy equation, together with constitutive relations and formulations for viscoplastic-elastic deformation. The full description is beyond this chapter; the reader is referred to Burov et al. (2003), Moresi & Solomatov (1998) and O'Neill et al. (2006) for further details of these style models.

The results of Burov et al. (2003), for a fully dynamic diapiric formulation, suggest that the ascent rate and final emplacement depth of plutons is controlled not so much by the density contrast between the pluton and the surrounding country rock, but by the heat budget of the diapir, which controls the viscosity and thus whether a diapir will stall or not. Only large diapirs (diameter $>\sim5$ km) will be able to penetrate into the brittle crust. Ascent rates within the ductile crust are relatively fast (0.1–5 m/yr), but slow within the brittle-ductile transition zone (5–10 km) to around 0.01–0.05 m/yr. Large diapirs within the brittle regime (<~5 km depth) typically rise very slowly, around a few mm per year. Interestingly, cold solid-state diapirs may continue to rise for over 1 Myr, in the models of Burov *et al.* (2003), by virtue of their buoyancy and viscosity. Surface effects such as erosion were found to be fairly second-order effects.

SUMMARY

Whilst models exist for various aspects of the melt ascent system, no fully consistent formulation exists encapsulating the full range of behaviors seen in Nature and outlined here. One of the main difficulties is the range of scales encompassed by the processes, both spatially and temporally. In terms of spatial scales, melt ascent involves constraining the wetting of grain boundaries by incipient melting, up to the movement of hundred kilometer-scale plutons through the crust. In terms of time, whilst melt production may be restricted by mantle velocities, melt ascent in island arcs can take <1000 years, based on U-series constraints (see Turner & Bourdon, this volume), and kimberlites may erupt through several hundred kilometers of lithosphere in just a few days (Sparks *et al.*, 2006). Despite these difficulties, individual processes in the sequence can be quantified as shown by the more common formulations presented here. The rates calculated from these relationships range from the order of cm/yr for porous melt migration from the source, to ~30 m/yr for the infilling velocity of small veins, to up to tens of kilometers per hour for dykes propagating by elastic fracturing – and down to several millimeters per year for plutons intruding into the brittle upper crust. One of the greatest challenges facing computational geodynamics in future years is the incorporation of the range of multiscale physical processes involved in magma ascent into existing solid-state geodynamic codes.

ACKNOWLEDGMENT

This is contribution 667 from the Australian Research Council National Key Centre for the Geochemical Evolution and Metallogeny of Continents (http://www.gemoc.mg.edu.au), supported by DP0880801.

REFERENCES

Barcilon V, Richter FM. 1986. Non-linear waves in compacting media. *Journal of Fluid Mechanics*, **164**: 429–448.

Bercovici D, Ricard Y, Schubert. G. 2001. A two-phase model for compaction and damage 1. General theory. *Journal of Geophyical Research – Solid Earth*, **106**(B5): 8887–8906.

Bourdon B, Elliott T. 2010. Melt production in the mantle: constraints from U-series. In: *Timescales of Magmatic Processes: From Core to Atmosphere*, Dosseto, A, Turner S, Van Orman JA (eds), Wiley-Blackwell, Oxford.

Buck WR, Su W. 1989. Focused mantle upwelling below mid-ocean ridges due to feedback between viscosity and melting. *Geophysics Research Letters* **16**: 641–644.

Burov E, Jaupart C, Guillou-Frottier L. 2003. Ascent and emplacement of buoyant magma bodies in brittle-ductile upper crust. *Journal of Geophysical Research* **108**: doi:10.1029/2002JB001904.

Cheadle MJ, Elliot MT, McKenzie D. 2004. Percolation threshold and permeability of crystallizing igneous rocks: The importance of textural equilibrium. *Geology* **32**: 757–760.

Daines MJ, Kohlstedt DL. 1993. A laboratory study of melt migration. *Philosophical Transactions* **342**: 43–52.

Etheridge MA, Wall VJ, Vernon RH. 1984. High fluid pressures during regional metamorphism and deformation: implications for mass transport and deformation mechanisms. *Journal of Geophysical Research* **89**: 4344–4358.

Forsyth DW, *et al.* 1998. Imaging the deep seismic structure beneath a mid-ocean ridge: the MELT experiment. *Science* **280**: 1215–1218.

Hart SR. 1993. Equilibration during mantle melting: A fractal tree model. *Proceedings of the National Academy of Sciences* **90**(24): 11,914–11,918.

Hier-Majumder S, Ricard Y, Bercovici D. 2006. Role of grain boundaries in magma migration and storage.

Earth and Planetary Science Letters **248**(3–4): 735–749.

Hirth G, Kohlstedt DG. 1995. Experimental constraints on the dynamics of the partially molten upper mantle: Deformation in the diffusion creep regime. *Journal of Geophysical Research* **100**(2): 1981–2002.

Holness MB. 1993. Temperature and pressure dependence of quartz-aqueous fluid dihedral angles: The control of absorbed H_2O on the permeability of quartzites. *Earth and Planetary Science Letters* **117**: 363–377.

Ito G, Martel, SJ. 2002. Focusing of magma in the upper mantle through dike interaction. *Journal of Geophysical Research* **107**(B10): 139: 67–80.

Katz R, Spiegelman M, Holtzman B. 2006. The dynamics of melt and shear localization in partially molten aggregates. *Nature* **442**: 676–679.

Kelemen PB, Braun M, Hirth G. 2000. Spatial distribution of melt conduits in the mantle beneath oceanic spreading ridges: Observations from the Ingalls and Oman ophiolites. *Geochemistry, Geophysics, Geosystems* **1**: doi:10.1029/1999GC000012

Kelemen PB, Hirth G, Shimizu N, Spiegelman M, Dick HJB. 1997. A review of melt migration processes in the adiabatically upwelling mantle beneath oceanic spreading ridges. *Philosophical Transactions of the Royal Society of London A* **355**: 283–318.

Laporte D, Watson EB. 1995. Experimental and theoretical constraints on melt distribution in crustal sources: the effect of crystalline anisotropy on melt connectivity. *Chemical Geology* **124**: 161–184.

Lister JR. 1994. The solidification of buoyancy-driven flow in a flexible-walled channel, Part 1. Constant-volume. *Journal of Fluid Mechanics* **272**: 21–44.

McKenzie D. 1984. The generation and compaction of partially molten rock. *Journal of Petrology* **25**: 713–765.

Mei S, Bai W, Hiraga T, Kohlstedt D. 2002. Influence of melt on the creep behavior of olivine- basalt aggregates under hydrous conditions. *Earth and Planetary Science Letters* **201**: 491–507.

Moresi L, Solomatov V. 1998. Mantle convection with a brittle lithosphere: thoughts on the global tectonic styles of Earth and Venus. *Geophysics Journal International* **133**: 669–682.

O'Neill C, Moresi L, Muller RD, Albert RA, Dufour F. (2006). Ellipsis 3D: A particle-in-cell finite-element hybrid code for modelling mantle convection and lithospheric deformation. *Computers & Geosciences* **32**: 1769–1779.

O'Neill C, Lenardic A, Jellinek AM, Kiefer WS. 2007. Melt propagation and volcanism in mantle convection simulations, with applications for Martian volcanic and atmospheric evolution. *Journal of Geophysical Research* **112**: E07003, doi:10.1029/2006JE002799.

O'Reilly S, Griffin W. 2010. Xenolith transport in ascending magma. In: *Timescales of Magmatic Processes: From Core to Atmosphere*, Dosseto, A, Turner S, Van Orman JA (eds), Wiley-Blackwell, Oxford.

Petford N, Kerr RC, Lister JR. 1993. Dike transport of granitoid magmas. *Geology* **21**: 845–848.

Ribe NM. 1986. Melt segregation driven by dynamic forcing. *Geophysics Research Letters* **13**: 1462–1465.

Ricard Y, Bercovici D, Schubert G. 2001. A two-phase model for compaction and damage 2. applications to compaction, deformation, and the role of interfacial surface tension. *Journal of Geophysical Research* **106**(B5): 8907–8924.

Rubin AM. 1995. Propagation of magma-filled cracks. *Annual Reviews of Earth and Planetary Science* **23**: 287–336.

Scott DR, Stevenson J. 1984. Magma solitons. *Geophysics Research Letters* **11**: 1161–1164.

Scott DR, Stevenson DJ. 1986, 1989. Magma ascent by porous flow. *Journal of Geophysical Research* **91**: 9283–9296.

Scott DR, Stevenson DJ. 1989. A self-consistent model of melting, magma migration and buoyancy-driven circulation beneath mid-ocean ridges. *Journal of Geophysical Research* **94**: 2973–2988.

Sleep NH. 1988. Tapping of melt by veins and dikes *Journal of Geophysical Research* **93**: 102,55–10,272.

Sparks W, Parmentier EM. 1991. Melt extraction from the mantle beneath spreading centers, *Earth and Planetary Science Letters* **105**: 368–377.

Sparks RSJ, Baker L, Brown RJ, *et al.* 2006. Dynamical constraints on kimberlite volcanism. *Journal of Volcanology and Geothermal Research* **155**: 18–48.

Spiegelman M. 1993a. Physics of Melt Extraction: Theory, Implications and Applications. *Philosophical Transactions* **342**(1663): 23–41.

Spiegelman M. 1993b. Flow in deformable porous media. part 1. Simple analysis. *Journal of Fluid Mechanics* **247**:17–38.

Spiegelman M. 1993c. Flow in deformable porous media. Part 2. Numerical analysis – The relationship between shock waves and solitary waves. *Journal of Fluid Mechanics* **247**: 39–63.

Spiegelman M. 1996. Geochemical consequences of melt transport in 2-D: The sensitivity of trace elements to mantle dynamics. *Earth and Planetary Science Letters* **139**: 115–132.

Spiegelman, M. 2003. Linear analysis of melt band formation by simple shear. *Geochemistry, Geophysics, Geosystems* **4**(8615): doi:10.1029/2002GC000499

Spiegelman M, McKenzie D. 1987. Simple 2-D models for melt extraction at mid-ocean ridges and island arcs. *Earth and Planetary Science Letters* **83**:137–152.

Spiegelman M, Kelemen P, Aharonov E. 2001. Causes and consequences of flow organization during melt transport: The reaction infiltration instability in compactible media. *Journal of Geophysical Research* **106**(B2): 2061–2077.

Stevenson DJ. 1989. Spontaneous small-scale melt segregation in partial melts undergoing deformation. *Geophysics Research Letters* **16**(9): 1067–1070.

Turcotte DL, Schubert G. 1982. *Geodynamics: Applications of Continuum Physics to Geological Problems.* John Wiley, Hoboken, NJ, 450 pp.

Turner S, Bourdon B. 2010. Melt transport from the mantle to the crust – U-series isotopes. In: *Timescales of Magmatic Processes: From Core to Atmosphere*, Dosseto, A, Turner S, Van Orman JA (eds), Wiley-Blackwell, Oxford.

Weinberg RF. 1999. Mesoscale pervasive felsic magma migration: alternatives to dyking. *Lithos* **46**: 392–410.

Weinberg RF, Podladchikov Y. 1994. Diapiric ascent of magmas through power law crust and mantle. *Journal of Geophysical Research* **99**: 9543–9559.

Wickham SM. 1987. The segregation and emplacement of granitic magma. *Journal of the Geological Society of London* **144**: 281–297.

Wiggins C, Spiegelman M. 1995. Magma migration and magmatic solitary waves in 3-D. *Geophysics Research Letters* **22**(10):1289–1292.

5 Melt Transport from the Mantle to the Crust – Uranium-Series Isotopes

SIMON P. TURNER[1] AND BERNARD BOURDON[2]

[1]GEMOC ARC National Key Centre, Department of Earth and Planetary Sciences,
Macquarie University, Sydney, NSW, Australia
[2]Institute of Geochemistry and Petrology, ETH Zurich, Switzerland

SUMMARY

Once melting commences within the Earth's mantle, melt movement commences by porous flow. Whether or not the entire path to the surface occurs via porous flow or whether channeled flow takes over is not well understood. Since channeled flow will be much faster than porous flow, time scale information provides one of the most powerful means of distinguishing the mechanism involved. U-series isotopes are the principle source of this time scale information. Studies of ^{238}U-^{230}Th and ^{235}U-^{231}Pa disequilibria in mid-ocean ridge, ocean island and island arc basalts all place minimum constraints on ascent rates of 1–20 m/yr. ^{226}Ra-^{230}Th disequilibria in island arc basalts (IAB) arguably provide the most stringent constraint suggesting ascent rates of ~70 m/yr. This requires channeled melt transport for the majority of source to surface path. It remains possible that ^{210}Pb disequilbria are a melting signature and if so this would require melt ascent rates of km/yr. Such estimates are consistent with available seismic and experimental constraints that melt transport is in general fast.

INTRODUCTION

How do mantle melts move from their source regions into the overlying crust or be erupted at the surface of the Earth? This is an important question because this process is the primary mechanism of differentiation of the Earth and an important, rapid means of heat transport (conductive heat transfer by comparison is a very slow process). There are a number of fundamental aspects of the physics of melt (or fluid) transport that need to be understood before any assessment of transport rates can be attempted. These were well synthesized by Spera (1980), who noted 'the elucidation of (melt) mobilization mechanics promises to be a complicated business.' Kelemen *et al.* (1997) provide a good synthesis of geochemical and physical constraints upon melt transport. The primary driving force for melt extraction is gravity. Because basaltic melt is less dense (ρ ~2,700 kg/m^3) than its surrounding solid peridotite matrix (ρ ~3,300 kg/m^3), it has positive buoyancy that causes it to rise. As we will discuss, there are a couple of alternative physical models for melt transport. In the absence of direct observation, perhaps the most convincing means to distinguish between them is to constrain the speed or timescale of melt transport, provided that the depth of melt production can also be established. In this way U-series isotope observations, in particular, can be used to infer minimum melt transport rates. Although this has proved difficult, a number of observations are beginning to emerge that may suggest some sort of consensus. It is also worth noting that most constraints on mantle to crust melt transport rates come from total source to surface transit times. Thus, this chapter implicitly incorporates, but will not discuss, transport times within the crust, since this is the subject of a subsequent chapter by Rushmer and Knesel.

INCEPTION OF MELTING

The mantle is not mono-mineralic and melting of peridotite occurs by reactions at grain boundaries (i.e. at the junction of olivine-orthopyroxene-clinopyroxene-spinel or olivine-orthopyroxene-clinopyroxene-garnet). Some mantle nodules contain evidence for melt being present at grain boundaries (Yaxley & Kamenetsky, 1999) and this has also been simulated experimentally (Vaughan & Kohlstedt, 1982). Melt transport or extraction from the matrix becomes possible when there is connectivity of the melt and a widely accepted theory is that this will occur if the dihedral angle (Figure 5.1) for melt-olivine is <60° (Beeré, 1975). However, this is less likely to strictly hold under conditions of deformation that will accompany compaction associated with melt extraction (Jin

et al., 1994) and that are likely to be prevalent within the Earth's convecting mantle. Furthermore, the dihedral angle does not specify the mechanism, and thus the rate, of melt transport. Some form of porous flow is necessary to extract melt when it forms along grain boundaries, but when considering the entire path to the surface, other melt transport mechanisms need to be considered. In principle, this mode of melt transport could be via porous flow all the way to the surface, channeled flow or a combination of the two. Much of this is covered in a separate chapter by O'Neill and Spiegelman, so we only present a summary of pertinent details here.

POROUS FLOW

Porosity (ϕ) is defined as a measure of the void spaces in a material and with regard to mantle melting this corresponds to the amount of melt present at grain boundaries and grain junctions. Assuming that melt is produced at grain boundaries, then the primary mechanism for, at least the inception of, melt transport should be by porous flow (Figure 5.2), most simply described by Darcy's law:

$$q = -k/\mu \, \nabla P$$

where q is the Darcy flux or filtration velocity, k is the permeability, μ is the viscosity and ∇P is the pressure gradient. The permeability is related to the porosity (ϕ) and crystal diameter (d) via:

$$k = \phi^3 \, d^2/\mu(1 - \phi)^2$$

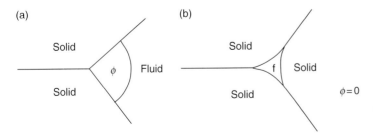

(a) Solid / Solid / ϕ / Fluid

(b) Solid / Solid / f / Solid / $\phi = 0$

Fig. 5.1 (a) The dihedral angle is that between two grain-melt interfaces. (b) If $\phi < 60°$, the pores are occupied by melt.

Fig. 5.2 Illustration of multi-phase porous flow in a melting medium. As the system undergoes compaction the grains (white spheres) compress and the fluid is expelled upwards. See Plate 5.2 for a colour version of this image.

McKenzie (1984) derived the basic equations describing the movement of the melt in the matrix of a partially molten material and showed that melt extraction will be controlled by the rate of deformation of the matrix. In the most straightforward situation this deformation is simply compaction and he calculated that the rate of melt extraction would be $\geq 10\,mm/yr$ for values of $\phi > 0.03$ (Figure 5.3). For $\phi > 0.1$ melt – matrix separation reaches $1\,m/yr^{-1}$ such that it is unlikely that such large melt fractions (porosities) are ever continuously present within the Earth. This further suggests that partial melting is best described as fractional (as opposed to batch), implying that melt is extracted rapidly once it is produced. Subsequent geophysical studies have confirmed that the melt beneath mid-ocean ridges must be <3% (Forsyth et al., 1998).

The porous flow treatment was subsequently expanded by Spiegelman & McKenzie (1987), who showed that the geometry of lithospheric plates and matrix advection at mid-ocean ridges and island arcs will lead to strong focusing of melt. In other words, melt will be extracted from a wide region of low porosity towards a zone of concentration or accumulation, where the total or combined melt fraction can reach tens of percent. Importantly, this reconciles the arguments for low porosity (instantaneous local melt fraction),

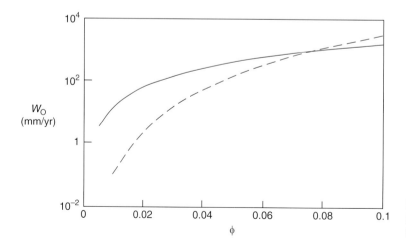

Fig. 5.3 Minimum melt velocity (w_0) as a function of porosity (modified from McKenzie, 1984).

with geochemical and experimental arguments for most basalts representing 5 to 20 % partial melts of peridotite. Additional complexity involves reactive porous flow where reaction between the melt and residual minerals occurs (Spiegelman *et al.*, 2001).

As noted by McKenzie (1984), the compaction model can only be applied to geological problems if the permeability of the matrix (i.e. the interconnectivity of the porosity) and the viscosity are known. These depend on the geometry of the melt that he assumed to be controlled by the dihedral angle. Results from static experiments lead to the conclusion that peridotite will only achieve sufficient permeability (i.e. basaltic melt will become extractable) when the melt fraction (porosity) exceeds 1% (Faul, 1997, 2000). However, the mantle undergoes ductile deformation, and dynamic experiments on partially molten peridotite show that melt geometry is strongly influenced by deformation, which enhances melt separation (Jin *et al.*, 1994; Holtzman *et al.*, 2003; Figure 5.4). Thus, melt separation may occur at very low porosity within the Earth but the limiting rate of melt transport by porous flow is likely to be of the order of 10 mm/yr (Figure 5.5).

CHANNEL FLOW

An alternative mechanism of transport is in channels where melt transport will be orders of magnitude faster (10–1,000's m/yr) than for porous flow. Channels can take the form of truly open channels, such as dykes, but also regions of locally high porosity/permeability where flow is relatively rapid but still porous (sometimes referred to as porous flow channels). Kelemen *et al.* (1997) concluded that channeled melt transport in the asthenosphere is permissible but requires optimal physical conditions. Note that if channeled flow occurs within the asthenosphere it must originate in, or be surrounded by, a zone of porous flow where the onset of melting occurs. In other words, porous flow would transition into channeled flow over some distance.

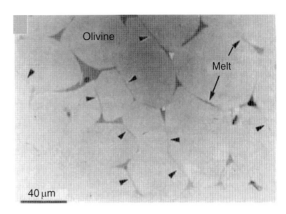

Fig. 5.4 The result of a dynamic partial melting experiment on peridotite. Melt films (indicated by arrows) can be seen on most of the grain boundaries and junctions (modified from Jin *et al.* (1994).

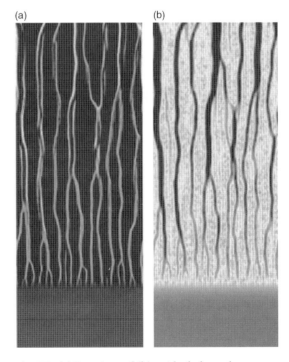

Fig. 5.5 (a) Porosity and (b) residual channel structure for the reactive flow model in Spiegelman *et al.* (2001). See Plate 5.5 for a colour version of these images.

Field observations of mantle sections

Dykes provide direct field evidence for melt transport in channels within the lithosphere and Kelemen *et al.* (1997) review evidence that dunite dykes in ophiolites represent former porous melt flow channels. The evidence is based on the observed structures and individual dykes can reach 200 m in thickness and 10 km in length. Although most are much smaller, they comprise 5 to 15% of the mantle section in the Oman ophiolite. They tend to be transposed, sub-parallel to the Moho, indicating they have undergone rotation by corner flow and thus may have originally formed in a vertical orientation in asthenosphere that subsequently became frozen onto the base of the lithosphere.

Geochemical arguments suggest that the dunite is derived from a melt reaction, in which the melt dissolves orthopyroxene and produces olivine (Kelemen, 1990). Based on the Mg in the olivines, the dunites are not believed to be residual cumulates but to have been in equilibrium with a mafic melt. Irrespective of the validity of these observations, a critical question remains as to whether a channeled flow mechanism also applies to melt transport in the convecting asthenosphere, or whether it simply reflects the response to the brittle behavior of the lithosphere?

Evidence from numerical simulations and experiments

In the absence of any direct means of observing channels in the asthenosphere, several groups have undertaken both numerical simulations and rheological experiments to assess their feasibility as a means of melt transport. To a large extent these are complementary approaches and both have implications that, in principle, can be tested with geochemical observables.

There is a general consensus that once channels exist, melt will be efficiently removed to the channels from surrounding regions of porosity – even in zones of active downwelling, such as beneath island arcs (Petford *et al.*, 2008). Thus, an important question is whether and how channels

might form in the first place and how long they might persist (Sleep, 1988). Stevenson (1989) has suggested that weak regions will have a slightly lower pressure than strong regions in viscously deforming media leading to mechanical instabilities. Thus, melt will tend to flow from the strong to the weaker regions, although the local pressure gradient between matrix and channel that drives melt flow is hard to assess. Hirth & Kohlstedt (1995) have shown that melt fraction is negatively correlated with strength in partially molten peridotite, so that this can unstably form melt-rich shear zones that initiate channel formation.

Dynamic experiments by Kohlstedt and co-workers have been used to investigate the rheological behavior of peridotite during partial melting. They have shown that deformation of partially molten rocks can produce melt segregation. An important result is that such stress-driven melt segregation may produce high-permeability pathways, allowing for rapid melt extraction along channels similar to the dunite dykes observed in ophiolites (Holtzmann *et al.*, 2003; Holtzmann & Kohldstedt, 2007), allowing for efficient melt-focusing beneath mid-ocean ridges (Katz *et al.*, 2006).

Aharonov *et al.* (1995) undertook numerical calculations to investigate the behavior of the mantle in the presence of partial melt and the tendency of such a system to self-organize into a coalescing network of high porosity channels (Figure 5.5). This work has been expanded by Spiegelman and co-workers in a number of studies, some of which have included reactive flow in deformable permeable media (Spiegelman *et al.*, 2001). In essence, if the dissolution of silicate minerals is sufficiently fast, the melt flow can lead to instabilities that lead to the formation of dissolution channels. As the channels are at a lower pressure than the surrounding regions, they suck melt into them, thus reducing the porosity of the interchannel region.

Other relevant observations

As noted above, observations relevant to melt transport are rare. Some geophysical data suggest that ascent rates could be extremely fast and values as high as 1.8 km/day have been estimated from seismic data in Vanuatu (Blot, 1972) but these are not common. As O'Reilly and Griffin discuss in the following chapter, the presence and size of mantle xenoliths can be used to calculate a minimum Stokes' law melt ascent velocity. For example, assuming a Bingham liquid, Sparks *et al.* (1977) calculated that a 25 cm diameter mantle nodule would settle unless the magma ascent velocity was ~0.8 km/day. Finally, Rutherford & Hill (1993) adopted an experimental approach to amphibole breakdown in lavas from Mt St Helens to estimate that magma ascent from 8 km depth to the surface occurred at between 15 and 66 m/hr or 131 to 578 km/yr.

Langmuir *et al.* (1992) used the Fe content in mid-ocean ridge basalts (MORB) to argue that the depth equilibration of melt was deep and accordingly suggested that melts do not equilibrate at shallow pressure and argued for a dynamic melting mechanism. That is, the mantle matrix is continuously melting whilst it is moving (typically ascending) through the melting region. Kelemen *et al.* (1997) reviewed evidence that in terms of both major and trace elements, MORB are out of equilibrium with the shallow upper mantle. Specifically, MORB should be saturated in orthopyroxene, but they are not. An attractive simple means by which to explain this observation is that the melts ascended in channels and never equilibrated with the shallow mantle.

With advances in analytical techniques, it has become possible to analyse trace element concentrations in melt inclusions. One of the outcomes has been the surprising observation that the incompatible element variation in melt inclusions at a single locality show huge variations that may reflect incomplete mixing of very different % melt fractions (Slater *et al.*, 2001). In addition, Sobolev & Shimizu (1993) have provided compelling evidence from melt inclusions that erupted melts are aggregates of melts from different depths. The corollary is that such observations are most compatible with very rapid melt transport and incomplete mixing.

Finally, Hauri (1997) has argued that chromatographic interaction between ascending melts and surrounding peridotite would quickly erase the unique isotope signatures of compatible elements such as Os that are observed in some oceanic plume basalts. His calculations suggest that the preservation of such signatures requires magma transport to the surface in channels that originate from within hundreds of meters of their source.

EVIDENCE FROM U-SERIES ISOTOPES

The key issue for our discussion is that different physical mechanisms for melt extraction, outlined briefly here and discussed in more detail in the chapter by O'Neill and Spiegelman, result in very different predicted rates of melt transport that can be tested geochemically. In particular, the short-lived isotopes of the uranium-series (hereafter U-series) decay chains afford the opportunity to test the relative importance of slow, porous flow and rapid, channeled melt ascent in the Earth.

U-series observations

In an earlier chapter, Bourdon and Elliott have introduced the principles of short-lived U-series isotopes and how these can provide timescale information for magmatic processes that is simply not available from other means (see also Bourdon *et al.*, 2003 for an extensive review). It is not our purpose here to review the details of the different in-growth melting models and the reader is referred to a recent summary by Elliott & Spiegelman (2003) for an exhaustive discussion.

Briefly, all of the elements in the U-series decay chains are highly incompatible (i.e. partition into the melt phase; $D < 1$) in mantle assemblages. Because fractionation between incompatible elements only occurs when the total extent of melting (F) is less than the bulk partition coefficient (D) of the (usually more compatible) parent element, the generation of disequilibria by fractionation alone requires $F \leq 0.05$ (Figure 5.6a). Since

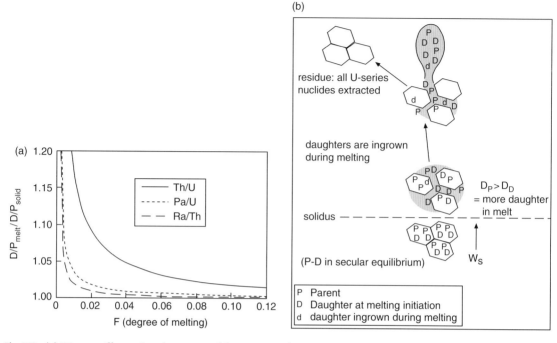

Fig. 5.6 (a) Diagram illustrating the extent of fractionation between U-series parent-daughter pairs versus degree of partial melting (in %). (b) Schematic representation of U-series daughter in-growth during partial melting that occurs over a timescale that is significant relative to the daughter half-life. Solid mantle approaching the solidus is in secular equilibrium. At the onset of melting, the U-series nuclides are 'instantaneously' partitioned according to their bulk partition coefficients (D), such that, in the case illustrated, the daughter isotope is enriched in the melt relative to the parent, following the fractionation illustrated in (a). This fractionation reflects the instantaneous or residual porosity (ϕ). However, as the solid continues to pass through the melting zone (at a matrix flow rate W_s) new daughter isotopes are produced (over time) by decay from their parent within the solid residue and these subsequently also become added to the melt leading to an additional excess of daughter in the melt that reflects the timescale of melting. The resultant disequilibria is controlled by the elemental partition coefficients, the porosity and the melting rate (a function of the rate of passage of matrix across the solidus) modified from Lundstrom (2003).

most MORB, ocean island basalts (OIB) and island arc basalts (IAB) are thought to reflect melting extents of several percent to tens of percent, no U-series disequilibria would be predicted by fractional (or batch) partial melting models. Thus, the observation of such disequilibria spawned the development of a number of variants of 'in-growth' models based on an initial treatment by McKenzie (1985). The fundamental feature of all of these models (illustrated in Figure 5.6b) is the same: namely, if melting occurs over a timescale that is comparable to the half-lives of the U-series

isotopes being considered, then daughter in-growth during melting can lead to significant parent-daughter disequilibria. So long as the melt can be extracted before such disequilibria decay back to secular equilibrium, observed disequilibria can provide constraints on the rates of melting and melt extraction (our interest here).

Whilst U-series disequilibria require in-growth during partial melting, the constraints on melt extraction rates are more model dependent. This boils down to *a priori* assumptions as to whether the melt is extract by channeled flow (often

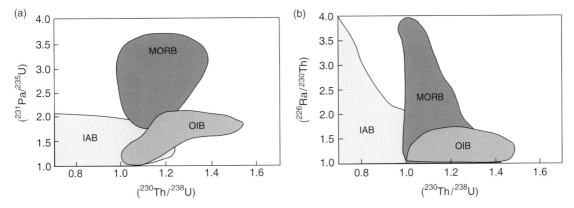

Fig. 5.7 (a) and (b) Observations of ^{238}U-^{230}Th, ^{235}U-^{231}Pa and ^{230}Th-^{226}Ra disequilibria in basalts from mid-ocean ridges, ocean islands and island arcs (after Lundstrom, 2003).

referred to at the 'dynamic melting models') as soon as the porosity is exceeded or whether the melt percolates upwards via porous flow and continues to interact with the matrix along the way. This model is often referred to as equilibrium porous flow (Spiegelman & Elliott, 1993). In this case, disequilibria for some short-lived and highly incompatible nuclides (e.g. ^{226}Ra) can be produced near the top of the melting column and therefore do not place major constraints upon melt ascent rates. Thus, the ability for U-series disequilibria to constrain the rate of melt ascent requires information that can independently demonstrate that the fractionation takes place at depth. Conversely, such information might dictate which models are more appropriate to the Earth.

The first-order observation from young volcanic rocks is that ^{238}U-^{230}Th-^{226}Ra and ^{235}U-^{231}Pa disequilibria are almost ubiquitous (Figure 5.7) and many also commonly preserve disequilibria between ^{226}Ra and ^{210}Pb (see Bourdon *et al.*, 2003 for a summary). Thus, at the simplest level, the mere observation of ^{238}U-^{230}Th-^{226}Ra disequilibria provide some minimum speed on melt ascent. Therefore, a minimum melt ascent rate is dictated by the need to preserve U-Th disequilibria. This requires ascent in less than a few half-lives of ^{230}Th (75 kyr), leading to inferred ascent rates of perhaps several 10's m/yr. The half-life of ^{231}Pa (33 kyr) is shorter than that of ^{230}Th, which might

further reduce the maximum ascent time allowed. However, this element is highly incompatible and so disequilibria could arguably be produced at shallower levels, by porous flow. Similarly, if observed ^{226}Ra disequilibria also source-related, then this would reduce the available time to <8,000 years and thus increase the melt ascent rate to arguably some 10's m/yr (see further discussion below).

^{210}Pb has a half-life of only 22 years and so ^{210}Pb-^{226}Ra disequilibria potentially provide even more robust constraints upon melt ascent rates. However, a major caveat is that ^{222}Rn is an intermediate product and so ^{210}Pb-^{226}Ra disequilibria (^{210}Pb deficits) could be produced by magma degassing (Gauthier & Condomines, 1999). On the other hand, Pb is much more compatible than Ra during mantle melting and so significant ^{210}Pb-^{226}Ra disequilibria (also ^{210}Pb deficits) are predicted during mantle melting. If such disequilibria could be shown to result from melting, this would imply very rapid melt ascent (~3 km/yr) indeed!

Constraints on melt velocities

As alluded to above, the key issue of importance to this chapter relies on the ability to constrain the depth of origin of U-series isotope signatures. If the signature is actually shallow, but assumed to be deep, then the speed of melt transport could be overestimated. On the other hand, even if assumptions about the depth of origin reflect reality, the rate could always be much faster than the U-series signature implies. U-series constraints always provide a minimum speed. So what are the robust constraints on the depth of origin of U-series disequilibria in the different tectonic environments – or indeed would we expect them to be similar?

Mid-ocean ridge basalts

A wide variety of geochemical arguments can be used to infer that it is likely that beneath ridges, melt is extracted from near the garnet-spinel transition zone around ~90 km (Robinson & Wood, 1999). These include the fact that it is only at these depths that U is significantly more compatible than Th (Blundy & Wood, 2003). Thus, in all 'in-growth' melting models, the bulk of ^{238}U-^{230}Th disequilibria originates near the base of the melting column because:

• it is here that, in the presence of residual high pressure clinopyroxene and/or garnet (~90 km depth), elemental fractionation between U and Th is maximized: and

• the overlying melting column is sufficiently long to allow for ^{230}Th in-growth during the course of melting in significantly less than a few half-lives of ^{230}Th (75 kyr).

Adding this rough depth of origin constraint to the arguments based on the first-order observations made above leads to increased inferred ascent rates of ~1.2 m/yr.

^{235}U-^{231}Pa and ^{226}Ra-^{230}Th disequilibria have the potential to place tighter constraints on melt ascent rates but suffer from the possibility that their origin is shallow. Possibilities include continuing interaction with the matrix via equilibrium porous flow and, in the case of ^{226}Ra, it has been suggested that observed disequilibria reflect shallow interaction with plagioclase in sub-axial magma chambers (Saal & Van Orman, 2004, Van Orman et al. 2006), whereby the faster diffusion of ^{226}Ra relative to ^{230}Th out of the plagioclase leads to ^{226}Ra excess in the melt. Thus, at present

these systems arguably do not offer much insight into the rate of melt ascent beneath ridges.

A more recent and highly provocative study by Rubin et al. (2005) measured significant ^{210}Pb-^{226}Ra disequilibria in zero-age MORB from the East Pacific Rise. Because MORB are erupted beneath a long water column, which will subdue degassing and the disequilibria did not increase with extent of differentiation, these authors attributed the disequilibria to partial melting. If correct, this would place by far the tightest constraints on the ascent rate of MORB (irrespective of the absolute depth of origin) requiring melt ascent in decades! Alternatively, Van Orman & Saal (2009) have recently suggested that the ^{210}Pb deficits might reflect interaction between melt and cumulate minerals. For MORB then, a crucial test of magma ascent rates will be assessing the validity of these data, particularly the inference that they reflect partial melting rather than degassing. Such a test is afforded by the ^{227}Ac-^{231}Pa system. ^{227}Ac also has a 22-year half-life but, unlike ^{210}Pb, there is no gaseous intermediate between it and its longer-lived parent ^{231}Pa (33 kyr). Ac is predicted to be reasonably compatible (similar to La) in the mantle compared with Pa (Blundy & Wood, 2003) and so mantle melting should produce significant ^{227}Ac-^{231}Pa disequilibria, whereas degassing will not. Unfortunately, analytical challenges have so far prevented the routine measurement of ^{227}Ac.

Ocean island basalts

Ocean island basalts (OIB) are generally thought to result from decompression melting in a manner analogous to MORB and the U-series observations are very similar (Figure 5.7). A difference is that the eruption of OIB through thick lithosphere and their inferred higher temperatures of partial melting suggest a greater average depth of melting (usually in the garnet-lherzolite facies), which is consistent with models for their combined ^{238}U-^{230}Th and ^{235}U-^{231}Pa disequilibria (Bourdon & Sims, 2003). This could perhaps be used to argue for slightly higher rates of melt ascent for OIB. Thus, the constraints on the depth of origin of these particular U-series disequilibria, and thus the rates of melt ascent, are not greatly different to those of MORB.

^{226}Ra-excesses tend to be smaller in OIB than in MORB (Figure 5.7b) and like MORB it remains possible that observed ^{226}Ra excesses are produced by shallow porous flow and so do not place strong constraints on melt ascent rates for OIB. However, some OIB exhibit deficits of ^{226}Ra relative to ^{230}Th and these have been interpreted to reflect melting in the presence of residual amphibole or phlogopite at the base of the lithosphere (Claude-Ivanaj et al. 1998; Bourdon et al., 1998; Bourdon & Van Orman, 2009). In this case, if the lithosphere is of a thermally equilibrated thickness of 12.5 km, the minimum speed of ascent is constrained to be ~15 m/yr in order to preserve the ^{226}Ra deficits.

There are currently very few measurements of ^{210}Pb in OIB but it is possible that, like MORB, these may reflect a melting signature rather than degassing. Again, the full implications for melt ascent rates must await verification from ^{227}Ac measurements.

Finally, Stracke et al. (2006) have argued that combined ^{238}U-^{230}Th, ^{235}U-^{231}Pa and ^{226}Ra-^{230}Th disequilibria is basalts from Iceland can be reconciled in a dynamic melting model with rapid (channeled) melt extraction without the need for shallower wall-rock interaction via porous flow.

Island arc basalts

Island arc basalts are erupted at volcanoes that lie, on average, ~110 km above subducting oceanic plates. These plates are the source of water, which induces partial melting by lowering the peridotite solidus. Thus, the cause of melting differs from MORB and OIB. IAB are also characterized by the reverse sense of U-Th disequilibria (i.e. excesses of ^{238}U over ^{230}Th) and this has long been attributed to addition of fluid mobile U from the subducting plate (see Turner et al., 2003 for a review). Since the depth of origin of the fluids is therefore reasonably well constrained, this immediately provides a maximum ascent rate of 0.3 m/yr. Inferred along-arc U-Th and U-Pa isochrons from the Mariana and

Tonga-Kermadec arcs have further reduced the transit time to 30–60 kyr (Elliott et al., 1997; Turner & Hawkesworth, 1997; Bourdon et al., 1999) and, accordingly, the ascent rate to ~4 m/yr. In fact, the most recent data from Tonga suggests horizontal U-Th isotope arrays within individual islands such that the time since fluid addition is not resolvable and could be <10 kyr (Caulfield, 2010) corresponding to an inferred ascent rate of 11 m/yr.

Probably the most important constraint on melt ascent rates comes from ^{226}Ra-^{230}Th disequilibria in IAB. Turner et al. (2000; 2001) showed that these can exceed 500% and that they are positively correlated with ratios such as Ba/Th (Figure 5.8a). Dosseto et al. (2003) showed that

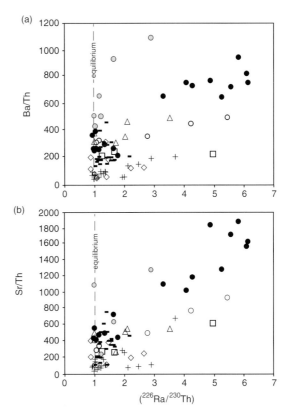

Fig. 5.8 (a) and (b) Correlations between ^{230}Th-^{226}Ra disequilibria and Ba/Th and Sr/Th in island arc basalts (after Turner et al., 2003).

there is also a positive correlation with Sr/Th (Figure 5.8b). (^{226}Ra/^{230}Th) ratios of >3 to 4 are extremely hard to reproduce by either dynamic melting or equilibrium porous flow models and require porosities <0.1% (Bourdon & Sims, 2003; Elliott & Spiegelman, 2003). Moreover, Th is more incompatible than Sr during mantle melting and so a positive correlation between (^{226}Ra/^{230}Th) and Sr/Th cannot be attributed to partial melting (Dosseto et al., 2003). Instead, Ba, Sr and, by analogy, Ra are all highly fluid mobile compared with Th (Brenan et al., 1995). Thus, it has been argued that the ^{226}Ra excesses observed in arc lavas reflect fluid addition (Turner et al., 2000, 2001), just like the ^{238}U excesses. Because fluid addition occurs at the base of the melting column at ~110 km depth, this appears to require very rapid melt ascent. The half-life of ^{226}Ra is only 1,600 years and theoretical calculations suggest that initial ^{226}Ra-^{230}Th disequilibria in the IAB source might be ~15 to 16 (Turner et al., 2003; Yokoyama et al., 2003). Thus, erupted lavas with (^{226}Ra/^{230}Th) ratios of 6 may have aged less than one half-life since fluid addition, suggesting melt ascent rates of ~70 m/yr.

The suggestion of such rapid melt ascent rates has been provocative because they would demand channeled melt ascent for much of the source to surface distance (at least beneath island arcs) and has been challenged by Feineman & DePaolo (2003) and Huang et al. (2008). Feineman & DePaolo (2003) suggested that ^{226}Ra excesses could result from the enrichment in ^{226}Ra in amphibole or phlogopite due to solid state diffusion out of the clinopyroxene where ^{226}Ra is initially produced from ^{230}Th decay. Subsequently, preferential melting of amphibole leads to a large ^{226}Ra excess in the melt. However, this model predicts that those IAB with the largest ^{226}Ra excesses should be low melt fractions (e.g. SiO_2 undersaturated), high in K_2O and have moderate Sr/Th ratios, all of which are the opposite of what is actually observed (see George et al., 2004 for a discussion). To the contrary, Bourdon & Van Orman (2009) argue that the presence of amphibole or phlogopite will lead to excesses of ^{226}Ra relative to ^{230}Th, only if melting is rapid and that

these would be accompanied by ^{230}Th excesses – again, this is contrary to observations since IAB are characterized by ^{238}U excesses.

Huang et al. (2008) pointed out that the positive correlations between (^{226}Ra/^{230}Th) and Ba/Th and/or Sr/Th might reflect the effects of time-dependent crystal fractionation. Although such an effect can lead to positive correlations on the Sr/Th diagram, it cannot produce the large observed ranges in Ba/Th, because both Ba and Th are highly incompatible. Moreover, it completely ignores the need to create a parental magma with high (^{226}Ra/^{230}Th), (^{238}U/^{230}Th), Ba/Th and Sr/Th in the first place. As discussed above, the most plausible means of this remains fluid addition to mantle peridotite and, as such, all arguments pertaining to melt ascent rates remain unchanged.

Many IAB exhibit excesses of ^{210}Pb over ^{226}Ra, although an equal number have the opposite sense of disequilibria (Turner et al., 2004). In principle, the deficits could form as a result of mantle melting (Rubin et al., 2005) or shallow level (sub-volcanic) magma degassing (Gauthier & Condomines, 1999). However, the characteristics of gases contained in IAB and the explosive nature of island arc volcanoes generally sway opinion towards the dominant ^{210}Pb signal being the result of differential gas-magma transport (Turner et al., 2004). Thus, at this stage, ^{210}Pb-^{226}Ra disequilibria do not appear to place additional constrains upon the rates of melt ascent through the mantle wedge.

One of the many intriguing aspects of the island arc setting is that induced convection in the mantle wedge will result in active downwelling rather than the passive upwelling that occurs beneath ridges and ocean islands. Whether or not melt velocity is sufficient to allow its extraction towards the surface is poorly understood. Nevertheless, Petford et al. (2008) undertook numerical calculations to estimate the timescales of melt extraction into channels in a downwelling matrix. The presence of channels was assumed because the actual mechanism of transition from porous flow to channel formation remains a major unsolved problem. Nevertheless, they showed that melt extraction could occur on timescales sufficiently fast to preserve ^{210}Pb (and [227]Ac) disequilibria. Thus, rather like MORB and OIB, the problem of whether any IAB ^{210}Pb disequilibria might reflect melting remains to be tested with [227]Ac measurements.

CONCLUDING REMARKS

The determination of melt ascent rates through the mantle is problematic and largely in its infancy. Intriguingly, estimates of crustal transit rates are remarkably consistent from a number of different sources of information. Seismic data, xenolith entrainment calculations and experiments on amphibole breakdown all suggest rates around 0.2 to 2 km/day, entirely consistent with channeled melt transport as evidenced by dykes in outcrop. Although dunite dykes in ophiolites provide evidence for channeled flow in the lithosphere and theoretical calculations indicate that channels could form in the asthenosphere, the primary constrains on the speed, and thus mechanism, of melt transport in the mantle come from U-series disequilibria. U-Th-Pa disequilibria in MORB, OIB and IAB all place minimum constraints on ascent rates of 1 to 20 m/yr. Ra disequilibria in IAB arguably provide the most stringent constraint suggesting ascent rates of 70 m/yr. This requires channeled melt transport for the majority of source to surface path. It remains possible (and testable with [227]Ac) that ^{210}Pb disequilbria are a melting signature and if so this would require melt ascent rates of km/yr similar to those estimated for the crust. This might be anticipated if the slowing effect of the lower density contrast in the crust is balanced by the increased tendency towards brittle behavior and easy dyke propagation. Perhaps the mantle and crust behave rheologically in the same way everywhere in the presence of partial melt.

This is GEMOC publication #669.

REFERENCES

Aharonov E, Whitehead JA, Kelemen PB, Spiegelman M. 1995. Channeling instability of upwelling melt in the mantle. *Journal Geophysical Research* **100**: 20433–20450.

Beeré W. 1975. A unifying theory of the stability of penetrating liquid phases and sintering pores. *Acta Metallica* **23**: 131–138.

Blot C. 1972. Volcanisme et séismes du manteau supérieur dans l'Archipel des Nouvelles-Hebrides. *Bulletin of Volcanolology* **36**: 446–461.

Blundy J, Wood B. 2003. Mineral-melt partitioning of Uranium, Thorium and their daughters. In: *Uranium-series Geochemistry*, Bourdon B, Henderson GM, Lundstrom CC, Turner SP (eds), *Reviews in Mineralogy and Geochemistry* **52**: 59–123.

Bourdon B, Van Orman JA. 2009. Melting of enriched mantle beneath Pitcairn seamounts: unusual U-Th-Ra systematics provide insights into melt extraction processes. *Earth and Planetary Science Letters* **277**: 474–481.

Bourdon B, Sims KWW. 2003. U-series constraints on intraplate basaltic magmatism. In: *Uranium-series Geochemistry*, Bourdon B, Henderson GM, Lundstrom CC, Turner SP (eds), *Reviews in Mineralogy and Geochemistry* **52**: 215–254.

Bourdon B, Joron J-L, Claude-Ivanaj C, Allègre CJ. 1998. U-Th-Pa-Ra systematics for the Grande Comore volcanics: melting processes in an upwelling plume. *Earth and Planetary Science Letters* **164**: 119–133.

Bourdon B, Turner S, Allègre C. 1999. Melting dynamics beneath the Tonga-Kermadec island arc inferred from [231]Pa-[235]U systematics. *Science* **286**: 2491–2493.

Bourdon B, Henderson G, Lundstrom C, Turner S. 2003. Uranium-series geochemistry. *Reviews in Mineralogy and Geochemistry* **52**: 656 pp.

Brenan JM, Shaw HF, Ryerson FJ, Phinney DL. (1995). Mineral-aqueous fluid partitioning of trace elements at 900°C and 2.0 GPa: Constraints on the trace element chemistry of mantle and deep crustal fluids. *Geochimica et Cosmochimica Acta*, **59**: 3331–3350.

Caulfield J. 2010. Nature and timing of magma genesis at Tofua volcano, Tonga. PhD thesis, Macquarie University.

Claude-Ivanaj C, Bourdon B, Allègre CJ. 1998. Ra-Th-Sr isotope systematics in Grande Comore Island: a case study of plume-lithosphere interaction. *Earth and Planetary Science Letters* **164**: 99–117.

Dosseto A, Bourdon B, Joron JL, Dupré B. 2003. U-Th-Pa-Ra study of the Kamchatka arc: new constraints on the genesis of arc lavas. *Geochimica et Cosmochimica Acta* **67**: 2857–2877.

Elliott T, Spiegelman M. 2003. Melt migration in oceanic crustal production: A U-series perspective. *Treatise on Geochemistry* **3**: 465–510.

Elliott T, Plank T, Zindler A, White W, Bourdon B. 1997. Element transport from slab to volcanic front at the Mariana arc. *Journal of Geophysical Research* **102**: 14991–15019.

Faul UH. 1997. Permeability of partially molten upper mantle rocks from experiments and percolation theory. *Journal of Geophysical Research* **102**: 10299–10311.

Faul UH. 2001. Melt retention and segregation beneath mid-ocean ridges. *Nature* **410**: 920–923.

Feineman MD, DePaolo DJ. 2003. Steady-state $^{226}Ra/^{230}Th$ disequilibrium in mantle minerals: Implications for melt transport rates in island arcs. *Earth and Planetary Science Letters* **215**: 339–355.

Forsyth DW, *et al.* 1998. Imaging the deep seismic structure beneath a mid-ocean ridge: the MELT experiment. *Science* **280**: 1215–1218.

Gauthier P-J, Condomines M. 1999. ^{210}Pb-^{226}Ra radioactive disequilibria in recent lavas and radon degassing: inferences on the magma chamber dynamics at Stromboli and Merapi volcanoes. *Earth and Planetary Science Letters* **172**: 111–126.

George R, Reagan M, Turner S, Gill J, Bourdon B. 2004. Comment on 'Steady-state $^{226}Ra/^{230}Th$ disequilibrium in mantle minerals: Implications for melt transport rates in island arcs' by Feineman MD DePaolo DJ [*Earth Planet. Sci. Lett.* 215 (2003) 339–355]. *Earth and Planetary Science Letters* **228**: 563–567.

Hauri EH. 1997. Melt migration and mantle chromatography, 1: simplified theory and conditions for chemical and isotopic decoupling. *Earth and Planetary Science Letters* **153**: 1–19.

Hirth G, Kohlstedt DL. 1995. Experimental constraints on the dynamics of the partially molten upper mantle: deformation in the diffusion creep regime. *Journal of Geophysical Research* **100**: 1981–2001.

Holtzman BJ, Groebner NJ, Zimmerman ME, Ginsberg SB, Kohlstedt DL. 2003. Stress-driven melt segregation in partially molten rocks. *Geochemistry, Geophysics, Geosystems* **4**: doi:10.1029/2001GC000258.

Holtzman BJ, Kohlstedt DL. 2007. Stress-driven melt segregation and strain partitioning in partially molten rocks: effects of stress and strain. *Journal of Petrology* **48**: 2379–2406.

Huang F, Gao L, Lundstrom CC. 2008. The effect of assimilation, fractional crystallization, and ageing on U-series disequilibria in subduction zone lavas. *Geochimica et Cosmochimica Acta* **72**: 4136–4145.

Jin Z-M, Green HW, Zhou Y. 1994. Melt topology in partially molten mantle peridotite during ductile deformation. *Nature* **372**: 164–167.

Katz RF, Spiegelman M, Holtzman B. 2006. The dynamics of melt and shear localization in partially molten aggregates. *Nature* **442**: 676–679.

Kelemen PB. 1990. Reaction between ultramafic wall rock and fractionating basaltic magma: Part I, Phase relations, the origin of calc-alkaline magma series, and the formation of discordant dunite. *Journal of Petrology* **31**: 51–98.

Kelemen PB, Hirth G, Shimizu N, Spiegelman M, Dick HJB. 1997. A review of melt migration processes in the adiabatically upwelling mantle beneath oceanic spreading ridges. *Philosophical Transactions of the Royal Society of London, A* **355**: 283–318.

Langmuir C, Klein E. Plank T. 1992. Petrological systematics of mid-ocean ridge basalts: Constraints on melt generation beneath ocean ridges. *AGU Monograph* **71**: 183–280.

Lundstrom CC. 2003. Uranium-series disequilibria in mid-ocean ridge basalts: observations and models of basalt genesis. In: *Uranium-series Geochemistry*, Bourdon B, Henderson GM, Lundstrom CC, Turner SP (eds), *Reviews in Mineralogy and Geochemistry* **52**: 175–214.

McKenzie D. 1984. The generation and compaction of partially molten rock. *Journal of Petrology* **25**: 713–765.

McKenzie D. 1985. ^{230}Th-^{238}U disequilibrium and the melting processes beneath ridge axes. *Earth and Planetary Science Letters* **72**: 149–157.

O'Neill C, Spiegelman, M. 2010. Melt transport from the mantle to the crust – Numerical modelling. In: *Timescales of Magmatic Processes: From Core to Atmosphere*, Dosseto, A, Turner S, Van Orman JA (eds), Wiley-Blackwell, Oxford.

O'Reilly S, Griffin W. 2010. Xenolith transport in ascending magma. In: *Timescales of Magmatic Processes: From Core to Atmosphere*, Dosseto, A, Turner S, Van Orman JA (eds), Wiley-Blackwell, Oxford.

Petford N, Koenders MA, Turner S. 2008. Channelised melt flow in downwelling mantle: implications for ^{226}Ra-^{210}Pb disequilibria in arc magmas. *Journal of Geophysical Research* **113**: doi:10.1029/2007JB005563.

Robinson JAC, Wood BJ. 1999. The depth of the spinel to garnet transition at the peridotite solidus. *Earth and Planetary Science Letters* **164**: 277–284.

Rubin KH, van der Zander I, Smith MC, Bergmanis EC. 2005. New speed limit for ocean ridge magmatism. *Nature* **437**: 534–538.

Rushmer T, Knesel K. 2010. Crustal melting. In: *Timescales of Magmatic Processes: From Core to*

Atmosphere, Dosseto, A, Turner S, Van Orman JA (eds), Wiley-Blackwell, Oxford.

Rutherford MJ, Hill PM. 1993. Magma ascent rates from amphibole breakdown: an experimental study applied to the 1980–1986 Mt St Helens eruption. *Journal of Geophysical Research* **98**: 19667–19686.

Saal AE, Van Orman JA. 2004. The ^{226}Ra enrichment in oceanic basalts: evidence for melt-cumulate diffusive interaction processes within the oceanic lithosphere. *Geochemistry, Geophysics and Geosystems* **5**: doi:10.1029/2003GC000620.

Slater L, McKenzie D, Gronvold K, Shimizu N. 2001. Melt generation and movement beneath Theistareykir, NE Iceland. *Journal of Petrology* **42**: 321–354.

Sleep NH. 1988. Tapping of melt by veins and dikes. *Journal of Geophysical Research* **93**: 10255–10272.

Sobolev AV, Shimizi N. 1993. Ultra-depleted primary melt included in an olivine from the Mid-Atlantic ridge. *Nature* **363**: 151–154.

Sparks RSJ, Pinkerton H, Macdonald R. 1977. The transport of xenoliths in magmas. *Earth and Planetary Science Letters* **35**: 234–238.

Spera FJ. 1980. Aspects of magma transport. In: *Physics of Magmatic Processes*, Hargraves RB (ed.), Princeton University Press, Princeton NJ, pp. 265–324.

Spiegelman M, McKenzie D. 1987. Simple 2-D models for melt extraction at mid-ocean ridges and island arcs. *Earth and Planetary Science Letters* **83**: 137–152.

Spiegelman M, Elliott T. 1993. Consequences of melt transport for uranium series disequilibrium. *Earth and Planetary Science Letters* **118**: 1–20.

Spiegelman M, Kelemen PB, Aharonov E. 2001. Causes and consequences of flow organization during melt transport: the reaction infiltration instability in compactable media. *Journal of Geophysical Research* **106**: 2061–2077.

Stevenson DJ. 1989. Spontaneous small-scale segregation in partial melts undergoing deformation. *Geophysical Research Letters* **16**: 1067–1070.

Stracke A, Bourdon B, McKenzie D. 2006. Melt extraction in the Earth's mantle: constraints from U-Th-Pa-Ra studies in oceanic basalts. *Earth and Planetary Science Letters* **244**: 97–112.

Turner S, Hawkesworth C. 1997. Constraints on flux rates and mantle dynamics beneath island arcs from Tonga-Kermadec. *Nature* **389**: 568–573.

Turner S, Bourdon B, Hawkesworth C, Evans P. 2000. ^{226}Ra-^{230}Th evidence for multiple dehydration events, rapid melt ascent and the time scales of differentiation beneath the Tonga-Kermadec island arc. *Earth and Planetary Science Letters* **179**: 581–593.

Turner S, Evans P, Hawkesworth C. 2001. Ultra-fast source-to-surface movement of melt at island arcs from ^{226}Ra-^{230}Th systematics. *Science* **292**: 1363–1366.

Turner S, Bourdon B, Gill J. 2003. Insights into magma genesis at convergent margins from U-series isotopes. In: *Uranium-series Geochemistry*, Bourdon B, Henderson GM, Lundstrom CC, Turner SP (eds), *Reviews in Mineralogy and Geochemistry* **315**: 255–315.

Turner S, Black S, Berlo K. 2004. ^{210}Pb-^{226}Ra and ^{232}Th-^{228}Ra systematics in young arc lavas: implications for magma degassing and ascent rates. *Earth and Planetary Science Letters* **227**: 1–16.

Van Orman JA, Saal AE. 2009. Influence of crustal cumulates on ^{210}Pb disequilibria in basalts. *Earth and Planetary Science Letters* **284**: 284–291.

Van Orman JA, Saal AE, Bourdon B, Hauri EH. 2006. Diffusive fractionation of U-series radionuclides during mantle melting and shallow level melt-cumulate interaction. *Geochimica et Cosmochimica Acta* **70**: 4797–4812.

Vaughan PJ, Kohlstedt DL. 1982. Distribution of the glass phase in hot-pressed olivine basalt aggregates: an electron microscope study. *Contributions to Mineralogy and Petrology* **81**: 253–261.

Yaxley GM, Kamenetsky V. 1999. *In situ* origin for glass in mantle xenoliths from southeastern Australia: insights from trace element compositions of glasses and metasomatic phases. *Earth and Planetary Science Letters* **172**: 97–109.

Yokoyama T, Kobayashi K, Kuritani T, Nakamura E. 2003. Mantle metasomatism and rapid ascent of slab components beneath island arcs: evidence from ^{238}U-^{230}Th-^{226}Ra disequilibria of Miyakejima volcano, Izu arc, Japan. *Journal of Geophysical Research* **108**: doi:10.1029/2002JB002103.

6 Rates of Magma Ascent: Constraints from Mantle-Derived Xenoliths

SUZANNE Y. O'REILLY AND W.L. GRIFFIN

GEMOC ARC National Key Centre, Department of Earth and Planetary Sciences,
Macquarie University, Sydney, NSW, Australia

SUMMARY

The rates of ascent of magmas through the lithosphere can be estimated by a variety of methods using physical and compositional parameters of lithospheric mantle fragments brought to the surface in these magmas. The xenoliths are relatively dense, and the magma must be ascending more rapidly than the xenoliths can sink through it. These methods suggest an average ascent rate through the whole lithosphere (mantle and crust) in the range of 0.2 to 0.5 m s^{-1} (about 0.5 to 2 km/hour). The ascent rates through the shallow crust may be much higher: >20 m s^{-1} and up to supersonic speeds (>300 m s^{-1}) in the uppermost crust. Residence times derived from microstructural observations in recrystallised minerals in the xenoliths and element diffusion profiles in xenolith minerals suggest ascent rates in the range of 0.2 to 0.4 m s^{-1}. These methods all provide minimum velocities ranging from about 0.2 to 2 m s^{-1} for relatively low-volume melts such as alkali basalts, to ~4 to 40 m s^{-1} (and up to supersonic) for volatile-charged ultramafic melts such as kimberlites.

In summary, magmas carrying mantle xenoliths much reach the surface within a maximum of about 8 to 60 hours of picking up these dense fragments from depths of about 200 to 80 km depth.

Timescales of Magmatic Processes: From Core to Atmosphere, 1st edition. Edited by Anthony Dosseto, Simon P. Turner and James A. Van Orman.
© 2011 by Blackwell Publishing Ltd.

INTRODUCTION

The rates at which magmas rise through the crust and upper mantle cannot be measured directly, but can be estimated by several indirect means, based on observations from xenoliths in alkali basalts, kimberlites and related rocks.

Xenolith ('foreign rock') is a general term applied to a fragment of obviously foreign material enclosed in a magmatic rock; it implies that the xenolith was accidentally entrained, then transported by the magma from some depth to the place where the magma solidified. The mineral assemblages and mineral chemistry of many xenoliths can be compared with experimental studies to derive the depth and temperature at which they were picked up by the ascending magma. The presence of such xenoliths therefore can provide constraints on the shallowest depths of magma generation and segregation, and on how rapidly magmas reach the surface. In this chapter, we review the types of information that can be derived from xenoliths, to constrain the rates of magma ascent.

THE SIGNIFICANCE OF XENOLITHS

For the estimation of magma ascent rates, the most interesting xenoliths are those derived from the upper mantle, which include a range of ultramafic to mafic rock types, dominantly peridotites (olivine-dominant ultramafic mantle wall-rocks). Upper-mantle xenoliths are extremely rare in tholeiitic and fractionated basaltic rocks (both continental and oceanic), but relatively common in alkali basalts from both ocean islands and continental settings. They also are common in a range of more ultramafic intrusive and eruptive rocks, ranging from kimberlite to lamproite and ultramafic lamprophyres. The host magmas for xenoliths are generally generated in the asthenosphere (O'Reilly & Griffin, 2010) or at least at deeper levels than the lithospheric mantle regions from which the xenoliths were sampled. *Kimberlites and related rocks* typically erupt in intraplate continental (usually cratonic) settings, and carry xenoliths of harzburgite (olivine + orthopyroxene), lherzolite (olivine + orthopyroxene + clinopyroxene ± spinel ± garnet) and eclogite (clinopyroxene + garnet: the high-pressure form of basalt), derived from maximum depths of 180 to 220 km (O'Reilly & Griffin, 2006 and references therein). *Alkali basalts and related rocks* typically erupt in off-craton settings (mobile belts, young tectonic terranes, rifts, ocean islands) with relatively thin lithosphere; they commonly carry xenoliths of spinel (± garnet) lherzolite and pyroxenites (± garnet) derived from maximum depths of 60 to 80 km (Nixon, 1987). All of these rock types have densities significantly greater than their host magmas (Kushiro *et al.*, 1976; Spera, 1980, 1984), so their transport to the surface provides some minimum constraints on the ascent velocities of the magmas from mantle depths. Geochemical and microstructural evidence for the thermal history of the xenoliths can provide further constraints.

ENTRAINMENT OF XENOLITHS – BRITTLE FRACTURE IN THE UPPER MANTLE

At upper mantle temperatures and pressures (depths of 50–250 km and temperatures of 1,000–1,300°C), mantle rocks have been shown to deform plastically (Raleigh & Kirby, 1970; Mercier & Nicolas, 1975; Nicolas & Poirier, 1976; Mainprice *et al.*, 2005), at least in response to long-term stresses. However, the entrainment of xenoliths in an ascending magma requires that fragments of rock can be broken off the walls of magma chambers or conduits, a process that implies brittle fracture in response to short-term stresses.

Xenoliths, and particularly those derived from the upper mantle, are commonly referred to as 'nodules', implying a rounded shape. However, a study of over 4,000 mantle-derived peridotite and pyroxenite xenoliths in basaltic flows, dykes, sills and cinder cones from Australia and Spitsbergen (O'Reilly, unpublished data; Skjelkvåle *et al.*, 1989) shows that they are typically angular to subangular and many have polygonal and facetted shapes (Figure 6.1a, b, c). Significant rounding is restricted to xenoliths from breccias or those associated with the more explosive eruptions such as maars, diatremes and some cinder cones. Xenoliths in kimberlites are typically more rounded than those in basalts, but many display planar faces (Figure 6.1d). This rounding appears to reflect tumbling of xenoliths during turbulent ascent and eruption; we infer that at the time of their entrainment, most xenoliths are angular, and originated by brittle fracture.

Most spinel lherzolite xenoliths from alkali basalts, when subjected to mild hydraulic stress, split along planar surfaces in up to three different orientations, suggesting that pre-existing planes of weakness were present in the original upper-mantle wall rock. Basu (1980) described similar jointing in spinel lherzolites from San Quintin, Baja California, and concluded that these planar features formed because of high differential stress in the mantle, predisposing the wall-rock to be entrained in ascending magma.

The occurrence of common planar dykes, joints and faults in the upper mantle was shown by Wilshire & Kirby (1989) to be widespread, using examples from xenolith suites worldwide and also by reference to observations of straight-sided dykes cross-cutting other structures in exposed peridotite massifs.

The presence of fluids is probably a significant factor in generating brittle fracture ('hydrofracting'

Fig. 6.1 (a) and (b) Different views of angular xenoliths of spinel peridotite in alkali basalt from Batchelor Crater, Queensland, Australia. (c) Xenolith of a composite xenolith showing spinel peridotite veined by basaltic magma that crystallized within the mantle, within a volcanic bomb from Mt Shadwell, Victoria, Australia. Note the originally straight sides (emphasized by white dashed lines) and the rounded edges of the xenolith due to abrasion in the magma. The spinel peridotite areas appear rounded due to reaction with the invading basaltic magma. (d) Xenolith from kimberlite showing polygonal shape with planar surfaces. (e) Spinel peridotite xenolith showing straight-sided vein of garnet pyroxenite (basaltic magma crystallized within the mantle). (f) Photomicrograph of thin section of a peridotite showing an amphibole-rich vein with thin offshoot. Note the straight sides on both the large and small veins. Field of view is 2.5 cm across. See Plate 6.1 for a colour version of these images.

or 'shear fracture' (Wilshire & Kirby, 1989)) at the depths of magma generation and along magma conduits. Spera (1984) suggested that where the overpressure of a volatile-rich magma exceeds a critical value, crack propagation ensues in the overlying wall-rock, and proceeds at a velocity of centimeters to meters per second. Shaw (1980) demonstrated that localized shear fractures can result, showing a geometry that is consistent with the orientations of planar faces on polygonal xenoliths. There is abundant evidence that fluids in the mantle produce such fracturing. Many metasomatized xenoliths contain planar veins of amphibole or mica (<0.1mm to >10cm wide; Figure 6.1e). Cross-cutting planar veins oriented in up to three directions are reported from several localities worldwide (Wilshire *et al.*, 1980, 1991; Skjelkvåle *et al.*, 1989; Figure 6.1f). Other evidence for fluids within the mantle is provided by the presence of fluid inclusions within mineral grains. These inclusions typically define planar arrays, representing planar fractures; the fractures filled with fluids, which were trapped as inclusions with shapes that proceeded from amoeboid to negative-crystal forms as the host grain recrystallized to reduce its surface free energy (Roedder, 1984).

Detailed studies on volatiles and fluids in mantle-derived xenoliths and megacrysts have been carried out on samples from basalts in eastern Australia, particularly western Victoria (Andersen *et al.*, 1984, 1987; Porcelli *et al.*, 1986; O'Reilly, 1987; O'Reilly *et al.*, 1990). Xenoliths from western Victoria contain up to 3% by volume of crystal-lined cavities and fluid inclusions, all inferred to have been filled mainly with CO_2 at high pressure. Andersen *et al.* (1984) surveyed samples from xenolith-bearing localities worldwide and showed that xenoliths containing significant volumes of fluid cavities and fluid inclusions are confined to explosive-eruption vents such as Bullenmerri-Gnotuk (Victoria; Andersen *et al.*, 1984); Anakie (Victoria; Wass & Hollis, 1983); Spitsbergen (Amundsen *et al.*, 1987; Skjelkvale *et al.*, 1989); Craters 160 and 387 (Arizona; Andersen *et al.*, 1987); and Salt Lake Crater, (Hawaii; unpublished data). E. Roedder (personal communication, 1983) has pointed out

that such volumes of fluids represent high fluid pressure within the xenoliths relative to the magma, and that the direction of any volatile flow would be from the wall-rock – and the xenolith – outwards into adjacent magma. The presence of large volumes of fluid in some deep-seated wall-rocks therefore may favor brecciation and preferential entrainment of that horizon in the ascending magma, when heat from the magma leads to rapid outgassing of the wall rocks.

ASCENT VELOCITY: CONSTRAINTS FROM THE TRANSPORT OF XENOLITHS

In the 1960s, there was considerable controversy over the origin of the common peridotite fragments found in some basaltic to kimberlitic rocks worldwide; do they represent crustal cumulates from mafic magmas (O'Hara & Yoder, 1967) or are they actual samples of the mantle (Green & Ringwood, 1967)? Subsequently, after a mantle origin for the majority of such peridotite xenoliths was confirmed, there was debate over the degree to which mantle-derived xenoliths represent pristine samples of the upper mantle, or had re-equilibrated (thermally and/or compositionally) on their way to the surface. This debate led directly to the first attempts to quantify magma ascent rates. Most of the studies on this topic therefore derive from the 1970s and 1980s because of the very topical questions involved (then solved) at that time. Kushiro *et al.* (1976) carried out laboratory experiments on the static settling rates of olivine crystals in different magma compositions; the experiments simulated laminar-flow conditions. The results imply maximum ascent times of 60 hours for xenolith-bearing basalts, corresponding to minimum ascent rates of ~1 km/hour (0.28 ms^{-1}). Spera (1980, 1984) used Stokes' law, rheological parameters derived from numerous sources, and the assumption of Bingham behavior (in which the fluid has a finite shear strength) for the magma. For a Newtonian fluid (no shear strength), the ascent velocity can be calculated from the relationship:

$$V = (2g.r^2(d_{xen} - d_{liq}))/9\mu$$

where g is the gravitational acceleration, d is density, r is the radius of the xenolith and μ is magma viscosity. Magma densities can be estimated from bulk composition using relationships given by Bottinga & Weill (1970), whilst viscosities can be estimated at a given T by the methods of Shaw (1972) and Scarfe et al. (1987). If Bingham behavior is assumed, in order to account for the presence of other solids in the magma, μ must be replaced by a more complex term depending on the proportion of solids (Marsh, 1981).

As an example, Spera (1980) found that the transport of a 20 cm-diameter spherical xenolith of density 3.3 g/cc (a typical peridotite) in a basaltic magma containing 15 volume percent of phenocrysts would require a minimum ascent rate of ~0.1 ms^{-1}.

This methodology has since been applied to a range of magma types and xenolith types. Most estimates for the ascent rates of alkali basalts thus fall into the range of 0.1 to 0.5 ms^{-1}, with the highest rates corresponding to peridotite xenoliths ~40 cm in diameter. Wass & Pooley (1982) described lherzolite xenoliths up to 60 cm across from alkali basalts at Wallabadah (New South Wales, Australia), which had undergone concentric zonal alteration by carbonate-rich fluids during an extended residence time at mantle depths; the resulting decrease in density probably allowed the transport of these unusually large mantle fragments. Morin & Corriveau (1996) calculated the ascent rate of a minette magma as ~0.5 ms^{-1}, based on the presence of 40 cm pyroxenite xenoliths. The ascent rates for kimberlites are difficult to estimate because of the uncertainties in estimating viscosities; the intrinsic melt viscosity is probably low, but the common presence of abundant xenoliths and xenocrysts would raise the effective viscosity. Some kimberlites carry xenoliths of eclogite (d ≈ 3.5 gcm^{-3}) with diameters up to 60 cm, suggesting ascent rates of several tens of ms^{-1}. Sparks et al. (2006), recently summarised some of the other evidence (mainly fluid-flow calculations) for kimberlite ascent rates and suggested generalized dyke speeds (within the crustal regime) of >4 to 20 ms^{-1}, in agreement with ascent rates inferred from xenolith data.

These rates probably represent minimum estimates, because of the effects of volatiles in the magma. As noted above, many xenoliths contain evidence for the presence of volatiles (mainly CO_2) at depth. Other evidence for the important role of mantle-derived CO_2 includes: common cementation of basaltic breccias by carbonate (Emerson & Wass, 1980); vesicle linings of Fe-Mg-rich carbonate (Andersen et al., 1984); carbonation of some high-pressure xenoliths (Wass & Pooley, 1982); the high carbonate contents of kimberlites; and observations from recent Alaskan maars (Barnes & McCoy, 1979) that primary CO_2 is outgassing from the underlying mantle.

Spera (1984) carried out detailed fluid-dynamic modeling to test the hypothesis that high volatile contents in solution in magmas may act as propellants, enhancing velocity of ascent; he concluded that the propellant effect is relatively small. However, the exsolution of volatile components during magma ascent has a more important implication for the estimation of ascent velocities. Exsolution of volatiles (the generation of 'bubbles') decreases magma viscosity, and the magma-bubble mixture also has lower density (O'Donnell, 1984); both effects mean that a greater ascent velocity is required to prevent xenoliths from dropping out. An additional significant effect of volatile content is that exsolution of CO_2 will result in heating of the magma, rather than the cooling that accompanies exsolution of H_2O (Spera & Bergman, 1980). This means that ascending CO_2-rich magmas such as some alkali basalts, kimberlites and lamproites are unlikely to freeze (experience 'heat death') during ascent (Spera, 1984).

GEOCHEMICAL AND MICROSTRUCTURAL CONSTRAINTS

Geochemical and microstructural approaches to evaluating ascent rates are based on estimates of the rates at which xenoliths have been heated during transport in the host magmas. Since some of this heating, at least in the deepest-seated xenoliths, may have occurred before the initiation of magma ascent, these methods define minimum values for ascent rates.

Kimberlites commonly contain highly sheared xenoliths of garnet lherzolite; P-T estimates typically place these near the deepest levels (180–200 km) sampled by the individual kimberlites. In these mylonite-like rocks, porphyroblasts of olivine, pyroxenes and garnet sit in a very fine-grained foliated matrix. In detail, the boundaries of olivine grains in the matrix range from finely interdigitated to planar, reflecting continuous recrystallization during deformation. Mercier (1979) measured the grain sizes of olivine neoblasts in a number of these xenoliths and used experimental data on the kinetics of grain coarsening to estimate the maximum length of time that the xenoliths could have resided at high T. The estimated residence times in the kimberlite hosts were 4 to 6 hours, equivalent to velocities of 10 to 20 ms^{-1}.

Wanamaker (1990) studied secondary planes of CO_2 fluid inclusions in olivine grains of peridotite xenoliths from alkali basalts; these planes were assumed to have formed and healed during ascent. Using experimental measurements of the rates at which fractures in olivine anneal at high T, he estimated magma residence times of 80 to 170 hours, corresponding to ascent rates of 0.5 to 1 ms^{-1}.

Patches of glassy or partially crystallized melts, formed by the infiltration of magmas, are common in mantle-derived xenoliths, especially those in alkali basalts. These melt patches commonly crystallize daughter minerals, resulting in zoning of the melt composition. Hofmann & Magaritz (1977) measured the diffusion rate of Ca in melts and concluded that xenoliths with glassy patches that retained CaO heterogeneity could not have resided in the magma for >~3 days, equivalent to minimum ascent rates (from 60 km depth) of 0.2 ms^{-1}.

Selverstone (1982) calculated that xenoliths 10 cm in diameter will be heated to the temperature of the host magma within three hours of contact. Such heating will induce diffusional exchange of elements between grains of different minerals, tending toward equilibrium at the imposed higher T. This provides a basis for using the zoning of adjacent minerals to calculate the total magma-residence time. Ozawa (1983, 1984) used experimental data on the diffusion of Mg and Fe between olivine and spinel to develop a 'geospeedometry' method based on Mg/(Mg + Fe^{2+}) zoning in spinel in Cr-diopside lherzolite xenoliths. Modeling diffusion profiles in individual spinel grains, and correlating these with grain size, he estimated that spinel-lherzolite xenoliths in alkali basalts from the Ichinomegata crater (northeast Japan) had resided in the magma for <24 hours, giving ascent rates of >1.4 ms^{-1}. Bezant (1985) applied this technique to a range of lherzolites in alkali basalts from eastern Australia. He calculated heating durations ranging from 4 to 24 hours in different xenoliths, corresponding to ascent rates of 0.4 to 2 ms^{-1}.

Peslier & Luhr (2006) used a similar approach; they used FTIR measurements to establish zoning profiles of OH abundance in olivine grains in spinel peridotite xenoliths from several olivine basalts. They found high H contents in the cores of olivine grains, and gradients to lower contents near the rims of grains. They interpreted these profiles in terms of H diffusing out of olivine grains upon heating in the host magma. Using experimental diffusion data, they calculated magma residence times of 18 to 65 hours, corresponding to ascent rates of 0.2 to 0.5 ms^{-1}. A subsequent analogous study on peridotite xenoliths from kimberlites (Peslier *et al.*, 2008) indicates an ascent rate for kimberlites of 5 to 37 ms^{-1} although they caution that this estimation depends on the volatile and degassing behavior of the particular kimberlite.

Cherniak & Liang (2007) have shown experimentally that orthopyroxene can preserve significant grain-scale compositional variation in rare-earth elements (REE) and predict that new data for REE diffusion rates and mineral-melt and mineral-mineral partition coefficients will make it possible to develop a REE-in-orthopyroxene geospeedometer for ultramafic rocks. Van Orman *et al.* (2001, 2002) and Tirone *et al.* (2005) provide analogous REE diffusion data for clinopyroxene and garnet, respectively. Gallagher & Elliott (2009) have demonstrated lithium-isotope zoning in clinopyroxene due to the temperature dependence of lithium partitioning and diffusivity, which 'may be useful for tracing sub-volcanic processes'. At the time of writing, these possible geospeedometers for xenolith transport have not been tested, but may ultimately provide additional methods for assessing ascent rates. The diffusion profiles for the REE in mantle minerals would be only about 100 nm long for timescales of xenolith ascent, and thus with present-day analytical constraints, are difficult to measure, so currently are useful for longer-duration processes. However, as Li has faster diffusion rates, it could prove applicable to xenolith ascent estimation.

DISCUSSION

Most of the calculated ascent rates for alkali basalts ($0.2–2\,ms^{-1}$) are reasonably consistent, considering the varying precision of the parameters used and the range of approaches. The estimates for kimberlitic and lamproitic magmas are scarcer, but suggest higher speeds, on the order of tens of meters per second (generally $\sim4–40\,ms^{-1}$). Wilson & Head (2007) have modeled the eruption of kimberlites, based on the propagation of a CO_2-filled crack tip extending above the magma; they adopt an ascent rate of several tens of ms^{-1} at depth, accelerating rapidly to near-supersonic speeds ($\geq300\,ms^{-1}$) when the crack breaks through the upper crust.

These rapid ascent rates reinforce the validity of recognizing xenoliths as essentially unaltered fragments of deep-seated lithologies. They also are a strong argument against the concept that kimberlitic and undersaturated basaltic melts may rise slowly as kilometer-sized diapers. Rather, once a small percentage of partial melt creates a sufficiently high magma pressure, it surges up at high velocities (cf. Wilson & Head, 2007). The general absence of deep-seated xenoliths in tholeiitic magmas thus is probably related to their low volatile contents, and to the greater degrees of partial melting involved in their gene-

sis. These factors may reduce the potential for hydrofracture of the surrounding hot, weak, mantle wall-rocks, and the initiation of ascent is probably by slow percolation. In the absence of fluid-driven crack propagation, ascent velocities also will be lower and will allow settling of any entrained xenoliths.

CONCLUSIONS

The ascent rates of deep-seated magmas can be estimated using calculations based on the size and density of mantle-derived xenoliths, and from temperature-related diffusion phenomena (chemical zoning, grain-boundary adjustment) observed in xenoliths. These methods provide only minimum values, which range from 0.2 to $2\,ms^{-1}$ for relatively low-volume melts such as alkali basalts. Modeling of the eruption of kimberlites and similar volatile-charged ultramafic melts suggests rates of several meters per second ($\sim4–40\,ms^{-1}$) increasing to supersonic values as the propagating crack reaches the surface. High contents of CO_2 in magmas do not provide a significant 'propellant effect' to assist xenolith transport as the density and viscosity of the mixed magma-fluid system decrease, but the exsolution of CO_2 both provides heat that helps to prevent the magma from cooling during ascent, and enhances crack propagation. Magmas representing significant proportions of partial melting, such as tholeiitic basalts, rarely carry mantle-derived xenoliths; this probably reflects ascent rates that are too slow to hold (or possibly incorporate) high-density xenoliths entrained in the magma.

ACKNOWLEDGMENTS

The ideas summarised here represent the synthesis of a wealth of information gathered since the authors began working on mantle-derived xenoliths and their host magmas. The concepts have evolved from innumerable stimulating and question-provoking discussions with colleagues over many years that are greatly valued, including those with some of the early pioneers in xenolith and related studies, including Dale Jackson, Ikuo Kushiro and Howard Wilshire, as well as with those more recently developing relevant front-line experimental, geochemical and modeling studies too numerous to mention. Throughout all of the authors' relevant work, the Australian Research Council has consistently provided funding for projects that helped to add background information. This is GEMOC publication #666.

REFERENCES

Amundsen HEF, Griffin WL, O'Reilly SY. 1987. The lower crust and upper mantle beneath north-western Spitsbergen: evidence from xenoliths and geophysics. *Tectonophysics* **139**: 169–185.

Andersen T, Griffin WL, O'Reilly SY. 1987. Primary sulphide melt inclusions in mantle-derived megacrysts and pyroxenites. *Lithos* **20**: 279–294.

Andersen T, O'Reilly SY, Griffin WL. 1984. The trapped fluid phase in upper mantle xenoliths from Victoria, Australia: implications for mantle metasomatism. *Contributions to Mineralogy and Petrology* **88**: 72–85.

Barnes I, McCoy GA. 1979. Possible role of mantle-derived CO_2 in causing two 'phreatic' explosions in Alaska. *Geology* **7**: 434–435.

Basu AR. 1980. Jointed blocks of peridotite xenoliths in basalts and mantle dynamics. *Nature* **284**: 612–613.

Bezant C. 1985. Geothermometry and seismic properties of upper mantle lherzolites from eastern Australia. BSc. Honours thesis, Macquarie University, Sydney.

Bottinga Y, Weill, DF. 1970. Densities of liquid silicate systems calculated from partial molar volumes of oxide components. *American Journal of Science* **269**: 169–182.

Cherniak DJ, Liang Y. 2007. Rare earth element diffusion in natural enstatite. *Geochimica et Cosmochimica Acta* **71**: 1324–1340.

Emerson DW, Wass SY. 1980. Diatreme characteristics – evidence from the Mogo Hill intrusion, Sydney Basin. *Bulletin of the Australian Society of Exploration Geophysicists* **11**: 121–133.

Gallagher K, Elliott T. 2009. Fractionation of lithium isotopes in magmatic systems as a natural consequence of cooling. *Earth and Planetary Science Letters* **278**: 286–296.

Green, DH, Ringwood, AE. 1967. The genesis of basaltic magmas. *Contributions to Mineralogy and Petrology* **15**: 103–190.

Hofmann AW, Magaritz M. 1977. Diffusion of Ca, Sr, Ba, and Co in a basalt melt. Implications for the geochemistry of the mantle. *Journal of Geophysical Research* **82**: 5432–5440.

Kushiro I, Yoder HS, Mysen BO. 1976. Viscosities of basalt and andesite melts at high pressures. *Journal of Geophysical Research* **81**(B35): 6351–6356.

Mainprice D, Tommassi A, Couvy H. *et al.* 2005. Pressure sensitivity of olivine slip systems and seismic anisotropy of Earth's upper mantle. *Nature* **433**: 731–733.

Marsh BD. 1981. On the crystallinity, probability of occurrence and rheology of lava and magma. *Contributions to Mineralogy and Petrology* **78**(1): 85–98.

Mercier J-CC. 1979. Peridotite xenoliths and the dynamics of kimberlite intrusion. In: *The Mantle Sample: Inclusions in Kimberlites and Other Volcanics*, Boyd FR, Meyer HOA (eds), American Geophysical Union, Washington, DC, pp. 197–212.

Mercier J-CC, Nicolas A. 1975. Textures and fabric of upper mantle peridotites as illustrated by xenoliths from basalts. *Journal of Petrology* **16**: 454–487.

Morin D, Corriveau L. 1996. Fragmentation processes and xenolith transport in a Proterozoic minette dyke, Grenville Province, Quebec. *Contributions to Mineralogy and Petrology* **125**: 319–331.

Nicolas A, Poirier JF. 1976. *Crystalline Plasticity and Solid State Flow in Metamorphic Rocks*. Wiley-Interscience, New York, 444 pp.

Nixon PH. 1987. *Mantle Xenoliths*. Wiley, London, 844 pp

O'Donnell P. 1984. *The Night of Morningstar*. MacMillan, London. ISBN: 0-330-28168-2.

O'Hara MJ, Yoder HS Jr. 1967. Formation and fractionation of basic magmas at high pressures. *Scottish Journal of Geology* **3**: 67–117.

O'Reilly SY. 1987. Volatile-rich mantle beneath eastern Australia. In: *Mantle Xenoliths*, Nixon PH (ed.), Wiley, Chichester, 661–670.

O'Reilly SY, Griffin WL. 2006. Imaging chemical and thermal heterogeneity in the sub-continental lithospheric mantle with garnets and xenoliths: Geophysical implications. *Tectonophysics* **416**: 289–309.

O'Reilly SY, Griffin WL. 2010. The Continental lithosphere-asthenosphere boundary: Can we sample it? *Lithos* doi:10.1016/j.lithos.2010.03.016.

O'Reilly SY, Griffin WL, Segelstad TV. 1990. The nature and role of fluids in the upper mantle: evidence in xenoliths from Victoria, Australia. In: *Conference on Stable Isotopes and Fluid Processes in Mineralisation,*

Herbert HK, Ho SE (eds), Geology Department and University Extension, The University of Western Australia, Queensland, 315–323.

Ozawa K. 1983. Evaluation of olivine-spinel geothermometry as an indicator of thermal history for peridotites. *Contributions to Mineralogy and Petrology* **82**: 52–65.

Ozawa K. 1984. Olivine-spinel geospeedometry: Analysis of diffusion-controlled Mg-Fe^{2+} exchange. *Geochimica Cosmochimica Acta* **48**: 2597–2611.

Peslier AH, Luhr JF. 2006. Hydrogen loss from olivines in mantle xenoliths from Simcoe (USA) and Mexico: Mafic alkalic magma ascent rates and water budget of the sub-continental lithosphere. *Earth and Planetary Science Letters* **242**: 302–319.

Peslier AH, Woodland AB, Wolff JA. 2008. Fast kimberlite ascent rates estimated from hydrogen diffusion profiles in xenolithic mantle olivines from southern Africa. *Geochimica et Cosmochimica Acta* **72**: 2711–2722.

Porcelli D, O'Nions RK, O'Reilly SY. 1986. Helium and strontium isotopes in ultramafic xenoliths. *Chemical Geology* **54**: 237–249.

Raleigh CB, Kirby SH. 1970. Creep in the mantle. *Mineralogical Society of America Special Paper*, **3**: 113–121.

Roedder E. 1984. Fluid inclusions. Reviews in Mineralogy 12. *Mineralogical Society of America*, 644 pp.

Scarfe CM, Mysen BO, Virgo M, et al. 1987. Pressure dependence of the viscosity of silicate melts. In: *Magmatic Processes: Physiochemical Principles*, Mysen BO (ed.), *Geochemical Society Special Publication Number* **1**: 59–67.

Selverstone J. 1982. Fluid inclusions as petrogenetic indicators in granulite xenoliths, Pali-Aike volcanic field, Chile. *Contributions to Mineralogy and Petrology* **79**: 28–36.

Shaw HR. 1972. Viscosities of magmatic silicate liquids: an empirical method of prediction. *American Journal of Science* **272**: 870–893.

Shaw HR. 1980. The fracture mechanisms of magma transport from the mantle to the surface. In: *Physics of Magmatic Processes. Princeton University Press*, Hargraves EC (ed.), Princeton University Press, Princeton NJ, 201–264.

Skjelkvåle B, Amundsen HEF, O'Reilly S, Griffin WL, Gjelsvik T. 1989. A primitive alkali basaltic stratovolcano and associated eruptive centers, Northwestern Spitsbergen: volcanology and tectonic significance. *Journal of Volcanological and Geothermal Research* **37**: 1–19.

Sparks RSJ, Baker L, Brown RJ, et al. 2006. Dynamical constraints on kimberlite volcanism. *Journal of Volcanology and Geothermal Research* **155**: 18–48.

Spera FJ. 1980. Aspects of magma transport. In: *Physics of Magmatic Processes*, Hargraves EB (ed.), Princeton University Press, Princeton NJ, 265–323.

Spera FJ. 1984. Carbon dioxide in petrogenesis III: Role of volatiles in the ascent of alkaline magma with special reference to xenolith-bearing mafic lavas. *Contributions to Mineralogy and Petrology* **88**: 217–232.

Spera FJ, Bergman SC. 1980. Carbon dioxide in igneous petrogenesis I. Aspects of the dissolution of CO_2 in silicate liquids. *Contributions to Mineralogy and Petrology* **74**: 55–66.

Tirone M, Ganguly J, Dohmen R, et al. 2005. Rare earth diffusion kinetics in garnet: Experimental studies and applications. *Geochimica et Cosmochimica Acta* **69**: 2385–2398.

Van Orman JA, Grove TL, Shimizu N. 2001. Rare earth element diffusion in diopside: influence of temperature, pressure and ionic radius, and an elastic model for diffusion in silicates. *Contributions to Mineralogy and Petrology* **141**: 687–703.

Van Orman JA, Grove TL, Shimizu N, et al. 2002. Rare earth element diffusion in a natural pyrope single crystal at 2.8 GPa. *Contributions to Mineralogy and Petrology* **142**: 416–424.

Wanamaker BJ. 1990. Decrepitation and crack healing of fluid inclusions in San Carlos olivine. *Journal of Geophysical Research* **95**(B10): 15623–15641.

Wass SY, Hollis JD. 1983. Crustal growth in southeastern Australia – evidence from lower crustal eclogitic and granulitic xenoliths. *Journal of Metamorphic Geology* **1**: 25–45.

Wass SY, Pooley GD. 1982. Fluid activity in the mantle – evidence from large Iherzolite zenoliths. *Terra Cognita* **2**: 229.

Wilshire HG, Kirby, SH. 1989. Dikes, joints and faults in the upper mantle. *Tectonophysics* **161**: 23–31.

Wilshire HG, Pike JEN, Meyer CE, et al. (1980) Amphibole-rich veins in lherzolite xenoliths, Dish Hill and Deadman Lake, California. *American Journal of Science* **280A**: 576–593.

Wilshire HG, McGuire AV, Noller JS, et al. 1991. Petrology of lower crustal and upper mantle xenoliths from the Cima volcanic field, California. *Journal of Petrology* **32**: 169–200.

Wilson L, Head JW. 2007. An integrated model of kimberlite ascent and eruption. *Nature* **447**: 53–57.

7 Time Constraints from Chemical Equilibration in Magmatic Crystals

FIDEL COSTA[1,2] AND DANIEL MORGAN[3]

[1]Institut de Ciencies de la Terra 'Jaume Almera', CSIC, Barcelona, Spain
[2]Earth Observatory of Singapore, Nanyang Technological University, Singapore
[3]School of Earth and Environment, University of Leeds, UK

SUMMARY

The timescales of magmatic processes are some of the main physical variables that are still poorly-constrained in igneous petrology. This hinders our ability to quantify the rates at which many important processes operate, from the specifics of a single magma formation event at depth, to eruption, to the larger-scale Earth dynamics of crustal formation, differentiation, and recycling. Determination of the rates of magmatic processes is a complicated matter because a large variety of times and events are commonly involved in the petrogenesis of a given magma or rock suite. The paucity of information and complexity of the topic make it challenging and exciting. New findings and real advances can be expected in the following years. In this chapter we show that modelling of the compositional heterogeneities of crystals can be used to determine the time scales of a variety of magmatic processes. We first introduce the idea of using the zoning features in crystals to disentangle the magmatic processes recorded in given rock, and later show how such zoning can be re-equilibrated by diffusion. The diffusion equations and processes are dis-

cussed in some detail, and we also provide a list of diffusion coefficients that have been used in the literature. We illustrate the determination of time scales using three examples that exploit zoning in quartz, olivine, and plagioclase. In the final part of the chapter, we first show that most timescales derived from diffusion studies are in the range of days to hundreds of years and this is in contrast with the thousands of years or longer found by most studies employing radioisotopic methods. We propose several possibilities to explain such differences bearing in mind that some of the data, theoretical knowledge, and interpretations presented in this book are still evolving.

INTRODUCTION

The presence of chemical zoning in crystals has fascinated amateur mineralogists and scientists for over a century. Early workers recognized that crystals from magmatic rocks are zoned both at the hand specimen scale (e.g. crystals in granites and pegmatites) and at the thin section scale, shown by the variation of optical properties of minerals under the polarizing microscope (Figure 7.1). Crystals were employed very early as archives of magmatic processes (Larsen *et al.*, 1938) because they record changes in their environments as growth and dissolution zones. When *in-situ* chemical analyses of minerals

Timescales of Magmatic Processes: From Core to Atmosphere, 1st edition. Edited by Anthony Dosseto, Simon P. Turner and James A. Van Orman.
© 2011 by Blackwell Publishing Ltd.

(a)

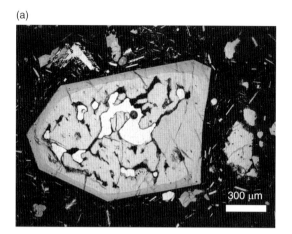

(b)

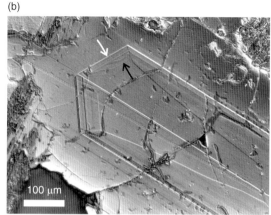

(c)

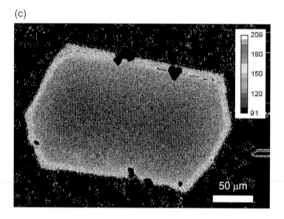

Fig. 7.1 Examples of zoning types in crystals and how they can be seen by different techniques. (a) Optical microscope photomicrograph of a zoned clinopyroxene with a clear rim of a different composition that the rest of the crystal. (b) Nomarski interference contrast image of plagioclase crystal taken with the optical microscope under reflected light. Note that details of the zoning include dissolution surfaces (black arrow). (c) X-ray distribution map of Fe in olivine obtained with the electron microprobe.

became possible with the development of the electron microprobe (Castaing, 1951), major element compositions were quantified and very quickly used to determine conditions of formation of magmatic rocks through thermodynamic calculations, resulting in the establishment of geothermobarometry (Buddington et al., 1955). With further development of in-situ analytical techniques, such as the ion microprobe (Slodzian, 1964), laser ablation mass spectrometry, and microdrilling, it became apparent that zoning in minerals occurs also in trace elements and isotopic ratios (Figure 7.1; Ginibre et al., 2007; Davidson et al., 2007).

Chemical zoning in crystals may result from a variety of magmatic processes, but it can be best interpreted as due to changes in the main intensive variables (e.g. pressure, temperature, chemical potential; Figure 7.2). It then follows that mineral zoning can be treated as an invaluable source of information about the physical and chemical processes (e.g. temperature fluctuations, additions of new melts) that the magma has experienced prior to and during solidification (Streck, 2008). However, chemical gradients that make up the crystal zoning are rarely the stable, equilibrium configuration, and they tend to decay due to the kinetic process of diffusion. The extent to which the zoning will be

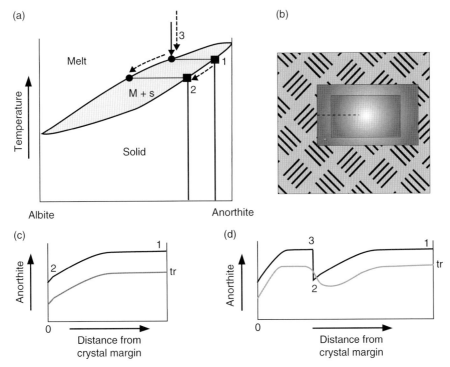

Fig. 7.2 Schematic illustration of development and re-equilibration of chemical zoning (anorthite and trace elements = tr) in plagioclase. (a) Plagioclase phase diagram, (b) plagioclase crystal as may be seen in thin section, (c) and (d) composition traverse of anorthite and a trace element (tr, in gray color) across the crystal at different stages of growth. Starting with a given bulk composition, a plagioclase of composition #1 will grow at a given temperature. If temperature decreases, the equilibrium composition will change (lower anorthite content) and the crystal will be zoned in anorthite down to composition #2. The trace element is shown here for simplicity to behave in the same manner during crystallization (part C of the figure). If at this moment we imagine that a large amount of new melt arrives at the system with the same composition (#3) as the original one (e.g. a replenishment event in the magma reservoir), the plagioclase may at first dissolve, but later as temperature decreases it will develop another zoning profile. Since the anorthite diffuses very slowly (Grove *et al.*, 1984) the zoning will retain the magmatic history, whereas a trace element such as Sr that diffuses much more rapidly (Cherniak & Watson, 1994) will tend to re-equilibrate. Thus, the crystal contains information about petrogenetic processes but also about their characteristic times.

equilibrated by diffusion depends on the mineral, the elements under consideration, temperature and time; however, many zoning patterns are the combined record of crystal growth (and dissolution) plus diffusion (Figure 7.2). This complicates a straightforward interpretation but has a huge added value; because of this dependence, the zoning patterns we observe cannot only unravel the processes but also the time that passed between different processes or events (Costa *et al.*, 2008).

RE-EQUILIBRATION OF CHEMICAL ZONING IN CRYSTALS BY DIFFUSION

Re-equilibration of chemical gradients in crystals is analogous to relaxation of gradients in topography, due to rock movement and erosion (Martin & Church, 1997). In the case of zoning in crystals, the gradients are of concentration (or more rigorously, chemical potentials), whereas in the case of topography, the gradient is of potential energy. Here we will concentrate on re-equilibration mediated by volume diffusion through crystals. Diffusion can be described in terms of the processes that occur at the atomic scale (e.g. potential energy wells and atom arrangements in solids) and/or in terms of the macroscopic measurements done in the laboratory (e.g. equilibration of concentration gradients). Both formulations are described below sequentially but are complementary (Manning, 1968). Here we present a relatively simplified view and equations of the diffusion processes that occur in silicate or oxide minerals, but we encourage the readers to take a look at Lasaga (1998), Watson & Baxter (2007), Chakraborty (2008), Costa *et al.* (2008) and references cited therein for a more in-depth treatment.

Atomistic description of diffusion

Diffusion in a mineral can be thought of as the relative movement of ions within the crystal lattice. This definition excludes flow of the crystal lattice, where all atoms move coherently. Atoms in the crystal move randomly and spontaneously due to their thermal energy. In other words, they move between lattice sites, even if there is no gradient in concentration. Crystal lattice sites, where ions tend to sit, can be thought of as regions of a lower, more stable energy, and therefore these form energy wells that hold the ions in place (Figure 7.3a). Some ions may have less energy than the average, leading to slower and less extensive vibrations, whilst others may have more energy and may vibrate faster and over longer distances, moving further up the energy well. If the energy increases further, eventually an energetic ion may have enough energy to jump to an adjacent energy well.

If we think about how the diffusion may proceed, it becomes apparent that an atom cannot jump into another site unless it is empty, or if another atom leaves. So either the two atoms exchange their positions, which is an extremely unlikely and energetically unfavorable situation, or there has to be a vacant site. Thus, diffusion

may occur in solids because there are defects in the crystal structure. These can be relatively complex defects such as dislocations, or point defects such as vacancies or interstitial atoms (Figure 7.4). A vacancy is a charge-balanced defect within a crystal where an ion is missing. It enables diffusion, as there is no need to form a very high-energy transition state as in the case of direct exchange of two atoms between sites; if an ion possesses sufficient energy to escape into the vacancy, diffusion is likely to occur. Crystals have vacancies for two main reasons. First, a few missing atoms increases the number of possible arrangements over the crystal sites compared to a 'pure' crystal, and so the creation of vacancies is favored as they increase the configurational entropy.

The free energy (G) can be described with a relation between the enthalpy (H), entropy (S) and temperature (T), $G = H - TS$. Thus, it is favorable to create a certain number of vacancies at all temperatures above absolute zero, and as the TS term increases, so will the number of vacancies as the free energy of the system is lowered (Flynn, 1972). These types of vacancies are called *intrinsic* and dominate at high temperature. In natural crystals, point defects are in many cases associated with the incorporation of trace elements (although they can also be associated with major elements) that have a different ionic charge than that of the regular atom. For example, we can consider the substitution of Eu^{3+} into plagioclase. Eu^{3+} has a similar radius to Ca^{2+} and substitutes into the structure in much the same manner as Sr^{2+}; however, the charge difference may lead to a cation vacancy through the following equivalence:

$$3\,Ca^{2+} \leftrightarrow 2\,Eu^{3+} + \blacksquare$$

where $\blacksquare$ denotes a vacant calcium site. These are called *extrinsic* vacancies.

Whereas the extrinsic vacancies are fixed in number due to crystal composition, the total number of vacancies in a crystal increases with increasing temperature as the increased thermal energy allows the formation of more intrinsic

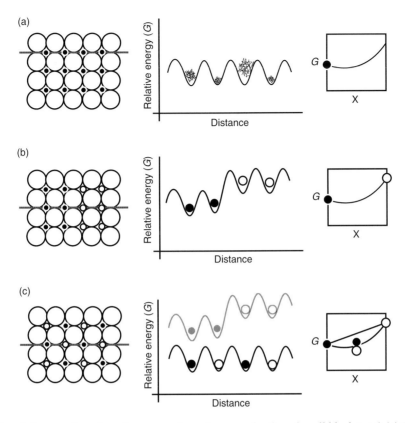

Fig. 7.3 Energies of sites as a function of traverses through a row of cations (small black ions). (a) Crystal sites in a pure homogeneous solid (e.g. MgO). All sites have the same depth of energy well, although the ions occupying the sites have variable energy states and may oscillate higher up the energy well at higher energies. (b) Crystal sites either side of an interface. Here, the interface is assumed to be sharp, with two end-members separated by a sharp plane. One set of ions (black circles) are of a mineral composition with a lower free energy (G) than the pale ions (white circles), which have a higher value of G reflecting the pure end-member composition. (c) Crystal sites following homogenization. Here, the filled and unfilled circles are distributed to show mixing. The system as a whole has lowered its free energy relative to a mechanical mixture by the act of mixing, due to the entropy increase. Note that all crystal sites have lowered their energies to the same level.

vacancies. In addition to more vacancies, increasing temperature also increases the number of ions of sufficient energy to jump lattice site, further increasing diffusion. The exact concentration of vacancies or point defects in crystals is difficult to determine experimentally as it is highly temperature-dependent but in general it is rather low (e.g. <10 lattice sites per million at 1,200 K; Dohmen & Chakraborty, 2007a). The low concentration of point defects in silicates is reflected in the low diffusivities that have been experimentally measured for many elements.

So far, we have described the process of diffusion simply driven by random oscillations of the atoms energies, without the involvement of a chemical gradient. Such diffusion is key in isotopic homogenization, but it is not useful for major or trace element purposes, because it does not lead to a net flux of ions. The presence of a chemical gradient encourages mixing in any

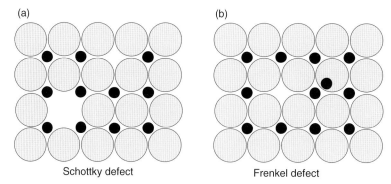

Fig. 7.4 Example of point defects in crystals. (a) Schottky defects are charge-compensated vacancies on different lattice sites, which may migrate independently but retain overall charge neutrality of the crystal. (b) A Frenkel defect is created by migrating an ion from a lattice site and either 'doubling up' in an adjacent site or by keeping that ion in an interstitial position, creating a vacancy on the original site. This typically is easiest in the case of ions with small ionic radii and low charge, such as Li.

miscible system, as the free energy of the system as a whole can be lowered through the mixing process. If two miscible, end-member minerals are brought together, this is a high-energy state (Figure 7.3b) that is unstable relative to a mixed system (Figure 7.3c), which has a lower bulk free energy (G). The change in energy between states shown in Figures 7.3b and 7.3c, also referred to as the *chemical potential*, is the driving force for homogenization. The chemical potential leads to complexities as it determines the equilibrium organizational state of the ions. In particular, various crystallographic sites can be occupied by different elements that may have different charges; there start to appear interactions between several different ions. Very quickly we have a multi-component diffusion process with a diffusion coefficient characteristic for each component (Lasaga, 1979). A very common application of multi-component diffusion in our context occurs when trying to model zoning in garnets (Spear, 1993; Faryad & Chakraborty, 2005).

Macroscopic description of diffusion and equations

The other description of diffusion is much more familiar to everyday experience, since it can be seen as part of the chemical homogenization

process of any heterogeneous mixture (e.g. coffee and milk). The equations to describe the chemical diffusion were proposed by Adolf Fick in 1855 after a series of experiments of diffusion of salt in water by Thomas Graham in 1833. The equation was derived for the flux of salt taking as analogy the laws of diffusion of heat and electricity, which were already proposed at that time (cf. Philibert, 2005). These laws state that the flux depends on the gradient of the potential and on the type of material. Then the so-called Fick's first law takes the form of a partial differential equation:

$$J = -D\frac{\partial C}{\partial x}$$

where J is the flux (e.g. mol m^{-2}s^{-1}), D is the diffusion coefficient, C is the concentration and x is the distance. A mass balance allows us to obtain Fick's second law that is used for modeling the homogenization of chemical gradients (in one dimension):

$$\frac{\partial C(x,t)}{\partial t} = \frac{\partial}{\partial x}\left(D\frac{\partial C(x,t)}{\partial x}\right)$$

where here we do not assume that D is independent of position or concentration. This equation describes how the concentration changes with time at every position of a concentration profile. With this equation we can see that to obtain robust timescales of magmatic processes, we need to consider the following points:

1 how well we know the values of the diffusion coefficient;
2 how well we have characterized the concentration gradients, paying special attention to the precision and spatial resolution of the analytical technique that we are using;
3 how we determine or infer the initial shape of the profile that we use to start the calculation (e.g. homogenous, ladder type, oscillatory). This is called the *initial condition*;
4 the type of boundary between the crystal and the surrounding media (e.g. at the crystal rim). These are called the *boundary conditions*.

These four points are developed in much more detail in the following sections.

Solutions to the diffusion equation

The solution to the diffusion equation can be *analytical* or *numerical*. Analytical solutions are typically in the form of an error function, Bessel function or exponentials and it is necessary to have a different solution for given initial conditions, boundary conditions and geometry (plane sheet, cylinder or sphere). Many common or simple situations have been solved and can be found in Crank (1975) or Carslaw & Jaeger (1986). Analytical solutions are very efficient because they provide the exact answer with little computational time, but they are difficult to apply in many geological situations due to the complexities of variable diffusion rate, chemical potential and multi-component diffusion. Figure 7.5 provides an example of analytical solutions and clarifies the roles of initial and boundary conditions. There are only a few analytical solutions for the diffusion equation that involve a composition-dependent diffusion coefficient. This limits their application because many elements in many minerals show compositional dependence (e.g. Fe-Mg interdiffusion in olivine; Dohmen *et al.*, 2007). Numerical solutions to the diffu-

sion equation can be found using the techniques of finite differences (Press *et al.*, 2007). The technique uses an approximate solution to the partial derivatives with a Taylor series approximation and algebraic manipulations (see for example Costa *et al.*, 2008). Numerical solutions have the disadvantage that the algorithms need to be written and that depending on the scale of the problems and the desired resolution, a significant amount of computer time is necessary. The advantage is that the solutions are very versatile and can be adapted to almost any initial and boundary conditions or dependencies of the diffusion coefficient.

THE DIFFUSION COEFFICIENT

As we noted during the inspection of the diffusion equation, it is important to note what determines the values of the diffusion coefficient. There are at least three definitions that are used to describe the diffusion coefficient, depending on the situation (Watson & Baxter, 2007; Costa *et al.*, 2008). The simplest coefficient is that of *self-diffusion*, and applies only to the situations of atoms migrating without a chemical gradient. This would be related to the random motions of atoms that we alluded to in the first part of this chapter but it is mainly a theoretical definition. A related type of diffusion coefficient is used to describe migration in a homogenous solid, but where we add an isotope of one element to make the migration of the element measurable in the laboratory. This applies to the diffusion coefficient of Fe in a fixed olivine composition, for example, 80% forsterite. This coefficient is commonly referred to as the *tracer diffusion coefficient* and the symbol $\dot{D}$ is commonly used. Note that this coefficient can depend on composition, for example, D^*_{Fe} is different for olivine with different forsterite contents (Chakraborty, 1997). This type of coefficient applies when considering the generation and re-equilibration of isotopic ratio disequilibria in crystals, for example, the equilibration of oxygen isotopes. The last type of coefficient is used for situations where a chemical

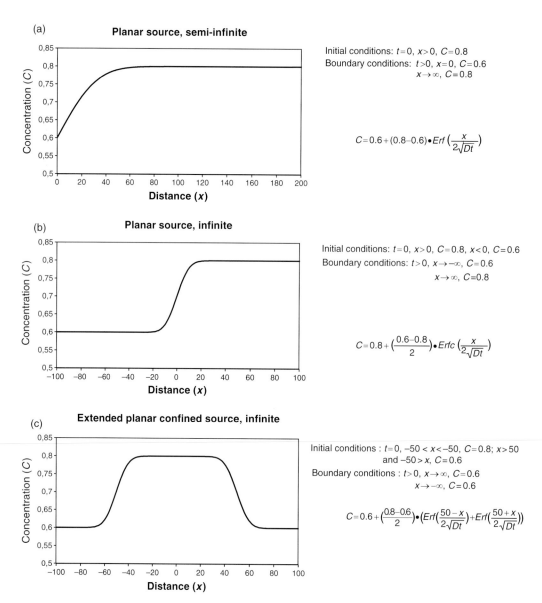

Fig. 7.5 Three analytical solutions of the diffusion equation to typical initial and boundary conditions encountered when modeling natural crystals (1 dimension and without any dependence of D on concentration or position). (a) This situation could be applied when a homogenous xenocryst is incorporated in a liquid (e.g. mantle xenolith in a basalt). (b) This configuration could be used to model abrupt changes in composition that occur in the middle of a crystal, far from the rim or when diffusion distance is small. It has been applied, for example, to Ba zoning in sanidine or Ti zoning in quartz. (c) Configuration for modeling a zone of a distinct composition in the middle of the crystal, or at least far from the rims or when limited diffusion has occurred. This has been applied in Fe-Mg modeling in clinopyroxene. Solutions are from Crank (1975).

potential gradient exists, which is commonplace in natural rocks. There are different types of formulations to describe *chemical diffusion*, depending on the situation. For example, if we want to study equilibration of the forsterite component we can use the *Fe-Mg binary interdiffusion coefficient* without the need to consider what other elements are doing (e.g. Si or O). In systems where there are more than two elements or components that are diffusing, for example Ca, Mn, Fe or Mg in garnet, we need to use a multicomponent formulation (Spear, 1993; Faryad & Chakraborty, 2005). There is a strong link between the tracer and chemical diffusion coefficients through theoretical models that allow calculating the chemical diffusion coefficients from experimentally measured tracer diffusivities (Lasaga, 1979). Moreover, the equilibration of trace elements can be sometimes described using tracer diffusion coefficients because of their low concentrations (Costa *et al.*, 2008). In general, it is a good practice to start by studying the multicomponent formulations and decide which simplifications apply to describe the diffusion process that we have at hand.

Parameters that control the diffusion coefficient and experimental determinations

The values of diffusion coefficients are mainly controlled by any variable that affects the concentration of the point defects (e.g. vacancies) in the crystal, and temperature (T) is the major player. Pressure (P), water fugacity (fH_2O), oxygen fugacity (fO_2), composition and crystallographic direction also play a role, depending on their values and the mineral and elements of interest. The manner in which the different variables affect the diffusion coefficient is expressed with an Arrhenius-type equation:

$$D = D_0 \cdot \exp\left(\frac{-Q - \Delta V(P - 10^5)}{R \cdot T} \right)$$

where Q is the activation energy at 10^5 Pa, ΔV is the activation volume, P is the pressure in Pascals,

R is the gas constant and D_o the pre-exponential factor. The 10^5 Pa can be left out for many practical purposes, because the effect of P on D is typically only significant for $P > 10^8$ Pa (see below). The effects of oxygen or water fugacities, the composition or crystallographic orientation, are typically incorporated in the pre-exponential factor D_0, although they may also have an effect in the activation energy (Costa *et al.*, 2008).

There are different protocols that can be used to determine the values of D, depending on the element, mineral or conditions (e.g. atmospheric or high pressure) in which we are interested (Holloway & Wood, 1988; Baker, 1993; Watson, 1994). The crystals used in the experiments are of gem-like quality, free of inclusions, deformations or cracks, and of a well-defined and homogeneous composition. Although there are no real standards for diffusion studies, authors try to work with material from the same locality (e.g. olivine from San Carlos peridotite) or source (e.g. a given Museum of Natural History standard). This is important because different crystals may have different trace elements and this could lead to different point defect concentrations (*extrinsic* vacancies) and thus different diffusion rates. However, theoretical analysis of point defects in simple minerals such as olivine shows that this is typically not the case, because the incorporation of trace elements with different charges may lead to antagonistic diffusion effects that largely compensate each other (Dohmen & Chakraborty, 2007a; Costa *et al.*, 2008). It has also been experimentally shown that within uncertainty, the same results for diffusivity are obtained when crystals with the same major element composition but coming from different localities are used (e.g. Fe-Mg in olivine; Dohmen *et al.*, 2007). Sneeringer *et al.* (1984) found that natural samples of diopsidic clinopyroxene had diffusivities of strontium two orders of magnitude greater than synthetic crystals of the same major element composition. However, the synthetic crystals were virtually Fe-free, which may have had a large effect on the defect concentration, from what is now known about the effect of Fe on cation diffusion in olivine (Dohmen & Chakraborty, 2007a).

The initial chemical gradient to drive diffusion into or out of the crystal can be created in the laboratory in several ways. Immersing the crystal in a liquid or even a fine powder of the same (or similar) major element composition but doped with an isotope or trace element of interest (e.g. Mg in plagioclase; LaTourrette & Wasserburg, 1998; REE in plagioclase; Cherniak, 2003), or putting in contact two crystals of the same mineral but of a different major or trace elements composition (e.g. Fe-Mg in olivine; Chakraborty, 1997), are common approaches. A major advance in recent years is the possibility of making high-quality thin films (e.g. pulsed laser or plasma deposition) of a doped mineral in crystallographic continuity on a prepared mineral substrate (Dohmen et al., 2002).

In the experiment, the diffusion couple is annealed, i.e. brought to the desired P, T and fO_2 conditions, and left there for a well-controlled period of time. In the best cases, only one parameter is varied (e.g. T) and the rest are kept constant (e.g. P and fO_2). However, this is not always possible, for example, in many experiments oxygen fugacity is buffered by a mineral reaction and thus both T and fO_2 co-vary. Disentangling the effects of the different variables is sometimes possible (Costa & Chakraborty, 2008) but in many cases diffusion data is reported along a certain oxygen buffer only. The times of the experiments are on the order of hours to a month and given the relatively slow diffusivities in solids, the length of the diffusion profiles are in general very short (e.g. some tens of nm to some tens of μm). However, the difference in experimental time and diffusion length scale between those used in experiments to characterize diffusion processes and those occurring in Nature does not really matter, as long as in both cases the main mechanism of transport is the same (e.g. volume diffusion).

With the diffusion profile and the time of the experiments, we can obtain the diffusion coefficient using the appropriate solution to the diffusion equation. Experiments are carried out for different temperatures, and the pre-exponential factor (D_0) and activation energy (Q) are obtained from a plot of Log D vs $1/RT$ at a given P (Figure 7.6; the equation above). The uncertainty of experimental data vary depending on the type of diffusion couple and analytical technique, and are typically in the range of 0.1 to 0.2 log units (1-sigma). This error is reflected in the values of D_0 and Q, which have uncertainties on the order of 50% relative for D_0, and <10% relative for Q. As the figure shows, the values of D_0 and Q are linked by a straight-line relationship regressed through the experimental data, where $\log_{10}D_0$ is the intercept on the y-axis and the gradient of the line is $-Q$. If these two uncertainties are combined improperly, we may get D with a large error, so it is necessary to combine the uncertainties in D_0 and Q carefully due to their innate relation.

Values for some commonly used diffusion coefficients

In Tables 7.1 and 7.2, we have compiled some diffusion coefficients and equations of elements in olivine, clinopyroxene, feldspars, quartz and garnet (Ganguly et al., 1998; Van Orman et al., 2002; Tirone et al., 2005) that have been well-characterized for main diffusion parameters and which have already been used, or have the potential to be used, to retrieve timescales of magmatic processes. It is not our purpose here to report a complete database of diffusion coefficients, but rather to give an introductory list of elements that can be used to model data obtained with the electron microprobe or LA-ICP-MS analyses. There is scarce and sometimes controversial information for many other elements and minerals but better quality data is necessary before it can be used to obtain timescales with confidence. The publications by Freer (1981) and Brady (1995) provide compilations of older data, and Cole & Chakraborty (2001) is, at the time of going to press, an up-to-date reference for oxygen. There is unfortunately no recent review of data that have been obtained since the year 2000, but an upcoming Reviews in Mineralogy and Geochemistry volume about diffusion is scheduled for publication in 2010 and should remedy this deficit.

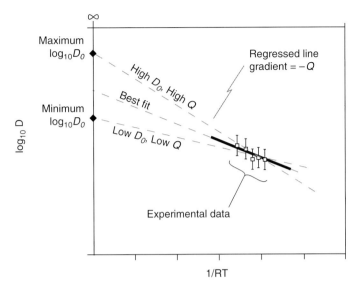

Fig. 7.6 Arrhenius-type of plot that is used to determine Q and D_0 from a series of experiments where D is obtained at various temperatures. An important point in considering the uncertainties for Q and D_0 is that they are not independent. This means that the high D_0 uncertainty should be processed with the high Q value to calculate one of the possible end-member diffusivities, and vice versa. As many diffusivity values are measured at relatively high temperatures, the rotation point about which the fitted lines rotate through the dataset (and at which the diffusivity is best controlled) may be close to the temperature of concern for the modeling, thus effectively decreasing the real uncertainties.

We encourage readers to read the source publication of any diffusion coefficient before applying it to modeling natural crystals. It is necessary to determine that the conditions of temperature, pressure, oxygen and water fugacities, anisotropy, mineral compositions and chemical activities at which the experiments were performed are close to the natural case under consideration. This has been overlooked in many applications and can lead to spurious results. A final point to note is that the recent experimental and analytical advances allow diffusion coefficients to be determined at the temperature and oxygen fugacity ranges relevant for most magmatic applications. It is therefore becoming less commonplace to extrapolate the data (and the consequent uncertainties of changes in activation energy) to temperatures strongly different from experimental conditions.

MEASURING THE CHEMICAL GRADIENT

In the previous sections we described how crystals may be chemically zoned by a series of processes and that such zoning is the combined result of crystal growth, dissolution and volume diffusion. Later, we saw that these diffusion processes can be described by partial differential equations that can be solved either numerically using iterations or analytically, typically using error functions. In this section, we address the question of how we move from these theoretical considerations to actually obtaining a time from natural materials. Historically, the major-element zonation of some minerals such as plagioclase was constrained with the optical microscope by manually measuring extinction angles of crystals in thin section. However, this is both subjective and

Table 7.1   Diffusion coefficients for different elements in olivine and clinopyroxene

	Crystal orientation	X	fO_2 (Pa)	fH_2O (Pa)	$\Delta V \cdot 10^{-6}$ ($m^3\,mol^{-1}$)	Do ($m^2\,s^{-1}$)	E ($Jmol^{-1}$)
Olivine							
Fe-Mg and Mn	// [001]	Fo: 0.86	1.26E–08		7	6.17E–10	201000
Fe-Mg	// [001]	Fo: 0.86	1.59E–09	1E+08	16	1.58E–15	220000
Ca	// [001]	Fo: 0.83–0.9	1.26E–08			9.55E–11	207000
	// [010]	Fo: 0.83–0.9	1.26E–08			3.47E–11	201000
	// [100]	Fo: 0.83–0.9	1.26E–08			1.66E–11	193000
Ni	// [001]	Fo: 0.9	1.26E–08		7	3.84E–09	216000
Cr	// [001]	Fo: 0.9	1.26E–08			2.24E–07	299000
	// [100]	Fo: 0.9	1.26E–08			4.90E–06	352000
Clinopyroxene							
(Fe,Mn)-Mg	// [001]	diopside	1.01E–08			3.39E–09	297000
Yb	⊥ [001]	diopside	QFM		9	2.29E–05	411000
Dy	⊥ [001]	diopside	QFM			4.90E–04	461000
Nd	⊥ [001]	diopside	QFM			1.12E–03	496000
Ce	⊥ [001]	diopside	QFM		9	7.94E–05	463000
Lu	⊥ [001]	diopside	QFM			6.03E–05	466000
Li	⊥ [010]	diopside	10^{-7} to 10^{-11}			2.90E–02	258000
Pb	⊥ [001]	Cr diopside	QFM			8.00E–07	351000
Pb	⊥ [001]	augite	QFM			3.80E–05	372000

* Equation is given if it is not $D = Do* \, Exp[-E/RT]$; Data sources: [1] = Dohmen and Chakraborty (2007a; 2007b); [2] = Holzapfel *et al* (2007); Wiedenbeck (2006); [8] = Van Orman *et al* (2001); [9] = Coogan *et al.* (2005a); [10] = Cherniak (2001).

T °C	P (Pa)	D (m²s⁻¹)	Observations	Ref	Equation*
1100	100000	1.30E−17	Parameters for $fO_2 > 10^{-10}$ Pa, see ref [1] for other parameters depending on fO_2	[1],[2]	$D_{[001]}^{Fe\text{-}Mg,Mn} = D_o \left(\dfrac{fO_2}{10^{-7}} \right)^{1/6} 10^{3(XFo-0.9)}$ $\exp\left[-\dfrac{E+(P-10^5)\Delta V}{RT} \right];$ $D_{[100]}^{Fe\text{-}Mg,Mn} \approx D_{[010]}^{Fe\text{-}Mg,Mn} \approx 1/6 D_{[001]}^{Fe\text{-}Mg,Mn}$
1100	1E+08	9.32E−17	measured along NNO fO_2 buffer	[3]	$D_{[001],NNO}^{Fe\text{-}Mg,Fo86} = D_o\, fH_2O^{0.9} \exp\left[-\dfrac{E+(P-10^5)\Delta V}{RT} \right]$
1100	100000	6.70E−19		[4]	$D_{[100],[010],[001]}^{Ca,Fo83\text{-}90} = D_{o[100],[010],[001]} \left(\dfrac{fO_2}{10^{-7}} \right)^{0.3}$ $\exp\left[-\dfrac{E_{[001],[010],[001]}}{RT} \right]$
1100	100000	4.11E−19		[4]	
1100	100000	3.97E−19		[4]	
1100	100000	1.43E−17		[5],[2]	$D_{[001]}^{Ni,Fo90} = D_o \left(\dfrac{fO_2}{10^{-7}} \right)^{1/4.2}$ $\exp\left[-\dfrac{E+(P-10^5)\Delta V}{RT} \right];$ $D_{[100]}^{Ni} \approx D_{[010]}^{Ni} \approx 1/6 D_{[001]}^{Ni}$
1100	100000	9.44E-19	measured along IW fO_2 buffer	[6]	
1100	100000	1.99E−19	measured along IW fO_2 buffer	[6]	
1100	100000	1.70E−20	fO_2 dependence found	[7]	
1100	100000	5.30E−21		[8]	
1100	100000	1.42E−21		[8]	
1100	100000	1.51E−22		[8]	
1100	100000	1.93E−22		[8]	
1100	100000	1.13E−22		[8]	
1100	100000	4.44E−12		[9]	
1100	100000	3.55E−20	fO_2 dependence found	[10]	
1100	100000	2.68E−19		[10]	

[3] = Hier-Majumder *et al.* (2005); [4] = Coogan *et al.* (2005b); [5] = Petry *et al.* (2004); [6] = Ito and Ganguly (2006); [7] = Dimanov and

Table 7.2 Diffusion coefficients for different elements in feldspars and quartz

	Crystal orientation	X	fO_2(Pa)	fH_2O (Pa)	$\Delta V \cdot 10^{-6}$ ($m^3\ mol^{-1}$)	Do (m^2s^{-1})	E ($Jmol^{-1}$)
Plagioclase							
Mg	//[010]	An0.95				7.10E−08	254000
Mg	//[001]	An0.95				1.20E−06	278000
Sr	⊥[001]	An0.23				8.43E−07	273000
Sr	⊥[001]	An0.43				1.80E−07	265000
Sr	⊥[001]	An0.67				9.35E−08	268000
Sr		An0.02	$0.2*10^5$ to 10^{-17}	up to 10^8		8.10E−05	277000
Sr		An0.27	air			7.40E−07	261000
Sr		An0.64	air			1.10E−06	295000
Sr		An0.96	air			5.70E−09	267000
Li	⊥[010]	0	air			1.58E−04	146000
Li	⊥[010]	0.9–1	air			2.51E−04	151000
K	⊥[010]	0	air	up to 10^8		3.98E−04	296000
K	⊥[010]	An0.1–0.3	air			6.31E−06	264000
K	⊥[010]	An0.5–0.7	air			3.16E−06	278000
Rb	⊥[010]	0	air			1.58E−06	283000
Ba	⊥[001]	An0.23	air			1.70E−07	303000
Ba	⊥[001]	An0.67	air			2.30E−07	323000
Ba	⊥[010]	An0.23	air			1.10E−04	377000
Ba	⊥[010]	An0.67	air			1.10E−06	341000
La	⊥[010]	An0.67	air			1.10E−02	464000
Nd	⊥[010]	An0.23	air			2.30E−03	425000
Nd	⊥[010]	An0.67	air			2.40E−02	477000
Nd	⊥[010]	An0.93	air			5.90E−06	398000
Dy	⊥[010]	An0.67	air			7.10E−03	461000
Yb	⊥[010]	An0.67	air			3.20E−01	502000
NaSi-CaAl		An0.8	air			1.10E−03	516000
NaSi-CaAl		An0,7–0,9	MH	2.5E+07		4.00E−16	103000
NaSi-CaAl		An0,7–0,9	MH	2.5E+07		1.10E−05	371000
NaSi-CaAl		An0–0,26	MH	2.5E+07		3.00E−08	303000
Alkali feldspars							
Sr	//[001]	Or0.94		1.0E+08		1.00E−11	167000
Sr	//[001]	Ab0.98		1.0E+08		2.51E−06	247000
Sr	⊥[001]	Or0.93	air			5.97E−07	284000
Sr	⊥[001]	Or0.27	air			2.25E−02	373000
Sr	⊥[010]	Or0.27	air			4.50E−03	374000
Sr	⊥[001]	Or0.61	QFM			8.40E+00	450000
Ba	⊥[001]	Or0.61	air			2.90E−01	450000
Quartz							
Ti	//[001]					7.00E−08	273000

*Equation is given if it is not D = Do Exp[−E/RT]; Data sources: [1] = LaTourrette and Wasserburg (1998); [2] = Costa *et al.* (2003); [3] = Cherniak Grove *et al.* (1984); [9] = Liu and Yund (1992); [10] = Giletti (1991); [11] = Cherniak and Watson (1992); [12] = Cherniak (1996); [13] =

T°C	P (Pa)	D (m²s⁻¹)	Observations	Ref	Equation*
1000	1.0E+05	2.69E–18	dependence on An is taken from that of Sr	[1], [2]	$D_{calc}^{Mg} = 2.92\,10^{-4.1*XAn-3.1}\exp[-\dfrac{266000}{RT}]$;
1000	1.0E+05	4.70E–18			
1000	1.0E+05	5.30E–18		[3]	
1000	1.0E+05	2.41E–18		[3]	
1000	1.0E+05	9.42E–19		[3]	
1000	10⁵ to 10⁸	3.49E–16		[4]	
1000	1.0E+05	1.44E–17		[4]	$D^{Sr} = 10^{-8.18}\,10^{-4.1*XAn+4.1}\exp[-\dfrac{276000}{RT}]$;
1000	1.0E+05	8.65E–19		[4]	
1000	1.0E+05	6.31E–20		[4]	
1000	1.0E+05	1.62E–10		[5]	
1000	1.0E+05	1.60E–10		[5]	
1000	10⁵ to 10⁸	2.85E–16		[5]	
1000	1.0E+05	9.28E–17		[5]	
1000	1.0E+05	1.24E–17		[5]	
1000	1.0E+05	3.87E–18		[5]	
1000	1.0E+05	6.28E–20		[6]	
1000	1.0E+05	1.28E–20		[6]	
1000	1.0E+05	3.73E–20		[6]	
1000	1.0E+05	1.12F–20		[6]	
1000	1.0E+05	1.01E–21		[7]	
1000	1.0E+05	8.38E–21		[7]	
1000	1.0E+05	6.42E–22		[7]	
1000	1.0E+05	2.75E–22		[7]	
1000	1.0E+05	8.62E–22		[7]	
1000	1.0E+05	8.07E–22		[7]	
1000	1.0E+05	7.39E–25		[8]	
1000	1.5E+08	2.38E–20	1000–1050°C	[9]	
1000	1.5E+08	6.58E–21	900–975°C	[9]	
1000	1.5E+08	1.11E–20		[9]	
1000	1.5E+08	1.40E–18		[10]	
1000	1.5E+08	1.84E–16		[10]	
1000	1.0E+05	1.33E–18		[11]	
1000	1.0E+05	1.11E–17		[11]	
1000	1.0E+05	2.03E–18		[11]	
1000	1.0E+05	2.88E–18		[12]	
1000	1.0E+05	9.95E–20		[6]	
1000	1.0E+05	4.40E–19		[13]	

and Watson (1994); [4] = Giletti and Casserly (1994); [5] = Giletti and Shanahan (1997); [6] = Cherniak (2002); [7] = Cherniak (2003); [8] = Cherniak *et al*. (2007).

of too coarse a resolution to be useful in determining timescales accurately. Optimal techniques have to have good spatial resolution (of the order of μm to tens of μm) and also a moderately good analytical resolution as a percentage of the contrast being observed (of the order of a few percent or less). These two factors trade off against each other, and it is better to have a technique with moderately good spatial and analytical precision, than a technique that has one precision extremely good and the other poor (Figure 7.7). In order to fit a diffusion model to a profile, we need at least four data points within the profile, and to be certain that there are no overlaps between adjacent points. This allows us to account for the effects of convolution, where a data point is representative of a significant area and which may include overlap with the adjacent data points (see section below on 'The effect of convolution and multiple dimensions').

Analytical techniques

The most common method employed by geologists is that of the electron microprobe or Electron Probe Micro-Analyser (EPMA). For certain mineral systems, Back-Scattered Electron Microscopy (BSE, or Z-contrast imaging) is also very useful. The use of trace elements increases the range of diffusivities and magmatic processes that can be determined from a given crystal, and the Laser-Ablation Inductively Coupled Plasma Mass Spectrometry (LA-ICPMS) or more specialized equipment such as the ion microprobe (or Secondary Ion Mass Spectrometry; SIMS) are commonly used. Other techniques that can be used but which are not discussed here are Proton-Induced X-ray Emission (PIXE), Analytical Transmission Electron Microscopy (ATEM) and cathodoluminescence (CL; Wark et al., 2007). See, for example, Potts et al. (1995) for a more detailed description of all these in-situ analytical techniques.

Electron microprobe

The electron microprobe revolutionized the field of mineral analysis shortly after its invention by

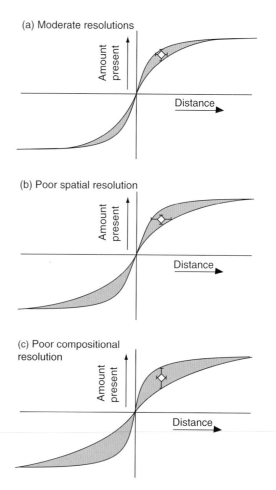

Fig. 7.7 Uncertainties in the compositional determination, or the size and position of the data point, can create uncertainty in the location of the true diffusion profile. This then translates into uncertainty relating to the timescale of diffusion. This is illustrated here by showing the problems of fitting a diffusion curve in the case where there is (a) moderate uncertainty on position and composition, (b) significant uncertainty on position and (c) significant uncertainty on composition. Note that the case with intermediate uncertainty is slightly better constrained, as shown by the smaller gray area between the curves.

Castaing in 1951, and is still the main tool for performing microanalysis for major elements in geological materials (Castaing, 1951). The electron microprobe uses a focused beam of electrons that interact with the sample and produce X-rays that are characteristic of the elements present (see Reed, 1975, for more details). All elements with an atomic weight higher than ~16 can be easily measured at concentrations higher than 0.01 wt.% and with precisions of 0.5 to 10% relative. Recent developments in spectrometer design allows measurement of light elements (B, C, N, O) and for some elements down towards the theoretical detection limit range of 10's of ppm (Ginibre *et al.*, 2002; Reed, 1975). With a focused beam, operating at 15 kV accelerating voltage, the minimum excited volume would be roughly a micrometer-diameter sphere below the surface of the sample. However, beam sizes used for analysis are typically larger than a tightly-focused beam, as such high energy densities tend to damage the sample surface and encourage migration of more volatile elements such as sodium. Therefore, beam diameters of 5 to 20 micrometers are not unusual. This is still the most versatile and accessible of the tools that we have available, however, and is necessary for the calibration of many other techniques, such as BSE, CL, LA-ICPMS and SIMS. Aside from one-dimensional traverses, we can also obtain X-ray maps and two-dimensional distributions of elements. There have been many applications of EMP to obtain timescales of magmatic processes, for example by using olivine (Klügel, 2001; Pan & Batiza, 2002; Costa & Chakraborty, 2004; Costa & Dungan, 2005, Shaw *et al.*, 2006).

Back scattered electron imaging

This technique relies on the elastic scattering of electrons from a focused electron beam as it is rastered across a sample in a Scanning Electron Microscope (SEM). The efficiency of the scattering depends on the number of electrons in the sample, which is effectively controlled by the mean atomic number (Z) of the sample. Regions with high mean atomic number show up as bright zones in the image, regions with low mean atomic number show up as dark regions. Whilst the image only gives intensity proportional to mean atomic mass, and not any element-specific infor-

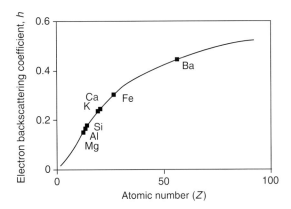

Fig. 7.8 The electron backscattering efficiency of some elements shown as a function of atomic number. Note that heavier elements are more efficient. The image shows that Fe substitution for Mg (i.e. in olivine or pyroxene) would increase the backscatter brightness of a crystal, as would substitution of Ba for Ca in feldspar. Calculated using equation 14.6 in (Reed, 1975). The Z-η diagram is useful in showing the magnitude of leverage that a substitution may have on the mean value of Z. Note that in a sense the Ba leverage in sanidine is anomalous; this is a case where a minor element has very large leverage, allowing the small amounts of Ba present to dominate the BSE signal.

mation, this can still be used if the range of substitutions in the mineral is restricted, since the electron backscattering efficiency varies depending on the atomic number (Figure 7.8).

The main advantage of BSE is the good spatial resolution (of ~0.1 µm), which may lead to a time resolution a factor of 100 shorter than equivalent EPMA analysis. The better spatial resolution of BSE compared to EMPA arises from the fact that the majority of backscattered electrons are from the upper 50 to 200 nm of the sample, whereas X-rays arise also from much deeper (Reed, 1975). BSE has the drawback that following calibration, the compositional precision of the analysis is typically poor. A nominal compositional resolution of ~0.1 mol.% per shade of grayscale is possible, but this will often have a random noise of ~0.3 mol.% (three shades of grayscale) superimposed. Examples where BSE imaging has been applied to determine compositions include the Anorthite-Albite Ca,Al – Na,Si exchange in plagioclase feldspar crystals (Ginibre *et al.*, 2002), the Fe-Mg exchange in olivine and clinopyroxene (Martin *et al.*, 2008; Morgan *et al.*, 2004), and also the Ba,Al–K,Si exchange in sanidine (Morgan *et al.*, 2006).

Laser-Ablation Inductively Coupled Plasma Mass Spectrometry (LA-ICP-MS)

This method of analysis uses a laser to excavate material from a sample, typically leaving a pit 20 to 200μm across and a few micrometers deep. The excavated material is carried into a plasma mass spectrometer; typically a quadrupole instrument, for trace element analysis (Perkins & Pearce, 1995). This then determines the abundances of a (potentially large) range of trace elements and the signal is calibrated to the abundance of a major element, such as aluminium or silicon, present in the sample. Compositional precision is of the order of a few percent, and spatial resolution is relatively poor, at 20 to 50μm. The relatively large size of the beam limits the application to natural samples with relatively long chemical gradients, but see, for example, Costa *et al.* (2010).

Secondary Ion Mass Spectrometry (SIMS and NanoSIMS)

In SIMS, a small pit is ablated into the sample surface using an ion beam, typically using either negatively-charged oxygen or positively-charged caesium ions. The material excavated in this process is then accelerated by a set of charged plates and fed through a magnetic sector mass spectrometer to a set of collectors (Slodzian, 1964). As for LA-ICPMS, the abundance of a variety of trace elements can be constrained, relative to the abundance of a major element, the amount of which has been determined through another method such as EPMA. One of the major problems is that there are very important matrix effects that need to be corrected for, and in many cases a period of standard characterization is necessary before unknowns can be measured with confidence (Hinton, 1995). The abundances of

elements can often be constrained to ~1%, with a good spatial resolution of down to 2μm. This method is possibly the most established for trace element analysis at the microscale, and whilst SIMS machines are certainly not common, there are usually established facilities at the national level to make these determinations. Recently there has been a development of the NanoSIMS, which can be used to measure profiles shorter than a micrometer and at low concentration levels (Hofmann *et al.*, 2009). This is certainly a very promising option, although the complexity of the measurements does preclude a systematic analysis of zoning and it is only warranted on a subset of samples. There have been quite a few applications of the ion probe to measure diffusion profiles for timescale determination (Zellmer *et al.*, 1999, 2003; Costa *et al.*, 2003).

The effect of convolution and multiple dimensions

It is very important when measuring the concentration gradients to be aware of the sample spot size of the analysis technique that is being used. The size of the analysis spots will influence the profile shape significantly if the size of the analysis is comparable to the size of the profile, leading to a convolution effect (Figure 7.9). There are several methods that can be used to deconvolve data (Ganguly *et al.*, 1988; Morgan *et al.*, 2004). A simple one is to consider a circular sample spot and, ignoring depth of field effects, to look at contributions from within that circular spot (Morgan *et al.*, 2004). This leads to a sine-wave contribution, dominated by the central point of the data analysis. Traditionally, convolution can be thought of as having a similar effect to a moving-window average of the dataset. If the true compositional profile is complex, and the sampling window large relative to the complexity, this creates a smoothing effect on the data that is almost indistinguishable from diffusion (Figure 7.9). This type of convolution should be avoided if at all possible, and requires a good awareness of the technique resolution and the complexity of the compositional variation in the sample.

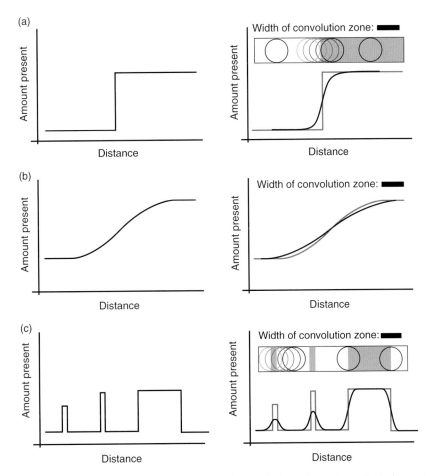

Fig. 7.9 This figure shows the most straightforward type of convolution where the analytical spot is large relative to the complexity of the compositional variation. (a) Sampling over a sharp interface produces a curved profile that looks to all intents and purposes like a diffusion profile, but which is actually produced by the analytical sampling. (b) When sampling across a diffusion profile, the effect of convolution is to broaden the diffusion profile appreciably, although to a lesser extent than in (a). (c) When convolution is applied to a profile of varying composition, it can be seen that the high-frequency and low-amplitude components of the profile will be most significantly affected by the convolution process, in the same way that they would suffer diffusion effects more strongly.

In cases where the analytical spot is relatively small compared to the length scale of diffusion, the diffusion profiles themselves are typically smooth, showing a monotonic decrease or increase in the abundance of an element along a sigmoidal trajectory. To some extent, this mitigates the effect of convolution. If we consider a simple diffusion profile, then a relatively small stretch is applied to the curve (Figure 7.9b).

Problems start to occur when the profile displays significant and varying curvature. In this case, the effect is to lower any peaks, fill in any troughs, and make the diffusion profile appear to be wider than it really is (Figure 7.9c), and to make it difficult to ascertain what the initial condition might have been. As the diffusion profile gets wider relative to the analysis spot, the effect of convolution gets proportionally smaller.

However, as a diffusion profile gets wider, small deviations in profile width represent larger uncertainties in time. These effects trade off such that convolution is typically a constant, positive uncertainty in diffusion time, but always yielding timescales that are longer than the true, deconvolved value.

The importance of multiple dimensions

Most chemical zoning data are acquired on thin sections of rock and thus two-dimensional data can be obtained using imaging techniques such as BSE microscopy and X-ray mapping. The advantages of two-dimensional data for diffusion modeling should not be neglected:

• It allows direct relationship to crystal textures and gives a much better appreciation of the context of the mineral zonation, both within a single crystal and between crystals.

• Many minerals are strongly anisotropic for diffusion, with some crystal axes showing faster diffusion than others. This means that the diffusion coefficient is different for different orientations of traverse line and this can be incorporated into two-dimensional models.

• Two-dimensional distributions can also be exploited to distinguish growth and diffusion if maps of elements with different diffusivities are compared to crystal morphology.

• Diffusion is occurring in three dimensions, and whilst a one-dimensional solution is approximately true when diffusion distances are small in the rims of relatively large crystals, as the extent of diffusion becomes greater, these solutions become less reliable, tending to strongly overestimate time. Two-dimensional modeling is more reliable than one-dimensional approximations. If two-dimensional distributions are not possible, it is a good practice to obtain two perpendicular traverses, which in turn are also perpendicular to the diffusion front to obtain profiles that approach the true diffusion lengths.

As we are constrained by sample preparation to a two-dimensional slice through a three-dimensional body, there are some geometric considerations that need to be incorporated into

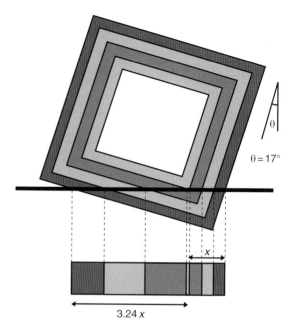

Fig. 7.10 Illustration the effective stretching that can be caused by shallow sectioning. Both the distance **x**, and the shallower section measuring 3.24x, have been stretched by shallow sectioning, as neither part of the zoning is orientated perpendicular to the sectioning line. The true distance is x cos (17°) or 0.956 x. Note however that the effect is much more pronounced for the shallower-dipping side of the crystal.

our model. The orientation of our grain edges with respect to the sectioning plane should be perpendicular, but this is rarely the case, particularly in a thin section (Figure 7.10). The effect introduces a stretch factor onto our traverses. If a crystal boundary is aligned vertically, there is no increase. We can express the measured traverse length as a function of the true traverse length through simple trigonometry:

$$measured = \frac{true}{\cos\theta}$$

where θ is the angle between the boundary and the vertical (the perpendicular to the sample surface). In most cases this angle has to be measured by hand

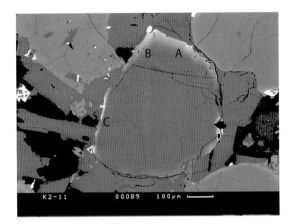

Fig. 7.11 Olivine crystal from the 1929 eruption of Nea Kameni, Santorini Greece. The crystal has been false-colored and the contrast increased to show the zonation more clearly. Three regions of the crystal were studied, A, B, C, and as the apparent diffusion profile length was different; these yield different timescales for the existence of that boundary from 167 to 49 days. When the boundary orientations were measured, it was found that they vary between 58 and 16 degrees away from the vertical. When this effect was corrected, the timescales collapsed into agreement and the three regions gave a time of 45 ± 2 days, as shown in Table 7.3. Image courtesy of V.M. Martin (Martin *et al.*, 2008).

Table 7.3 Data for the crystal regions A, B, and C shown in Figure 11, both before and after correction for the apparent versus true diffusion length

Region	Initial timescale	Boundary inclination to vertical	Strech factor $(1/\cos \theta)$	Corrected times
A	132 days	55°	1.74	43 days
B	167 days	58°	1.89	47 days
C	49 days	16°	1.04	45 days

using a universal stage to tilt the sample until the boundary is vertical and then the tilt can be determined from the stage axes. An example is shown with the olivine highlighted in Figure 7.11 and Table 7.3, which is from the Nea Kameni eruption of Santorini, (1929, Greece; Martin *et al.*, 2008).

MODELING THE ZONING PATTERNS OF MINERALS WITH CASE STUDIES

Now that we are equipped with knowledge regarding the diffusion equation, diffusion coefficients, and we have carefully measured a given chemical gradient in a mineral, we focus upon those additional ingredients required to obtain a meaningful time. The first thing is to consider what magmatic processes the crystals are recording. This seems like a superfluous or obvious consideration, but it is very important to think what process we are studying because the validity and significance of the initial and boundary conditions are entirely determined by our hypothesis. The same type of zoning in different samples can be modeled in different ways, depending on the geological context. This means that we have to conduct a detailed petrological study before we launch into the models. The other two important aspects are the initial and boundary conditions, which are discussed below. This is followed by a presentation of three cases studies of increasing complexity.

The initial conditions refer to the concentration distributions of the elements at the time we think that diffusion started to play a role. This is not always straightforward, and depending on the crystal growth and diffusion rates of the elements of interest, the observed profile is a combination of both growth and diffusion. In such situations it is probably better to either employ a more complex diffusion plus growth model (Costa *et al.*, 2008) or to accept the limitation that diffusion alone will overestimate timescales, and should be employed to place an upper bound. Such growth plus diffusion models have been rarely applied because they also introduce other uncertainties, such as how the melt concentration is changing at the growth surface. We need to care about the initial conditions because these effectively set up the zero time of the diffusion clock. Different initial profiles would yield different timescales as a result, although a detailed analysis of the most common possible end-member possibilities shows that the effects on absolute times are not as large as we would anticipate (Costa *et al.*, 2008). There have been different approaches to assessing the initial conditions; these fall mainly into three categories:

1 use the profile shapes of the slow-diffusing elements, combined with a consideration of melt chemistry and partitioning, as a proxy for the initial profile shape of faster-diffusing elements. A perfect candidate for this approach is plagioclase, where the slow diffusivities of the anorthite component involving coupled diffusion of NaSi-CaAl (Grove *et al.*, 1984) are a good indication of the profiles created during magmatic processes and thus can be used to model the initial condition for the faster diffusing elements such as Mg or Sr (Costa *et al.*, 2003. 2010).

2 assume a very sharp profile as the initial condition, in which case the time obtained is a maximum. This has been shown to be useful for modeling Fe-Mg in clinopyroxene (Morgan *et al.*, 2004) or Ti diffusion in quartz (Wark *et al.*, 2007).

3 use a homogenous initial profile and prove *a posteriori* that this is a good assumption by the consistent results obtained by modeling multiple elements and multiple traverses along different orientations. Olivine has been a good candidate for such an approach because of the diffusion anisotropy of many elements (Costa & Chakraborty, 2004; Costa & Dungan, 2005).

We stress that these are just three types of initial conditions and others can be found in Zellmer *et al.* (1999, 2003), Costa *et al.* (2008) or Morgan & Blake (2006). The other important decision that we need to make is the type of boundary that is applicable to the problem that we want to model. Two end-member cases are an *isolating boundary* where the crystal does not exchange matter or equilibrate with its surroundings (e.g. liquid), and an *open boundary* where re-equilibration of the crystal with its host liquid occurs. In the case of magmatic rocks and crystals that are freely set in a liquid of a much larger mass than the crystals that we are studying, it is likely that the boundary of the crystal will be open to exchange with the liquid. This open boundary condition has

been applied in many studies and is mainly of two types: the concentration is constant with time, or the concentration at the boundary changes because of crystallization, mixing or changes in temperature and the partition coefficients. It is difficult to imagine a natural situation, related to the topic we are discussing, where an isolating boundary would be applicable. In situations where the chemical gradient we are modeling is not very large and occurs far from the crystal margin (= the position of the boundary in our case) the results of using an isolating or open boundary would yield the same results and thus it is not important which one we choose. This situation is, for example, seen in the case study we describe below.

Titanium diffusion in quartz

In some respects, the diffusion of titanium in quartz is perhaps the simplest type of diffusion to model. Quartz has a single-component composition, with no real solid solution, and no compositional dependence for the diffusion coefficient. Al and Ti both occur as trace elements and due to their mutual low abundance, there is also likely to be no interference between diffusion of Al and Ti. Therefore, we can simply use the second equation above without any compositional dependence for the diffusion coefficient and an analytical solution (Figure 7.5). Titanium uptake in quartz is effectively controlled by the temperature and is useable as a geothermometer (Wark & Watson, 2006). Thus, Ti in quartz has been used to unravel the temperature fluctuations of granites and high-silica volcanic rocks (Cherniak *et al.*, 2007; Wiebe *et al.*, 2007), but in addition it is also useful to determine the timescales during which the temperature changes may occur.

The determination of the Ti diffusivity conducted by Cherniak *et al.* (2007), reports that there seems to be no noticeable anisotropy of Ti diffusion in quartz, with diffusion parallel and perpendicular to the [001] axis being comparable in rate. Ti diffusion is relatively slow, even at the magmatic temperature of silicic melts (Table 7.2). This means that the diffusion distances are typically small relative to the crystal size, and a simple, one-dimensional model is perfectly adequate to describe the diffusion accurately. Typical contents of Ti in quartz range from 10 to 600 ppm, which is close to the detection limit by electron microprobe (~50 ppm). Ti concentrations can be determined by using a calibrated cathodo-luminescence (CL) image, since in many (but not all) samples, Ti abundance is observed to be proportional to CL intensity (Cherniak *et al.*, 2007). With calibration, the intensity of CL emission can be used as a map of Ti concentration, provided that due care is taken regarding the CL decay time, because luminescence has a significant emission time relative to the beam raster rate, meaning that beams must scan the sample slowly to avoid strong convolution.

Once a CL image has been calibrated, the diffusion profile can be taken from the image and then modeled given a suitable magmatic temperature. As noted above, the peak Ti abundance can be used as a geothermometer if the Ti activity of the melt is known (Wark & Watson, 2006). However, this will be of dubious merit as the Ti content of the quartz crystal will have been modified by the diffusion process we are attempting to constrain, and is therefore likely to be somewhat unreliable. We could therefore use a different thermometer, and in silicic magmatic systems, other options include the ternary feldspar solvus (Elkins & Grove, 1990), plagioclase-melt (Putirka, 2005), zircon-saturation (Watson, 1979) or Ti-in-zircon (Ferry & Watson, 2007) thermometers. However, in this case we also need to consider if the closure temperature of the geothermometer is higher or lower than the closure temperature of the Ti diffusivity.

Figure 7.12 is representative of the type of image from which work can be carried out. Note that the concentration profiles are short (~40 μm), indicating relatively little diffusion has taken place. In this case the initial condition was considered a very sharp profile (type ii, above), and thus the entire length of the profile is due to diffusion. The boundary conditions for this example are not very important since the gradient occurs in the center of the crystal rather than at the crystal margin (our left boundary in this case) and the

diffusion length is short. We would obtain the same time if isolating or open boundary conditions were used. The highlighted area was selected to take a diffusion profile, as the boundaries in this area are straight, suggesting a planar dissolution surface, which is more reliable. Picking the shoulders of the profile, the central 86% of the variation spans the central 26 micrometers of the traverse. We can combine this with the diffusivity of Ti in quartz (Table 7.2) and fit the data to a profile described by the equation (Figure 7.5):

$$C_{(x,t)} = C_0 + \frac{C_1 - C_0}{2} erfc \frac{x}{2\sqrt{Dt}}$$

where $C_{(x,t)}$ is the composition at point x at time t, C_1 and C_0 are the initial compositions on each side of the boundary, x is the distance from the boundary, D the diffusivity and t the time. This yields a timescale of diffusion calculated at 200 to 900 years. In this particular example, this time would reflect the time spent at $789 \pm 22°C$, since the low Ti rim grew and the ensuing eruption.

Major and minor element zoning in olivine

Major element zoning in olivine has been used for some time to learn about magmatic processes and mantle xenoliths. The electron microprobe allows obtaining olivine major element zoning of Fe-Mg at standard conditions, and since current can be increased to 30 or 60 nA without damaging the crystal, it is also possible to determine the concentration of minor elements Ni, Ca and Mn with relatively good precision (Costa & Dungan, 2005). This makes zoning in olivine a very robust source of time information, since up to four time determinations can be obtained from a single traverse, allowing for a test of consistency of our hypothesis and models.

One of the main particularities of diffusion in olivine is the marked anisotropy of diffusion (Table 7.1). For example, diffusion of Fe-Mg parallel to the [001] direction is about six times faster than parallel to the [100] or [010] directions, which each have about the same diffusion rate (Dohmen & Chakraborty, 2007a). This means

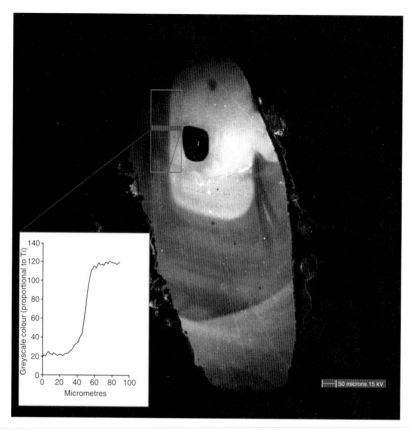

Fig. 7.12 Cathodoluminiscense image of quartz showing Ti zonation. The selected region was used to extract a diffusion profile (inset) and gave times ranging from 200 to 900 years. Figure and thermometry courtesy of Dr Kate Saunders.

that in addition to measuring the concentration gradients, we also need to measure the orientation of the crystallographic axis in the crystal of interest. This can be done *in situ* using electron backscatter diffraction patterns (Prior *et al.*, 1999). Once we know the orientation of the crystallographic axes we need to calculate the diffusion coefficient parallel to the electron microprobe traverse that we are interested in modeling and this can be readily done by using the following formula:

$$D_i^{traverse} = D_i^{[100]}(\cos\alpha)^2 + D_i^{[010]}(\cos\beta)^2 + D_i^{[001]}(\cos\gamma)^2$$

where $D_i^{traverse}$ is the diffusion coefficient parallel to the traverse direction in which composition is measured, D_i^a, D_i^b and D_i^c are the diffusion coefficients parallel to the [100], [010] and [001] principal axes, and α, β and γ are the angles between the traverse and each axis, respectively.

In addition to the anisotropy, the diffusion coefficients of several elements in olivine have been shown to depend on the oxygen fugacity, and the olivine major element composition (e.g. Fe-Mg ratio). We are fortunate that these effects have been experimentally calibrated (see Table 7.1 for equations). The concentration dependence of the diffusion coefficient precludes using an analytical solution for the modeling and thus

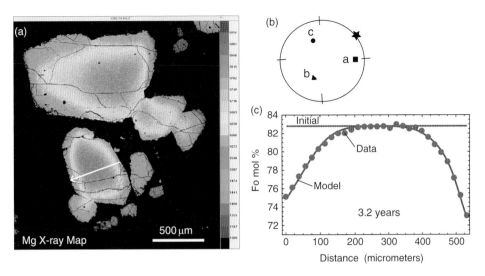

Fig. 7.13 Example of zoning pattern and diffusion modeling of Fe-Mg in olivine from a basalt of the Tatara-San Pedro volcanic complex. (a) X-ray map of Mg concentration. Red is high concentration and blue is low. (b) Crystallographic axis orientations for the same crystal and electron microprobe traverses projected onto the lower hemisphere. For this case, the angles between the [100], [010] and [001] axes are 29°, 63° and 81°, respectively, and these have been used with the above equation to calculate D parallel to the traverse (star). (c) Forsterite profile showing the measured data, initial conditions, model profiles, and calculated time for 1,125°C and oxygen fugacity at the Ni-NiO buffer. The diffusion model includes a compositional dependence for Fe-Mg the effect of oxygen fugacity, and the effect of anisotropy.

we need to rely on numerical models (see Appendix of Costa *et al.* 2008).

In Figure 7.13 we show the X-ray map distribution of Mg of some olivines from a basaltic lava of the Tatara-San Pedro volcanic complex (Dungan *et al.*, 2001). These crystals are thought to be xenocrysts, and they were derived from gabbroic plutons that were partially assimilated by the ascending basalts. Thus, modeling their zoning patterns provides some insight into the times of assimilation of the crust by basaltic magmas. In this case the mineral assemblage did not contain a geothermometer that was already calibrated, and so Costa & Dungan (2005) used the MELTS thermodynamic algorithm (Ghiorso & Sack, 1995) to obtain a range of temperature (1,125 ± 25°C) and oxygen fugacities around the Ni-NiO oxygen buffer. The crystal orientations were determined by EBSD and the diffusion coefficient for Fe-Mg within the sample was calculated using stereographic analyses and the equation above.

For this case, Costa & Dungan (2005) assumed a homogenous initial condition and open boundary conditions with constant concentration at the rim. This made sense within their interpretation that the crystals were recycled from mafic plutonic rocks and thus it could be expected to have been homogenized during cooling of the intrusion. The finding that different elements (with different diffusivities) in different crystals gave a coherent answer using these initial conditions was taken as confirmation that the initial conditions were appropriate. The open boundary was necessary to fit the profiles and also made sense in the context of crystals that are foreign to the magma and thus try to re-equilibrate with it.

Plagioclase trace element zoning

Chemical zoning in plagioclase has been used from the earliest times, since its zoning can be easily seen with a polarizing optical microscope in cross-polarized light. Major element zoning in plagioclase can be easily measured with the electron microprobe. Some minor elements such as K and Fe can in most cases also be analysed using standard protocols. Mg concentration in feldspars from mafic samples can also be obtained with relatively good precision by EPMA provided that long counting times of 100 seconds are employed in both peak and background. If the plagioclase is Ca-rich (e.g. anorthite content higher than 80 mol.%), the electron beam current can be increased by up to 30 nA without damaging the crystals. Plagioclase contains quite a few trace elements, which have been well characterized for their diffusivities (Table 7.2) and thus it is worth going to the ion microprobe or LA-ICP-MS to measure their concentrations.

The example of obtaining times from a zoned plagioclase is in most cases more complex than the previous two studies. This complexity arises because of two factors. First, the diffusion rates of many elements depend on the anorthite content (Table 7.2) and thus unless we have a crystal un-zoned in major elements, we need to introduce such dependency in a similar manner to that employed for olivine. The other is that the equilibrium distribution of trace elements in plagioclase is also dependent on the anorthite content and results from the activity coefficient for the trace element varying with the major element composition. Zellmer et al. (1999) and Costa et al. (2003) proposed to model this effect using the plagioclase/melt partition coefficients (Blundy & Wood, 1991; Bindeman et al., 1998) and derived the following expression (Costa et al., 2003):

$$\frac{\partial C_i}{\partial t} = \left(D_i \frac{\partial^2 C_i}{\partial x^2} + \frac{\partial C_i}{\partial x} \frac{\partial D_i}{\partial x} \right) - \frac{A}{RT} \left(D_i \frac{\partial C_i}{\partial x} \frac{\partial X_{An}}{\partial x} \right.$$
$$\left. + C_i \frac{\partial D_i}{\partial x} \frac{\partial X_{An}}{\partial x} + D_i C_i \frac{\partial^2 X_{An}}{\partial x^2} \right)$$

where X_{An} is the mol fraction of anorthite and A is a thermodynamic factor that relates the dependence of the partition coefficient on the anorthite content (Blundy & Wood, 1991; Bindeman et al., 1998). One of the simplifications is that the CaAl-

NaSi diffusion is so slow (Grove et al., 1984; Baschek & Johannes, 1995) that the anorthite content is considered constant for the time that the trace element is diffusing. The first term in parenthesis on the right-hand side of the equation is Fick's second law with a concentration dependence of the D, and is the same that is used, for example, in modeling the Fe-Mg diffusion in olivine. The second term accounts for the dependence of the activity coefficient on the anorthite. There is no analytical solution to this equation and thus it is necessary to solve it using numerical algorithms.

The example we show is from a crystal from a gabbroic xenolith from Volcan San Pedro (Costa & Singer, 2002). This crystal belongs to a suite of xenoliths that are thought to be part of a partially crystallized zone of from the margins of the San Pedro magma reservoir (Costa et al., 2002). The major elements were measured with an electron microprobe and the trace elements with a Cameca IMS-4f ion microprobe using an O- primary beam at the Edinburgh ion microprobe facility (Figure 7.14). What can be seen is that anorthite and all elements (Ti, Sr, Ba, La) except for Mg have a more or less homogenous core and then change abruptly at the rim. Costa et al. (2003) were interested in modeling the core to rim transition, as the presence of the rim indicates a melt migration event, and the diffusion is the result of re-equilibration following rim growth and prior to eruption. As an initial profile for Mg, they used the shape of the anorthite and the maximum Mg concentration measured in the plagioclase. This is a good approach because of the slow diffusivity of the anorthite content and because there are abrupt jumps in compositions that are seen in most elements. The left boundary was considered to be open. There were not many suitable geothermometers for this system and 850°C (±30°C) was used for the calculation, obtained by the hornblende-plagioclase geothermometer of Holland & Blundy (1994). Modeling of this crystal gave a time ~140 years between the event that metasomatized the xenoliths and the eruption (Costa et al., 2003).

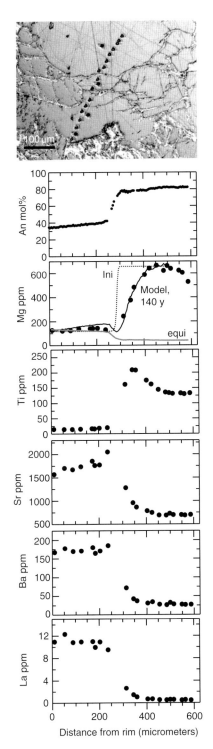

Fig. 7.14 Normarski image of plagioclase that has been analysed with the electron microprobe for the major elements (anorthite content), and with the ion microprobe for the trace elements (Mg, Ti, Sr, Ba, La). The pits made by the ion microprobe are clearly visible in the Normarski image and have diameters between 15 to 30μm. The 2-σ precision for most elements is under 5% relative. Only La is in some cases slightly above 10%. Note the sharp change in anorthite, which is also seen by Ti, Sr, Ba La, whereas Mg shows a much smoother profile. Given the higher diffusivity of Mg compared to these other trace elements (Table 7.2), its profile reflects equilibration between the core and the rim. The diffusion model gave a time of ~140 years since the formation of the rim associated by the metasomatic event and eruption (Costa *et al.*, 2003). Figure modified from Costa (2000) and Costa *et al.* (2003).

Uncertainty in timescale determinations

One drawback of using diffusion to obtain timescales is that the resulting time estimates may
have significant uncertainties. This is partly
related to the difficulty of precisely measuring
diffusion coefficients (although this will continu-
ally get better with improving experimental and
analytical techniques), but also to the fact that
diffusion rates vary exponentially with temperature (see the Arrhenium-type equation above).
This means that we need to know precisely the
temperature of our natural samples if we want to
have small uncertainties, and it makes sense to
spend some time obtaining more than one temperature determination from more than one type
of thermodynamic geothermometer. In addition,
other variables, such as P, fO_2 and fH_2O, of the
natural sample also need to be properly known.
Other uncertainties may arise if our samples have
experienced protracted temperature-time paths
during cooling or crustal residence. The convolution problem associated with detailed, small-
scale measurements also has to be considered.

But before we go into a detailed discussion of
the uncertainties, it is worth noting that this
technique allows obtaining many independent
timescale determinations from a single rock sample. Since we can obtain multiple profiles from a
single crystal, we have an internal check for consistency; the situation gets better if we measure
multiple elements with diffusivities measured
from different laboratories and different crystals
and protocols. With such a test we can have a
more realistic estimate of the uncertainties
involved for every case study. This exercise can
be further improved if multiple crystals are used
and if different minerals from the same samples
are used. It is not uncommon to have >20 time
determinations for a given sample. This allows
for statistical treatment of the results and a better
assessment of uncertainties.

Uncertainties to be considered

Many of the uncertainties have an exponential
effect on the diffusion coefficient, and therefore
on the determined time. This means that the

uncertainties are asymmetric, running to larger
values on larger timescales than the low-time
side. However, these uncertainties are near-
Gaussian in log-time, and so we can propagate
the uncertainties effectively in log-units before
converting back to absolute values.

The uncertainties on temperature, the pre-
exponential factor (D_0) and activation energy (Q)
values of the diffusion coefficients all have a logarithmic effect on the calculated times. The largest effect of all the uncertainties belongs typically
to the temperature determination. Typical diffusivities vary rapidly with temperature variation,
dependent on the activation energy and the temperature range. As an example, if we consider a
system with an activation energy of $200\,kJ\,mol^{-1}$,
the diffusivity increases by a factor of 4 between
1,000 and 1,100°C, and this factor increases to 16
if $Q = 400\,kJ\,mol^{-1}$. With typical good thermometry giving 2σ uncertainties of ~30°C, this can easily yield a factor of 2 or more uncertainty on our
timescale estimates. In cases where experiments
have been carried out to determine the equilibration conditions, temperature uncertainty can be
decreased to ~10°C.

There is a small correction to consider related
to the calibration of the spatial resolution of the
data. This could be the reproducibility and spatial
resolution of the XY stage in the EPMA, or the
quality of the magnification calibration (and
hence the length scale) in the SEM. Typically,
this is a relatively small effect, of the order of 1%,
and is thoroughly outweighed by the effects of
temperature, Q or D_0. However, under certain circumstances, it may become significant.

Combining uncertainties

The logarithmic uncertainties can be combined
together relatively straightforwardly. This requires that they are calculated as one-sigma logarithmic uncertainties, and then combined as a
root mean square log-scale uncertainty, which is
then doubled to give the 2σ uncertainty (roughly
95% confidence). The combination above considers that these contributions are independent.
This is an assumption that is not completely

appropriate. Oxygen fugacities are often buffered by a chemical reaction, which has temperature dependence that therefore links the oxygen and temperature uncertainties. For mineral systems where this linkage is a strong effect (i.e. Fe-Ti oxides; Ghiorso & Sack, 1995), the uncertainties in oxygen fugacity and temperature should be combined first taking this into account, before combination with the Q and D_0 uncertainty.

RELATION OF TIMESCALES DETERMINED FROM MODELING THE ZONING PATTERNS WITH OTHER TECHNIQUES

Modeling the zoning patterns of crystals or glasses provides relative time information about a given process occurring at a given temperature. The 'clock' of the diffusion system effectively stops, once the temperature drops below a given value, as the diffusion coefficient will decrease by several orders of magnitude. In a volcanic rock, this is when eruption occurs, and thus the time we obtain spans that of the process that created the zoning (e.g. magma mixing) and the eruption. It is a relative timescale, although if we know the time of eruption we can convert this to an absolute time. In contrast, methods based on radioactive decay provide absolute ages since a magmatic event such as crystallization, but in order to get the relative time before the eruption we will need to subtract the eruption age from the radiometric results. Despite this main difference, it is worth noting that methods based on radioactive decay are still dependent on diffusion rates through the concept of closure temperature (Dodson, 1973, 1986). The closure temperature is a means of determining when a given crystal will become effectively closed for exchange with the environment and thus when the radioactive system will become a reliable clock. For elements with fast diffusion rates such as Ar, this occurs at quite low temperatures, and thus isotopic systems with low closure temperatures are the way to obtain eruption ages in most cases. For systems with low diffusivities (e.g. Pb in zircon), the system

closes at magmatic temperatures yielding crystallization ages. Crystallization ages need not be close to eruption, and by comparing eruption ages with crystallization ages we can obtain the pre-eruptive crystal residence times (Reid, 2003; Costa, 2008).

Aside from the basic fundamental difference in approach between radioisotopic and diffusion methods noted above, analytical constraints also impose important differences. The radioisotopic disequilibria systems yield information for timescales comparable to the half-life. When investigating processes occuring over short timescales, a short-lived isotope therefore gives the best time resolution. This poses sample mass limitations, as the low abundance of short-lived isotopes demands either highly-enriched, or large-mass (i.e. multi-crystal) samples. Although possible, it is still far from routine to be able to employ U-series dating work with *in-situ* analytical techniques in most minerals, although great advances have been made with zircon (Charlier *et al.*, 2005). From this, it results that most U-series disequilibrium work is obtained either from a bulk crystal separate, multiple or whole single mineral grains. The potential for age differences between core and rim within single crystals gives a wealth of detail in the U-series measurement. In such cases, young rims may dominate short-lived isotopic disequilibria (i.e.^{230}Th-^{226}Ra disequilibrium), while older cores may dominate the longer-lived isotopes (e.g.^{234}U-^{230}Th disequilibrium). Understanding the results may require significant interpretation (for a full discussion, see Charlier & Zellmer, 2000). The single-crystal nature of diffusion studies gives a different perspective, where individual crystals yield different timescales as part of a spectrum that spans the time over which crystallization processes operated. As recent studies of U-Th disequilibria in multiple, single zircon grains have shown (Charlier *et al.*, 2005), the U-series data yield a similar kind of spectrum when treated at the scale of a population of single grains. This emphasises the need for both a population of measurements in diffusion studies, and for detailed interpretation of the complementary U-series data.

Summary of times obtained by diffusion

The number of studies that have used diffusion modeling to understand magmatic processes is relatively low compared to those obtained with radioactive isotopes, but at the time of going to press is increasing very rapidly. One of the general findings is that the times are relatively short, from a few days to a few years, and rarely extend to more than about a few hundred years. Among the shortest times that have been determined are those of time of magma transport from the mantle to the surface, which give several hours to days using H zoning in mantle olivines or Ar isotope re-equilibration in phlogopite (Kelley & Wartho, 2000; Demouchy et al., 2006; Peslier & Luhr, 2006). However, in other studies of mantle xenoliths and xenocrysts Klügel (2001), Shaw (2004), Shaw et al. (2006) and Costa et al. (2005) have used Fe-Mg zoning in olivine and obtain transport times that are longer, from some months to a few hundred years.

Most determinations have dealt with magma mixing or the time since magma replenishment (or assimilation) and eruption. This focus reflects that such magmatic events are important for understanding volcanic hazards but also they typically leave a clearly identifiable record in the crystals in the form of abrupt zoning patterns. This makes diffusion modeling more reliable, as the initial and boundary conditions can be identified through characterization of the end members involved in mixing or assimilation. Many studies have used Fe-Ti zoning in magnetite (Nakamura, 1995a; Venezky & Rutherford, 1999; Coombs et al., 2000; Chertkoff & Gardner, 2004) or the major and minor element zoning in olivine (Nakamura, 1995b; Coombs et al., 2000; Pan & Batiza, 2002; Costa & Chakraborty, 2004; Costa & Dungan, 2005). Fe-Mg zoning in clinopyroxene has been used to determine the time between magma intrusion and eruption (Costa & Streck, 2004; Morgan et al., 2004). Zoning of trace elements (Sr, Mg, Ba) in plagioclase and Ti zoning in quartz has also been applied (Costa et al., 2003, 2010; Wark et al., 2007).

Longer timescales have been obtained for the generation of silica-rich magmas by partial melting or remobilization of crustal rocks. Oxygen isotopes in quartz and zircon (Bindeman & Valley, 2001), and Sr and Ba zoning in plagioclase (Zellmer et al., 1999, 2003) give time ranges of between a few decades and a few thousand years. Morgan & Blake (2006) and Zellmer & Clavero (2006) have employed Sr and Ba zoning in sanidine with similar results of thousands to tens of thousands of years.

The majority of timescales that we report above range from a few hours to a few thousand years. These are in general significantly shorter than those obtained from radioactive isotopes that are in the range of a few 10's to 100's of kiloyears (Turner & Costa, 2007; although see Sigmarsson, 1996). This difference may have several origins. First, U-series reflects the entire growth history, where diffusion may only address the time of rim formation. Second, diffusion only operates at high temperatures, and so if the system has been cooled and re-heated, the diffusion clock does not record the cool phases while the radiometric clock was still ticking. At the same time, the radiometric clock may also be 'opened' and the age reset depending on the closure temperature. It may provide a complex picture if old crystals are recycled into the magma (e.g. classical discussion about residence times of silicic magmas). A third component to this difference is the half-life of the isotopic systems; these make it difficult to see very short-lived events immediately prior to eruption, except for the rare cases when ^{210}Pb-^{226}Ra disequilibria can be used (Condomines et al., 2003). We are not aware of any study to date that has applied diffusion modeling and radiogenic isotope dating to exactly the same crystals, although this is an area of active research.

CONCLUDING REMARKS

Chemical zoning in crystals is the main archive that we have about the magmatic and volcanic processes that magmas may experience between their formation and their solidification. In this

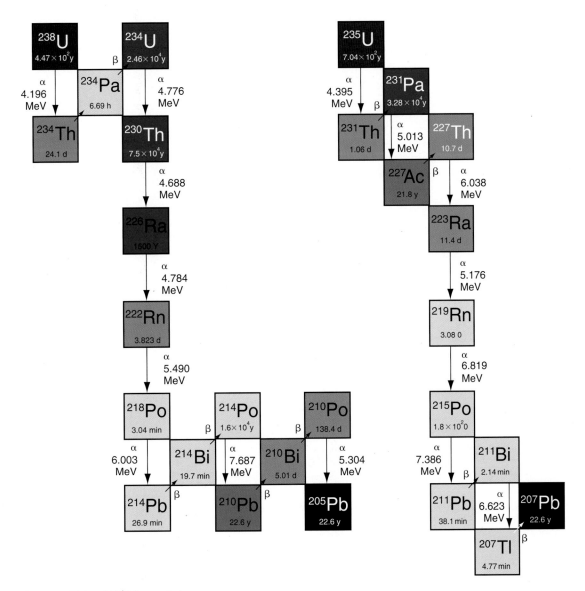

Plate 0.1 ^{238}U and ^{235}U decay chains.

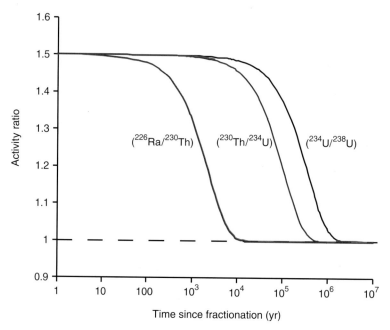

Plate 0.2 Decay of radioactive disequilibria for different systems. The timescale of return to secular equilibrium is directly a function of the daughter nuclide half-life in the system considered: ~8,000 years for ^{226}Ra-^{230}Th, 300,000 years for ^{230}Th-^{238}U.

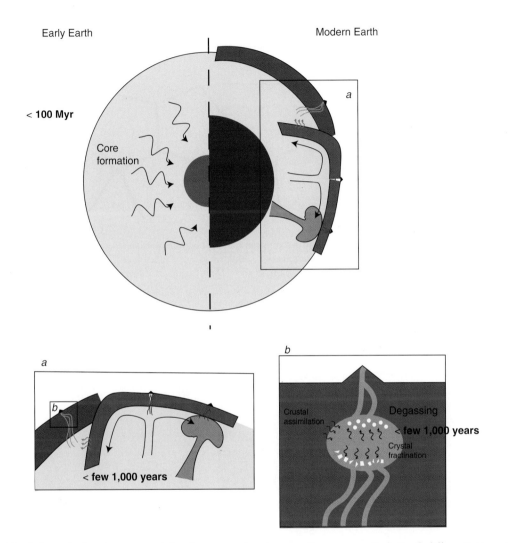

Early Earth

Modern Earth

< 100 Myr

Core
formation

a

a

b

< few 1,000 years

b

Crustal
assimilation

Degassing

< few 1,000 years

Crystal
fractination

< few 1,000 years

Plate 0.3   Schematic drawing summarizing the timescales of magmatic processes. Early Earth differentiation occurs shortly after the beginning of the solar system (less than a few 100 million years). Inset (a) shows melt production at different tectonic settings (from left to right: subduction, divergent plates and hotspot). Timescales of magma production and transfer towards the surface are inferred to be very short, of the order of a few 1,000 years or less. Inset (b) shows magma differentiation in the crust (regardless the tectonic setting). Crystal fractionation, crustal assimilation or magma degassing occur over short timescales, typically a few 1,000 years or less. Note however that some magma bodies can exist for several 100,000 years, but the actual process of differentiation of the magma into a wide range of composition is very short. Moreover, it takes 100,000's of years to heat up the crust to allow crustal assimilation, but once the required thermal regime is reached, production and assimilation of crustal melts into mantle-derived magmas is very rapid.

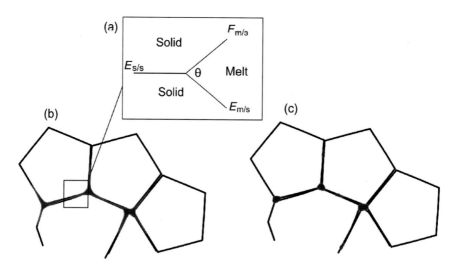

Plate 4.1 The effect of the dihedral angle on the distribution of melt on grain boundaries. (a) The dihedral angle set by the surface energies between the melt and the grains ($E_{m/s}$), and the grains themselves ($E_{s/s}$). It is defined as $\cos(\theta/2) = E_{s/s}/2E_{m/s}$. (b) Distribution of melt if the dihedral angle is <60°. Melt forms an interconnected network at low melt fractions, whereas (c) shows the distribution of melt for dihedral angle >60°. The dihedral angle for melt coexisting with olivine is <50°.

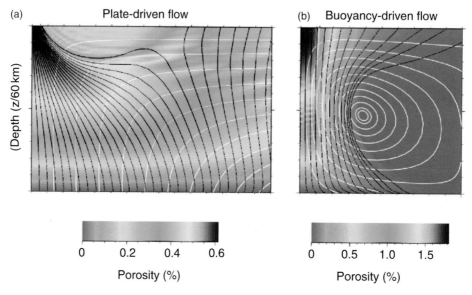

Plate 4.3 Porosity and flow-lines for two mid-ocean ridge models. (a) Assumes purely plate driven flow. White lines represent mantle flow lines, black lines represent melt flowlines. (b) For lower shear viscosity, buoyancy of the melt can dominate the flow-field, resulting in a focused higher viscosity beneath the ridge crest. The buoyancy number R, which denotes the relative contribution of buoyancy and plate-driven flow, is 0.26 is the plate-driven flow case, and 169 in the buoyancy driven flow case (modified from Spiegelman, 1996).

Porosity (simulation), vol. frac. gamma = 2.79

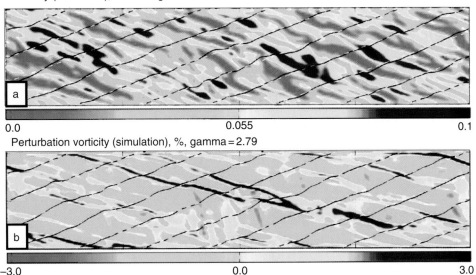

0.0   0.055   0.1

Perturbation vorticity (simulation), %, gamma = 2.79

−3.0   0.0   3.0

Plate 4.4 Porosity and perturbation vorticity (the vorticity minus the simple shear contribution) fields demonstrating the development of low-angle shear bands in a numerical experiment based on a laboratory core deformation experiment. In this experiment $n = 6$, $\alpha = -27$, and the shear strain is 2.79. Black lines are tracers which were initially vertical at commencement of shear, white dotted lines demonstrate the expected position due to only simple shear (modified from Katz *et al.*, 2006).

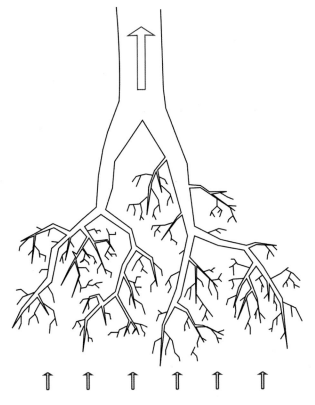

Plate 4.5 Schematic of a bifurcating vein network. A large number of smaller veins upstream tap the background porous flow, and coalesce to form one large transporting dyke.

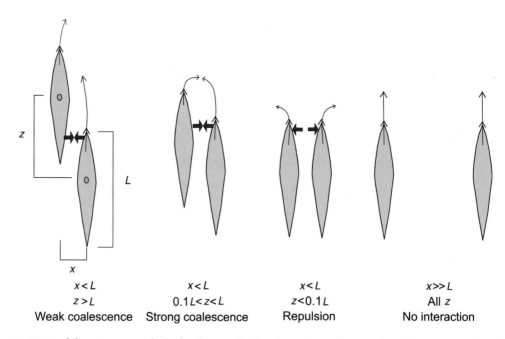

$x < L$	$x < L$	$x < L$	$x \gg L$
$z > L$	$0.1L < z < L$	$z < 0.1L$	All z
Weak coalescence	Strong coalescence	Repulsion	No interaction

Plate 4.6 Rising dykes interact with the local stress field, and can thus affect nearby dykes in a number of ways, depending on their spacing. Here L is the length of the dyke, z their vertical spacing, and x their horizontal spacing. For all x < L, dykes will interact. If they are at similar depths (z <0.1 L), they will tend to repulse each other. If their depths are different, they will tend to attract, and the strength of attraction depends on z.

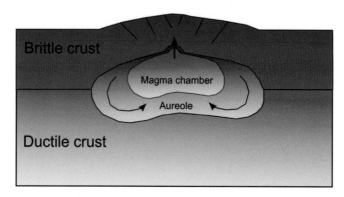

Plate 4.7 Schematic of the emplacement of viscoelastic-plastic magmatic pluton. A partially molten diapir may rise ductilely through the lower crust. It is more impeded by the cool, brittle upper crust but may induce uplift and fracturing of the brittle crust as it migrates upwards at rates of millimeters per year (based on the work of Burov *et al.*, 2003).

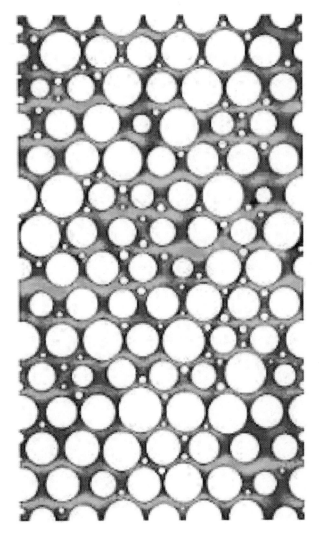

Plate 5.2 Illustration of multi-phase porous flow in a melting medium. As the system undergoes compaction the grains (white spheres) compress and the fluid is expelled upwards.

(a)   (b)

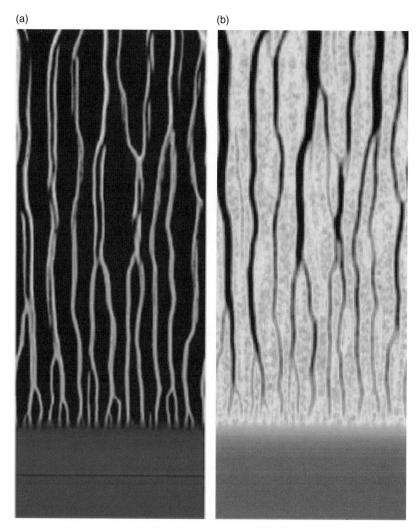

Plate 5.5   (a) Porosity and (b) residual channel structure for the reactive flow model in Spiegelman *et al.* (2001).

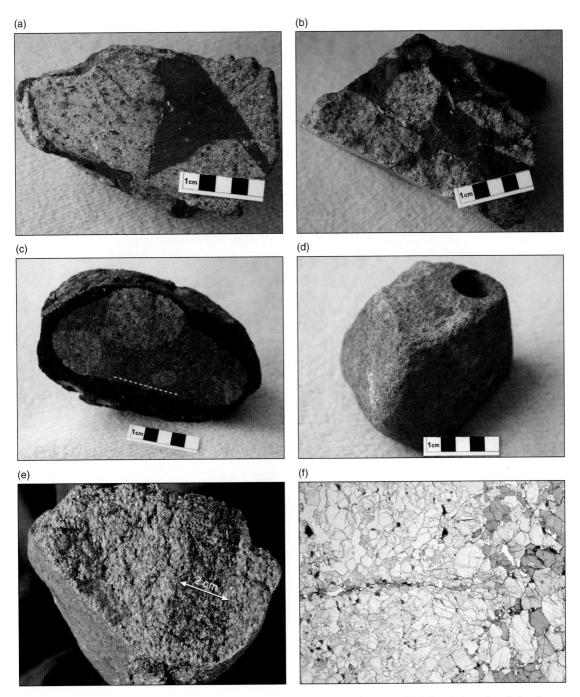

Plate 6.1   (a) and (b) Different views of angular xenoliths of spinel peridotite in alkali basalt from Batchelor Crater, Queensland, Australia. (c) Xenolith of a composite xenolith showing spinel peridotite veined by basaltic magma that crystallized within the mantle, within a volcanic bomb from Mt Shadwell, Victoria, Australia. Note the originally straight sides (emphasized by white dashed lines) and the rounded edges of the xenolith due to abrasion in the magma. The spinel peridotite areas appear rounded due to reaction with the invading basaltic magma. (d) Xenolith from kimberlite showing polygonal shape with planar surfaces. (e) Spinel peridotite xenolith showing straight-sided vein of garnet pyroxenite (basaltic magma crystallized within the mantle). (f) Photomicrograph of thin section of a peridotite showing an amphibole-rich vein with thin offshoot. Note the straight sides on both the large and small veins. Field of view is 2.5 cm across.

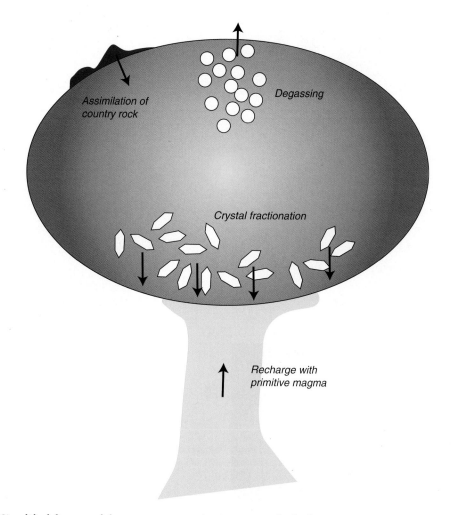

Plate 8.1 Simplified diagram of the processes occurring in a magma body that promote differentiation.

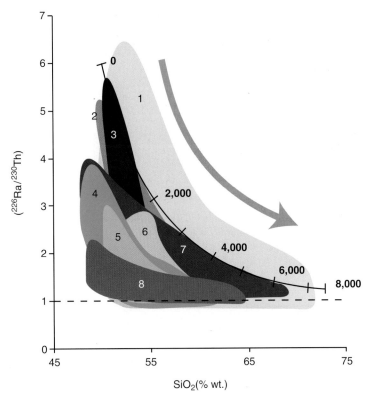

Plate 8.6 Compilation of whole rock $(^{226}Ra/^{230}Th)$ activity ratio vs SiO_2 content in island and continental arc volcanic rocks. The general decrease in $(^{226}Ra/^{230}Th)$ during differentiation (increasing SiO_2 content; gray arrow) can be used to infer duration of differentiation and crystallization rate, assuming closed system differentiation and little fractionation between Ra and Th during crystal fractionation. For instance, the thick curve shows the compositions during differentiation of a primary magma with $(^{226}Ra/^{230}Th) = 6$ and $SiO_2 = 50$ wt. %. The curve is labeled in years of differentiation. The crystallization rate is set at 10^{-4} g/g/yr and bulk solid/melt partition coefficients for Ra, Th and SiO_2 are taken to be respectively 0.001, 0.01 and 0.5. Fields show data from different arcs: 1: Tonga-Kermadec (Turner & Hawkesworth, 1997); 2: Vanuatu (Turner *et al.*, 1999); 3: Marianas (Elliott *et al.*, 1997); 4: Sunda (Turner & Foden, 2001); 5: Nicaragua (Reagan *et al.*, 1994); 6: Lesser Antilles (Turner *et al.*, 1996; Chabaux *et al.*, 1999); 7: New Britain/ Bismarck (Gill *et al.*, 1993); 8: Aleutians (George *et al.*, 1999).

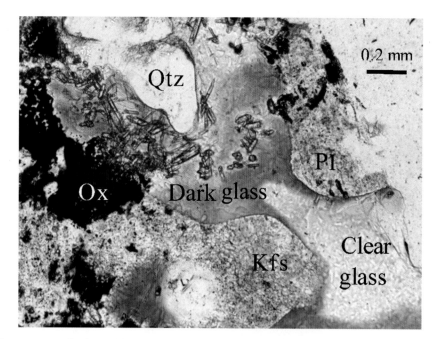

Plate 9.8 Photomicrograph of partially melted granitoid, showing development of dark and clear melt (quenched to glass) during near-surface melting along the margin of basaltic dyke in the Sierra Nevada Batholith, California. *Kfs* alkali feldspar, *Ox* is Fe-Ti oxide, *Pl* plagioclase, *Qtz* quartz.

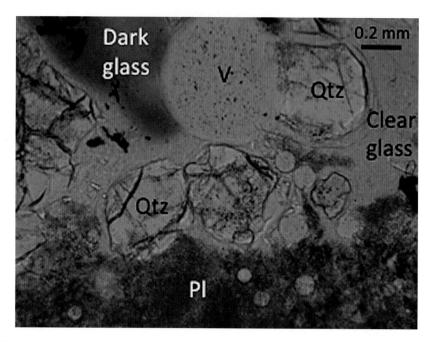

Plate 9.10 Photomicrograph of partially melted granite showing development of dark and clear melt generated by melting of solid granite blocks at 1 atm in the laboratory. *Pl* plagioclase, *Qtz* quartz, *V* vesicle.

Plate 10.4 World map showing locations (white stars) of some well-known large explosive volcanic eruptions (≥ 100 km^3) over the last 1 Ma (based on Self 2006 and Sparks *et al.* 2005). Note the abundance of supereruptions around the Pacific Ring of Fire. See Hughes and Mahood 2008 for a more exhaustive database. Base map from http://geology.com/nasa/world-topographic-map/

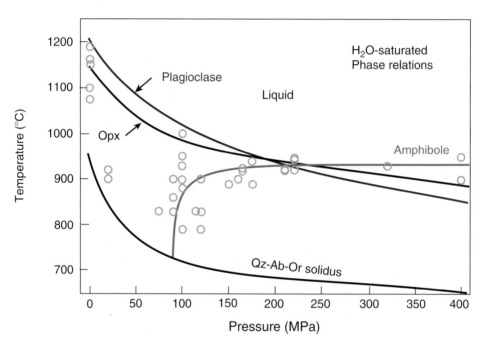

Plate 11.4 Phase relations of H_2O-saturated dacite from Mt St Helens, constructed by Blundy & Cashman (2001) from the experimental datapoints (shown as open circles) of Merzbacher & Eggler (1984), Rutherford *et al.* (1985), Rutherford & Devine (1988), Rutherford & Hill (1993). Appearance of different mineral phases is shown by the solid lines. The iron-titanium oxides have been omitted for clarity. The solidus is taken from the haplogranitic quartz-albite-orthoclase system (Johannes & Holtz, 1996).

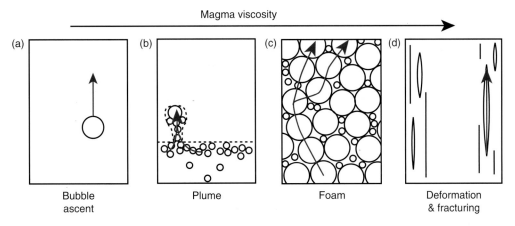

Plate 11.6 Schematic illustration of different ways in which gas can segregate from magma. (a) ascent of individual bubbles through low viscosity magma (e.g. basalt, carbonatite); (b) accumulation and coalescence of bubbles either at an interface (compositional, temporal) or by larger bubbles catching up with smaller ones; (c) gas flow through an interconnected network of bubbles; (d) deformation of bubbles and fracturing of magma in high stress regions such as along the conduit walls (high viscosity magma only).

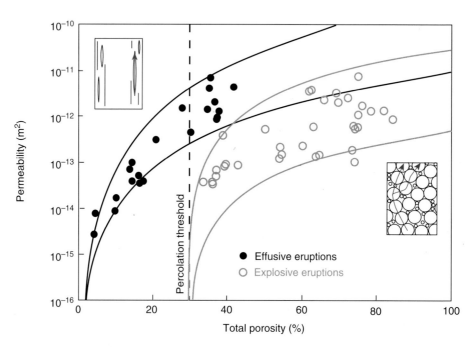

Plate 11.7 Porosity (vesicularity)-permeability relation such as measured in volcanic rocks (modified after Mueller *et al.*, 2005). In bubble foams connectivity is reached at ~30% bubble volume, the percolation threshold. Thus gas loss during explosive eruptions is late and bubbles can expand until they reach permeability (closed system degassing). Effusively erupted magma is permeable at much lower vesicularity and the nature of the connectivity differs, fracturing and deformation provide efficient pathways for gases to escape (open system degassing).

chapter, we have seen that the zoning patterns cannot only be used to obtain information about processes, but also about the duration of such processes. The increasing precision and spatial resolution of *in-situ* analytical techniques will enable this field to grow considerably in the future. Time determinations from multiple elements in a single crystal allow for very robust estimates; the only limitation that is starting to be apparent is that timescale determinations from natural samples require laboratory determinations of diffusion coefficients. This is a time-consuming process, especially if not only temperature, but also the roles of other variables such as oxygen fugacity or composition are to be determined. The good news is that thanks to recent analytical advances, the determination of diffusion coefficients no longer require significant extrapolations to magmatic conditions.

ACKNOWLEDGMENTS

Anthony Dosseto, Simon Turner and James Van Orman are thanked for giving us the opportunity to write this chapter and for their editorial work. Reviews by J. Van Orman and G. Zellmer are acknowledged and helped to clarify various parts of the chapter. FC acknowledges many discussions with S. Chakraborty and R. Dohmen on the topics presented in this chapter. DM acknowledges the support of the University of Leeds.

REFERENCES

Baker DR. 1993. Measurement of diffusion at high temperatures and pressures in silicate systems. In: *Experiments at High Pressure and Applications to the Earth's Mantle*. Mineralogical Association of Canada Short Course 21: 305–355.

Baschek G, Johannes W. 1995. The estimation of NaSi-CaAl interdiffusion rates in peristerite by homogenization experiments. *European Journal of Mineralogy* 7: 295–307.

Bindeman IN, Valley JW. 2001. Low δ¹⁸O rhyolites from Yellowstone: magmatic evolution based on analyses

of zircons and individual phenocrysts. *Journal of Petrology* 42: 1491–1517.

Bindeman IN, Davis AM, Drake MJ. 1998. Ion microprobe study of plagioclase-basalt partition experiments at natural concentration levels of trace elements. *Geochimica et Cosmochimica Acta* 62: 1175–1193.

Blundy JD, Wood BJ. 1991. Crystal-chemical controls on the partitioning of Sr and Ba between plagioclase feldspar, silicate melts, and hydrothermal solutions. *Geochimica et Cosmochimica Acta* 55: 193–209.

Brady JB. 1995. Diffusion data for silicate minerals, glasses, and liquids. In: *Mineral Physics and Crystallography: A Handbook of Physical Constants*, Ahrens TH. (ed.), AGU Reference Shelf 2, AGU, Washington DC, 269–290.

Buddington AF, Fahey J, Vlisdis A. 1955. Thermometric and petrogenetic significance of titaniferous magnetite. *American Journal of Science* 253: 497–532.

Carslaw HS, Jaeger JC. 1986. *Conduction of Heat in Solids*. Oxford University Press, New York.

Castaing R. 1951. Application des sondes electroniques aune methode d'analyse ponctuelle chimique et crystallographique. PhD thesis, University of Paris.

Chakraborty S. 1997. Rates and mechanisms of Fe-Mg interdiffusion in olivine at 980°C–1300°C. *Journal of Geophysical Research* 102: 12,317–12,331.

Chakraborty S. 2008. Diffusion in solid silicates: a tool to track timescales of processes comes of age. *Annual Reviews of Earth and Planetary Sciences* 36: 153–190.

Charlier BLA, Zellmer GF. 2000. Some remarks on U-Th mineral ages from igneous rocks with prolonged crystallisation histories. *Earth and Planetary Science Letters* 183: 457–469.

Charlier BLA, Wilson CJN, Lowenstern JB, Blake S, Van Calsteren PW, Davidson JP. 2005. Magma generation at a large, hyperactive silicic volcano (Taupo, New Zealand) revealed by U-Th and U-Pb systematics in zircons. *Journal of Petrology* 46: 3–32.

Cherniak DJ. 1996. Strontium diffusion in sanidine and albite, and general comments on strontium diffusion in alkali feldspars. *Geochimica et Cosmochimica Acta* 60: 5037–5043.

Cherniak DJ. 2001. Pb diffusion in Cr diposide, augite, and enstatite, and consideration of the dependence of cation diffusion in pyroxene on oxygen fugacity. *Chemical Geology* 177: 318–397.

Cherniak DJ. 2002. Ba diffusion in feldspar. *Geochimica et Cosmochimica Acta* 66: 1641–1650.

Cherniak DJ. 2003. REE diffusion in feldspar. *Chemical Geology* 193: 25–41.

Cherniak DJ, Watson EB. 1992. A study of strontium diffusion in K-feldspar, Na-K feldspar and anorthite using Rutherford Backscattering Spectroscopy. *Earth and Planetary Science Letters* **113**: 411–425.

Cherniak DJ, Watson EB. 1994. A study of strontium diffusion in plagioclase using Rutherford backscattering spectroscopy. *Geochimica et Cosmochimica Acta* **58**: 5179–5190.

Cherniak DJ, Watson EB, Wark DA. 2007. Ti diffusion in quartz. *Chemical Geology* **236**: 65–74.

Chertkoff DG, Gardner JE. 2004. Nature and timing of magma interactions before, during, and after the caldera-forming eruption of Volcán Ceboruco, Mexico. *Contributions to Mineralogy and Petrology* **146**: 715–735.

Cole DR, Chakraborty S. 2001. Rates and mechanisms of isotope exchange. *Reviews in Mineralogy and Geochemistry* **43**: 83–223.

Condomines M, Gauthier P-J, Sigmarsson O. 2003. Timescales of magma chamber processes and dating of young volcanic rocks. *Reviews in Mineralogy and Geochemistry* **52**: 125–174.

Coogan LA, Kasemann SA. Chakraborty S. 2005a. Rates of hydrothermal cooling of new oceanic upper crust derived from lithium-geospeedometry. *Earth and Planetary Science Letters* **240**: 415–424.

Coogan LA, Hain A, Stahl S, Chakraborty S. 2005b. Experimental determination of the diffusion coefficient for calcium in olivine between 900°C and 1500°C. *Geochimica et Cosmochimica Acta* **69**: 3683–3694.

Coombs ML, Eichelberger JC, Rutherford MJ. 2000. Magma storage and mixing conditions for the 1953–1974 eruptions of Southwest Trident volcano, Katmai National Park, Alaska. *Contributions to Mineralogy and Petrology* **140**: 99118.

Costa F. 2000. The petrology and geochemistry of diverse crustal xenoliths, Tatara-San Pedro Volcanic Complex, Chilean Andes. *Terre & Environnement* **19**: 120 pp. PhD. thesis University of Geneva.

Costa F. 2008. Residence times of silicic magmas associated with calderas. In: Caldera Volcanism: Analysis, Modelling and Response. Gottsmann, J, Marti J. (eds), *Developments in Volcanology* **10**: 1–55.

Costa F, Chakraborty S. 2004. Decadal time gaps between mafic intrusion and silicic eruption obtained from chemical zoning patterns in olivine. *Earth and Planetary Science Letters* **227**: 517–530.

Costa F, Chakraborty S. 2008. The effect of water on Si and O diffusion rates in olivine and their relation to transport properties and processes in the upper mantle. *Physics of the Earth and Planetary Interiors* **166**: 11–29.

Costa F, Singer BS. 2002. Evolution of Holocene dacite and compositionally zoned magma, Volcan San Pedro, Southern Volcanic Zone, Chile. *Journal of Petrology* **43**: 1571–1593.

Costa F, Streck M. 2004. Periodicity and timescales for basaltic magma replenishment at Volcán Arenal (Costa Rica). IAVCEI General Assembly 2004 abstracts.

Costa F, Dungan M. 2005. Short timescales of magmatic assimilation from diffusion modelling of multiple elements in olivine. *Geology* **33**: 837–840.

Costa F, Dungan M, Singer BS. 2002. Hornblende- and phlogopite-bearing gabbroic xenoliths from Volcan San Pedro (368S), Chilean Andes: Evidence for melt and fluid migrationand reactions in subduction related plutons. *Journal of Petrology* **43**: 219–241.

Costa F, Chakraborty S, Dohmen R. 2003. Diffusion coupling between trace and major elements and a model for calculation of magma residence times using plagioclase. *Geochimica et Cosmochimica Acta* **67**: 2189–2200.

Costa F, Chakraborty S, Fedortchouk Y, Canil D. 2005. Crystals in Kimberlites: Where are the Phenocrysts?. Eos Trans. AGU, **86**(52): Fall Meet. Suppl., Abstract V13B-0545.

Costa F, Dohmen R, Chakraborty S. 2008. Timescales of magmatic processes from modeling the zoning patterns of crystals. *Reviews in Mineralogy and Geochemistry* **69**: 545–594.

Costa F, Coogan L, Chakraborty S. 2010. The timescales of magma mixing and mingling involving primitive melts and melt-mush interaction at Mid-Ocean Ridges. *Contributions to Mineralogy and Petrology*, **159**: 371–387.

Crank J. 1975. *The Mathematics of Diffusion.* Oxford Science Publications, Oxford.

Davidson JP, Morgan DJ, Charlier LA, Harlou R, Hora JM. 2007. Microsampling and isotopic analysis of igneous rocks: implications for the study of magmatic systems. *Annual Reviews of Earth Planetary Sciences* **35**: 273–311.

Demouchy S, Jacobsen SD, Gaillard F, Stern CR. 2006. Rapid magma ascent recorded by water diffusion profiles in mantle olivine. *Geology* **34**: 429–432.

Dimanov A, Windenbeck M. 2006. (Fe,Mn)-Mg interdiffusion in natural diopside: effect of PO_2. *European Journal of Mineralogy* **18**: 705–718.

Dodson MH. 1973. Closure temperature in cooling geochronological and petrological systems. *Contributions to Mineralogy and Petrology* **40**: 259–274.

Dodson MH. 1986. Closure profiles in cooling systems. *Material Sciences Forum* **7**: 145–154.

Dohmen R, Chakraborty S. 2007a. Fe-Mg diffusion in olivine II: point defect chemistry, change of diffusion mechanisms and a model for calculation of diffusion coefficients in natural olivine. *Physics and Chemistry of Minerals* **34**: 409–430.

Dohmen R, Chakraborty S. 2007b. Fe-Mg diffusion in olivine II: point defect chemistry, change of diffusion mechanisms and a model for calculation of diffusion coefficients in natural olivine (vol 34, pg 409, 2007). *Physics of Chemistry of Minerals* **34**: 597–598.

Dohmen R, Becker H-W, Chakraborty S. 2007. Fe-Mg diffusion coefficients in olivine, Part I: Experimental determination between 700 and 1200°C as a function of composition, crystal orientation and oxygen fugacity. *Physics and Chemistry of Minerals* **34**: 389–407.

Dohmen R, Becker H-W, Meissner E, Etzel T, Chakraborty S. 2002. Production of silicate thin films using pulsed laser deposition (PLD) and applications to studies in mineral kinetics. *European Journal of Mineralogy* **14**: 1155–1168.

Dungan M, Wulff A, Thompson R. 2001. A refined eruptive stratigraphy for the Tatara–San Pedro Complex (36 S, Southern Volcanic Zone, Chilean Andes): Reconstruction methodology and implications for magma evolution at long-lived arc volcanic centers. *Journal of Petrology* **42**: 555–626.

Elkins LT, Grove T. 1990. Ternary feldspar experiments and thermodynamic models. *American Mineralogist* **75**: 544–559.

Faryad SW, Chakraborty S. 2005. Duration of Eo-Alpine metamorphic events obtained from multicomponent diffusion modeling of garnet: a case study from the Eastern Alps. *Contributions to Mineralogy and Petrology* **250**: 306–318.

Ferry JM, Watson EB. 2007. New thermodynamic models and revised calibrations for the Ti-in-zircon and Zr-in-rutile thermometers. *Contributions to Mineralogy and Petrology* **154**: 429–437.

Flynn CP. 1972. *Point Defects and Diffusion.* Clarendon Press, Oxford.

Freer R. 1981. Diffusion in silicate minerals and glasses: a data digest and guide to the literature. *Contributions to Mineralogy and Petrology* **76**: 440–454.

Ganguly J, Bhattacharya RN, Chakraborty S. 1988. Convolution effect in the determination of compositional zoning by microprobe step scans. *American Mineralogist* **73**: 901–909.

Ganguly J, Cheng W, Chakraborty S. 1998. Cation diffusion in aluminosilicate garnet: experimental determination in pyrope-almandine diffusion couples. *Contributions to Mineralogy and Petrology* **131**: 171–180.

Ghiorso MS, Sack RO. 1995. Chemical mass transfer in magmatic processes IV. A revised and internally consistent thermodynamic model for the interpolation and extrapolation of liquid-solid equilibria in magmatic systems at elevated temperatures and pressures. *Contributions to Mineralogy and Petrology* **119**: 197–212.

Giletti BJ. 1991. Rb and Sr diffusion in alkali feldspars, with implications for cooling histories of rocks. *Geochimica et Cosmochimica Acta* **55**: 1331–1343.

Giletti BJ, Casserly JED. 1994. Strontium diffusion kinetics in plagioclase feldspars. *Geochimica et Cosmochimica Acta* **58**: 3785–3797.

Giletti BJ, Shanahan TM. 1997. Alkali diffusion in plagioclase feldspar. *Chemical Geology* **139**: 3–20.

Ginibre C, Kronz A, Wörner G. 2002. High resolution quantitative imaging of plagioclase composition using accumulated back scattered electron images: new constraints on oscillatory zoning. *Contributions to Mineralogy and Petrology* **142**: 436–448.

Ginibre C, Wörner G, Kronz A. 2007. Crystal zoning as an archive for magmatic evolution. *Elements* **3**: 261–266.

Grove TL, Baker MB, Kinzler RJ. 1984. Coupled CaAl-NaSi diffusion in plagioclase feldspar: experiments and applications to cooling rate speedometry. *Geochimica et Cosmochimica Acta* **48**: 2113–2121.

Hier-Majumder S, Anderson IM, Kohlstedt DL. 2005. Influence of protons on Fe-Mg interdiffusion in olivine. *Journal of Geophysical Research* **110**: doi: 10.1029/2004JB003292.

Hinton R. 1995. Ion microprobe analysis in geology. In: *Microprobe Techniques in the Earth Sciences*, Potts PJ, Bowles JFW, Reed SJB & Cave MR (eds), Chapman & Hall, London.

Hofmann AE, Valley JW, Watson EB, Cavosie AJ, Eiler JM. 2009. Sub-micron scale distributions of trace elements in zircon. *Contributions to Mineralogy and Petrology* **158**: 317–335.

Holland T, Blundy J. 1994. Non-ideal interactions in calcic amphiboles and their bearing on amphibole-plagioclase thermometry. *Contributions to Mineralogy and Petrology* **116**: 433–447.

Holloway JR, Wood BJ. 1988. *Simulating the Earth: Experimental Geochemistry.* Springer-Verlag, Berlin.

Holzapfel C, Chakraborty S, Rubie DC, Frost DJ. 2007. Effect of pressure on Fe–Mg, Ni and Mn diffusion in $(Fe_xMg_{1-x})_2SiO_4$ olivine. *Physics of the Earth and Planetary Interiors* **162**: 186–198.

Ito M, Ganguly J. 2006. Diffusion kinetics of Cr in olivine and 53Mn-53Cr thermochronology of early solar system objects. *Geochimica et Cosmochimica Acta* **70**: 799–809.

Kelley SP, Wartho JA. 2000. Rapid kimberlite ascent and the significance of Ar-Ar ages in xenolith phlogopites. *Science* **209**: 609–611.

Klügel A. 2001. Prolonged reactions between harzburgite xenoliths and silica-undersaturated melt: Implications for dissolution and Fe-Mg interdiffusion rates of orthopyroxene. *Contributions to Mineralogy and Petrology* **141**: 1–14.

Larsen ES, Irving J, Gonyer FA, Larsen ES. 1938. Petrologic results of a study of the minerals from the Tertiary volcanic rocks of the San Juan region, Colorado. *American Mineralogist* **23**: 227–257.

Lasaga AC. 1979. Multicomponent exchange and diffusion in silicates. *Geochimica et Cosmochimica Acta* **43**: 455–469.

Lasaga AC. 1998. *Kinetic Theory in the Earth Sciences*. Princeton University Press, Princeton NJ.

LaTourrette T, Wasserburg GJ. 1998. Mg diffusion in anorthite: implications for the formation of early solar system planetesimals. *Earth and Planetary Science Letters* **158**: 91–108.

Liu M, Yund RA. 1992. NaSi-CaAl interdiffusion in plagioclase. *American Mineralogist* **77**: 275–283.

Manning JR. 1968. *Diffusion Kinetics for Atoms in Crystals*. Princeton University Press, Princeton NJ.

Martin Y, Church M. 1997. Diffusion in landscape development models: on the nature of basic transport equations. *Earth Surface Processes and Landforms* **22**: 273–279.

Martin VM, Morgan DJ, Jerram DA, Caddick MJ, Prior DJ, Davidson JP. 2008. Bang! month-scale eruption triggering at Santorini Volcano. *Science* **321**: 1178.

Morgan DJ, Blake S. 2006. Magmatic residence times of zoned phenocrysts: introduction and application of the binary element diffusion modelling (BEDM) technique. *Contributions to Mineralogy and Petrology* **151**: 58–70.

Morgan DJ, Blake S, Rogers NW, et al. 2004. Timescales of crystal residence and magma chamber volume from modeling of diffusion profiles in phenocrysts: Vesuvius 1944. *Earth and Planetary Science Letters* **222**: 933–946.

Morgan DJ, Blake S, Rogers NW, DeVivo B, Rolandi G, Davidson J. 2006. Magma recharge at Vesuvius in the century prior to the eruption AD 79. *Geology* **34**: 845–848.

Nakamura M. 1995a. Continuous mixing of crystal mush and replenished magma in the ongoing Unzen eruption. *Geology* **23**: 807–810.

Nakamura M. 1995b. Residence time and crystallization history of nickeliferous olivine phenocrysts from the northern Yatsugatake volcanoes, Central Japan: Application of a growth and diffusion model in the system Mg-Fe-Ni. *Journal of Volcanology and Geothermal Research* **66**: 81–100.

Pan Y, Batiza R. 2002. Mid-ocean ridge magma chamber processes: Constraints from olivine zonation in lavas from the East Pacific Rise at 9°30′N and 10°30′N. *Journal of Geophysical Research* **107**: doi 10.1029/2001JB000435

Perkins WT, Pearce NJG. 1995. Mineral microanalysis by laserprobe inductively coupled plasma mass spectrometry. In: *Microprobe Techniques in the Earth Sciences*, Potts PJ, Bowles JFW, Reed SJB, Cave MR (eds), Chapman & Hall, London.

Peslier AH, Luhr JF. 2006. Hydrogen loss from olivines in mantle xenoliths from Simcoe (USA) and Mexico: Mafic alkalic magma ascent rates and water budget of the sub-continental lithosphere. *Earth and Planetary Science Letters* **242**: 302–319.

Petry C, Chakraborty S, Palme H. 2004. Experimental determination of Ni diffusion coefficients in olivine and their dependence on temperature, composition, oxygen fugacity, and crystallographic orientation. *Geochimica et Cosmochimica Acta* **68**: 4179–4188.

Philibert J. 2005. One and a Half Century of Diffusion: Fick, Einstein, before and beyond. *Diffusion Fundamentals* **2**: 1–10.

Potts PJ, Bowles JFW, Reed SJB, Cave MR. 1995. *Microprobe Techniques in the Earth Sciences*. Chapman & Hall, London.

Press WH, Teukolsky SA, Vetterling WT, Flannery BP. 2007. *Numerical Recipes*. Cambridge University Press, Cambridge.

Prior DJ, Boyle AP, Brenker F, et al. 1999. The application of electron backscatter diffraction and orientation contrast imaging in the SEM to textural problems in rocks. *American Mineralogist* **84**: 1741–1759.

Putirka K. 2005. Igneous thermometers and barometers based on plagioclase + liquid equilibria: Tests of some existing models and new calibrations. *American Mineralogist* **90**: 336–346.

Reed SJB. 1975. *Electron Microprobe Analysis*. Cambridge University Press. Cambridge.

Reid M. 2003. Timescales of magma transfer and storage in the crust. In: *Treatise on Geochemistry, Vol. 3: The Crust*. Holland HD, Turekian KK (eds), 167–193.

Shaw CSJ. 2004. The temporal evolution of three magmatic systems in the West Eifel volcanic field, Germany. *Journal of Volcanology and Geothermal Research* **131**: 213–240.

Shaw CSJ, Heidelbach F, Dingwell DB. 2006. The origin of reaction textures in mantle peridotite xenoliths from Sal Island, Cape Verde: the case for metasomatism by the host lava. *Contributions to Mineralogy and Petrology* **151**: 681–697.

Sigmarsson O. 1996. Short magma chamber residence time at an Icelandic volcano inferred from U-series disequilibria. *Nature* **391**: 440–442.

Slodzian G. 1964. Etude d'une methode d'analise locale chimique et isotopique utilisant l'emission ionique secondaire. *Annales de Physique* **9**: 591–648.

Sneeringer M, Hart SR, Shimizu N. 1984. Strontium and samarium diffusion in diopside. *Geochimica et Cosmochimica Acta* **48**: 1589–1608.

Spear FS. 1993. *Metamorphic Phase Equilibria and Pressure-temperature-time Paths*. Mineralogical Society of America, Washington DC.

Streck M. 2008. Mineral textures and zoning as indicators for open system processes. *Reviews in Mineralogy and Geochemistry* **69**: 595–622.

Tirone M, Ganguly J, Dohmen R, Langenhorst F, Hervig R, Werner H-W. 2005. Rare earth diffusion kinetics in garnet: experimental studies and applications. *Geochimica et Cosmochimica Acta* **69**: 2385–2398.

Turner S, Costa F. 2007. Measuring timescales of magmatic evolution. *Elements* **3**: 267–272.

Van Orman JA, Grove TL, Shimizu N, Graham, DL. 2001. Rare earth element diffusion in diopside: influence of temperature, pressure and ionic radius, and an elastic model for diffusion in silicates. *Contributions to Mineralogy and Petrology* **141**: 687–703.

Van Orman JA, Grove TL, Shimizu N, Layne GD. 2002. Rare earth element diffusion in a natural pyrope single crystal at 2.8 GPa. *Contributions to Mineralogy and Petrology* **142**: 416–424.

Venezky DY, Rutherford MJ. 1999. Petrology and Fe-Ti oxide reequilibration of the 1991 Mount Unzen mixed magma. *Journal of Volcanology and Geothermal Research* **89**: 213–230.

Wark DA, Hildreth W, Spear FS, Cherniak DJ, Watson EB. 2007. Pre-eruption recharge of the Bishop magma system. *Geology* **35**: 235–238.

Wark DA, Watson EB. 2006. TitaniQ: A titanium-in-quartz geothermometer. *Contributions to Mineralogy and Petrology* **152**: 743–754.

Watson EB. 1979. Zircon Saturation in Felsic Liquids: Experimental Results and Applications to Trace Element Geochemistry. *Contributions to Mineralogy and Petrology* **70**: 407–419.

Watson EB. 1994. Diffusion in volatile-bearing magmas. *Reviews in Mineralogy* **30**: 371–412.

Watson EB, Baxter EF. 2007. Diffusion in solid-Earth systems. *Earth and Planetary Science Letters* **253**: 307–327.

Wiebe RA, Wark DA, Hawkins DP. 2007. Insights from quartz cathodoluminescence zoning into crystallization of the Vinalhaven Granite, coastal Maine. *Contributions to Mineralogy and Petrology* **154**: 439–453.

Zellmer GF, Clavero JE. 2006. Using trace element correlation patterns to decipher a sanidine crystal growth chronology: an example from Taapaca volcano, Central Andes. *Journal of Volcanology and Geothermal Research* **156**: 291–301.

Zellmer GF, Blake S, Vance D, Hawkesworth C, Turner S. 1999. Plagioclase residence times at two island arc volcanoes (Kameni Islands, Santorini, and Soufriere, St. Vincent) determined by Sr diffusion systematics. *Contributions to Mineralogy and Petrology* **136**: 345–357.

Zellmer GF, Sparks RSJ, Hawkesworth C, Wiedenbeck M. 2003. Magma emplacement and remobilization timescales beneath Montserrat: Insights from Sr and Ba zonation in plagioclase phenocrysts. *Journal of Petrology* **44**: 1413–1431.

8 Magma Cooling and Differentiation – Uranium-series Isotopes

ANTHONY DOSSETO[1] AND SIMON P. TURNER[2]

[1]GeoQuEST Research Centre, School of Earth and Environmental Sciences,
University of Wollongong, Wollongong, Australia
[2]GEMOC ARC National Key Centre, Department of Earth and Planetary Sciences,
Macquarie University, Sydney, NSW, Australia

SUMMARY

Timescales of magma cooling and differentiation can be quantified by measuring and modelling the composition of uranium-series (U-series) isotopes in crystals and whole rock samples. The U-series isotope composition of crystals yields a wide range of timescales: from a few tens of years to several hundred thousand years. Magma differentiation is inferred to be generally rapid (a few 1,000 years) at mid-ocean ridges and oceanic islands compared to island, continental arcs and intra-continental volcanic centres (>1,000 to several 100,000 years). For large volume, highly silicic rocks (dacites, rhyolites), timescales of differentiation > 100,000 years are generally inferred, reflecting the time required to heat up and assimilate the crust. U-series timescale information can be used to distinguish between different models for magma differentiation. For example, a model of frequent magma recharge combined with crustal assimilation seems to best explain U-series isotope data in many instances (and is probably the most realistic physical model).

Timescales of Magmatic Processes: From Core to Atmosphere, 1st edition. Edited by Anthony Dosseto, Simon P. Turner and James A. Van Orman.
© 2011 by Blackwell Publishing Ltd.

It is shown that, once the thermal regime that allows melting of the country rock is met (often requiring hundreds of thousands years), it takes only a few 1,000 years to produce highly silicic compositions.

INTRODUCTION

Most magmas originate in the mantle, or require mantle heat input, and previous chapters have explored how they are produced and transferred to the crust. Some magmas can be transferred directly from their source region to the Earth's surface, but most of them are stored for various amounts of time in the crust before eruption at the surface. Differentiation of magmas during storage is the main process responsible for the diversity of igneous rock compositions in the crust and an understanding of magma differentiation is required to understand this diversity. This question is also economically significant as a large number of ore deposits are related to the differentiation and/or degassing of magma bodies. Here, we use the term *magma differentiation* to refer to the compositional evolution of a magma that may be the result of various processes such as crystal fractionation, magma mixing or crustal assimilation (Figure 8.1).

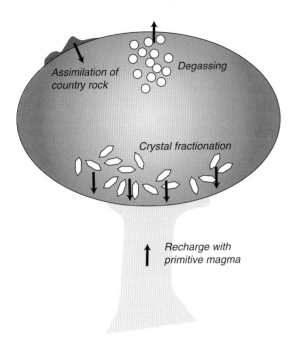

Fig. 8.1 Simplified diagram of the processes occurring in a magma body that promote differentiation. See Plate 8.1 for a colour version of this image.

Several mechanisms have been proposed to account for the differentiation of magma bodies but it is beyond the scope of this chapter to review them in detail and the reader is referred to Winter (2001). An important driving mechanism of magma differentiation is the formation of crystals and their physical separation from the magma body (*crystal fractionation*; Figure 8.1). Newly-formed crystals deplete the magma in some elements, termed *compatible elements*, such that their abundance in the magma progressively decreases during crystal fractionation. Conversely, the abundances of elements that do not enter a crystal structure, termed *incompatible elements*, increase in the magma during crystal fractionation. Thus, as a consequence of compatible or incompatible behavior, the process of magma differentiation, driven by crystal fractionation, causes the abundance of elements in the magma to decrease or increase, depending on which mineral phases crystallize. In geochemical analysis, major element abundances are generally reported as oxides and the composition of most minerals in igneous rocks can be described as a sum of oxides. For instance, in basaltic systems, the initial phases to crystallize are typically Fe-Mg rich phases such as olivine and pyroxene, and so the amounts of magnesium oxide (MgO) and iron oxide (FeO) in the magma tend to decrease as differentiation proceeds, whereas the oxide concentrations of potassium (K_2O), sodium (Na_2O) and silicon (SiO_2) increase, as the fractionated minerals are low in these components. Because of these types of differentiation trend, magma differentiation was first described geochemically using *x-y* diagrams of major element oxides against silica (Harker diagrams). Co-genetic volcanic rocks define trends in these diagrams termed *liquid lines of descent*. In some instances it has been argued that, instead of reflecting continuous magma differentiation by crystal fractionation, these trends result from mixing between a primitive (e.g. low SiO_2 and high MgO) and an evolved magma (high SiO_2 and low MgO). However, two end-member mixing cannot explain the inflections observed from some elements in many liquid lines of descent and a combination of crystal fractionation and assimilation of wall-rock is often inferred to be important, leading to Assimilation and Crystal Fractionation (AFC) models (DePaolo, 1981). Because the energetics of crystallization, crystal settling, mixing and wall-rock melting differ, it is likely that the timescales of these processes will also differ and so the timescales of differentiation observed for a magma system may allow us to determine the relative importance of each process.

For example, crystal fractionation can be driven by cooling of the magma body (by conduction and/or convection (Spera, 1980)) or by degassing of the magma (Cashman & Blundy, 2000). In each case, associated timescales will be very different: cooling of a large, deep magma body occurs over timescales >1,000 years, whereas degassing takes place over timescales of weeks to years. Consequently, a quantification of the characteristic timescales of magma differentiation is required to better understand what the driving mechanisms are.

As the fundamental process by which crystals and liquids separate, the rate of crystal settling can be modeled to gain an initial insight into the rate of differentiation. Assuming spherical olivine crystals (radius = 0.1 cm) settling in a basaltic liquid, we can use Stokes' Law to infer that they should settle by gravity at a rate of ~1 m per day. A 0.1-cm radius feldspar crystal would settle in a rhyolitic melt at a much slower rate of 2 cm/year. Gravity settling is driven by the difference in density, ρ, between crystals and the liquid, and resisted by the liquid viscosity. The density is partly controlled by the amount of iron (Fe) in the liquid, whilst viscosity is a function of silica content (increases polymerization), water and alkali contents (depolymerizing agents) and temperature (depolymerization). Interestingly, whilst most crystals sink because they are more dense than the liquid, feldspar crystals ($\rho = 2.7\,g/cm^3$) will float in a Fe-rich liquid ($\rho > 2.7\,g/cm^3$). Thus, it appears that crystal fractionation is possible both in basaltic and granitic magmas, but settling is faster in basaltic magmas because of their lower viscosity. However, these settling velocities should be taken as maximum values for several reasons:

• Crystals are not spherical. Non-spherical crystals (e.g. tabular, acicular or platy) will settle much slower;
• Stokes' Law assumes that the magma is a Newtonian fluid (i.e. a fluid with no yield stress, deforming as soon as a differential stress is applied).

However, only basaltic magmas near or above their liquidus temperature behave as Newtonian fluids (McBirney & Noyes, 1979). Once they begin to crystallize, they develop a significant yield strength that will dramatically reduce crystal settling rates. Consequently, it is difficult to predict the rate of magma differentiation by simply studying gravity settling of crystals and other approaches must be considered.

The development of uranium-series (U-series) isotope measurements in igneous rocks has given us the unique opportunity to greatly improve our understanding of the timescales of magmatic processes. Each daughter-parent system of the

U-series records magmatic processes on various timescales that are a function of the daughter half-life. For instance, ^{238}U-^{230}Th can record fractionation between these two isotopes up to 300,000 years old, ^{230}Th-^{226}Ra up to 8,000 years old and ^{226}Ra-^{210}Pb up to 100 years old. Previous chapters have described how U-series isotopes can be used to quantify how fast magmas are produced and migrate towards the Earth's surface (Bourdon & Elliott, 2010; Turner & Bourdon 2010). In this chapter, we are interested in the processes that occur when magmas stall in the Earth's crust. Here, magmas cool down, crystallize and degas (magma degassing is discussed in detail in a separate chapter). We focus on the timescale of crystal fractionation, magma differentiation and the production of evolved volcanic rocks (i.e. high SiO_2 content). The cooling and crystallization of plutonic rocks may be slower and in any case, by the time they are exposed at the surface, such rocks are too old to be amenable to U-series analysis.

Note that to gain access to the composition of the evolving magma, we have two possibilities:
1 erupted glass provides a direct sample of the magma composition. However, glass samples at sub-aerial volcanoes are rare and often limited to highly silicic eruptions.
2 sampling and analysis of the whole volcanic rock can also be used as a proxy for the composition of the magma. Although a volcanic rock often contains crystals, because U, Th, Pa and Ra are very incompatible, their budget in a volcanic rock will be dominated by the composition of the groundmass, which reflects the composition of the evolving magma.

The approach used here is to measure *radioactive disequilibrium* in mineral separates or whole rock samples. As explained in previous chapters, a radioactive disequilibrium represents the fractionation between a parent nuclide and its daughter. The radioactive disequilibrium is expressed as a daughter/parent activity ratio, for example ($^{230}Th/^{238}U$) activity ratio for the ^{238}U-^{230}Th system, and ($^{226}Ra/^{230}Th$) activity ratio for the ^{230}Th-^{226}Ra system (a common convention that we will follow in this chapter is that parentheses denote activity ratios). Primitive magmas, i.e. prior to differentiation, are assumed to record the radioactive disequilibrium acquired during magma generation and transfer, whereas minimal disequilibria are likely to be generated during crustal melting (Berlo *et al.*, 2004). As seen in previous chapters, $(^{230}Th/^{238}U)$ ratios can be >1 or <1, depending on the source and the conditions of melting, whereas $(^{226}Ra/^{230}Th)$ is, in general, >1 (because in most cases radium, Ra, is largely more incompatible than thorium, Th). In the following sections, we will see how processes related to magma differentiation (crystal fractionation, recharge and assimilation) can modify the activity ratios of a primitive magma and how this can be used to infer timescales of magma differentiation.

FRACTIONATION OF U-SERIES ISOTOPES DURING MAGMA DIFFERENTIATION

Crystal fractionation

In previous chapters, we have seen how U-series isotopes fractionate during magmatic processes and how they partition between minerals and melt (Bourdon & Elliott, 2010; Turner & Bourdon, 2010). The same concepts apply here, where U-series isotopes fractionate between a cooling magma and newly formed crystals. Whereas during partial melting, uranium (U), thorium (Th), protactinium (Pa) and radium (Ra) can fractionate significantly, depending on the mineralogy of the source and the conditions of partial melting (porosity, melting rate) (Bourdon & Elliott, 2010), these nuclides undergo little fractionation during crystallization as a result of their very low mineral/melt partition coefficients. An exception can arise when significant amounts of plagioclase accumulate in a differentiating magma, because Ra is enriched in plagioclase relative to Th, and this could potentially fractionate the $(^{226}Ra/^{230}Th)$ activity ratio. However, in a study of volcanic rocks from Ardoukoba volcano (Asal Rift, Eastern Africa), Vigier *et al.* (1999) have shown that up to 30% plagioclase accumulation affects $(^{226}Ra/^{230}Th)$

ratios by only 0.1 to 7%. Thus, except when U-Th-rich accessory phases such as allanite, sphene or zircon fractionate, we assume that the variation in daughter-parent activity ratios, for example $(^{230}Th/^{238}U)$ or $(^{226}Ra/^{230}Th)$, in a magma body evolving as a closed system (e.g. no recharge by primitive magma, and no assimilation of the host rock) is primarily the result of radioactive decay. Consequently, variations of activity ratios in the magma should be a direct function of time and provide information on the timescale of differentiation.

Magma mixing (recharge, assimilation)

In the previous section, we considered only crystal fractionation in a closed-system magma chamber. In this case, we see that most of the variation in activity ratios in the magma is related to the residence time of the magma body in the crust, during which it undergoes differentiation. In reality, magma bodies can behave as open systems where different types of 'contaminants' can be added to the differentiating magma body:

• *Primitive magma that recharges the differentiating magma body*. In this case, as the magma body differentiates, it is frequently or continuously recharged by a primitive melt, added to the magma body from the melting region. The primitive melt is likely to have higher activity ratios than the differentiating magma body, since activity ratios decrease with time during differentiation as a result of radioactive decay.

• *Melts derived from partial melting of the country rock*. Emplacement of magma bodies in the crust provides heat that promotes, in some cases, partial melting of the country rock. This crustal melt is likely to have a different U-series isotope composition than the differentiating magma body. Berlo *et al.* (2004) have shown that partial melting of the crust would generate contaminant melts with little or no radioactive disequilibrium (i.e. activity ratios close to unity). This implies that:

1 the radioactive disequilibrium observed in silicic volcanic rocks (e.g. dacites, rhyolites) cannot be *produced* by crustal assimilation;

2 mixing of mantle-derived magma with crustal melts will typically result in a 'dilution' of the source-derived radioactive disequilibria of the differentiating magma, effectively hastening the return to secular equilibrium.

TIMESCALES OF DIFFERENTIATION FROM U-SERIES ISOTOPES IN CRYSTALS

Because minerals have different atomic structures and compositions, partition coefficients for U-series isotopes of interest (^{238}U, ^{230}Th and ^{226}Ra) vary for different mineral phases. Thus, the U-series isotope composition of minerals can be used to derive their crystallization age. Early

studies used these crystal ages to constrain the eruption age of prehistoric eruptions (Kigoshi, 1967; Taddeucci *et al.*, 1967; Fukuoka, 1974; Peate *et al.*, 1996; Condomines, 1997). However, obtaining eruption ages using this approach requires that the time lag between crystallization and eruption is negligible (Taddeucci *et al.*, 1967; Allègre, 1968). If the eruption age is independently known, it can be subtracted from the crystallization age and yield a pre-eruptive age that will reflect that time elapsed between precipitation of minerals from the differentiating magma and eruption.

We first consider the ^{238}U-^{230}Th system. In this case, we use an *isochron diagram* that plots the daughter isotope, ^{230}Th, against the parent isotope, ^{238}U, both normalized to ^{232}Th (Figure 8.2). In this instance, because the ^{238}U-^{230}Th system records fractionations up to 300,000 years, ^{232}Th

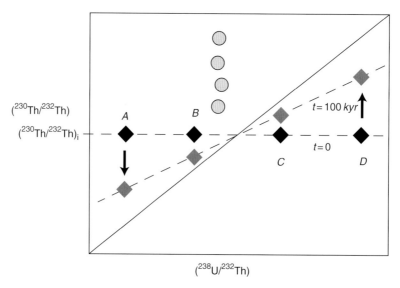

Fig. 8.2 Schematic representation of mineral dating using the ^{238}U-^{230}Th radioactive system. At t = 0, 4 minerals A, B, C and D crystallize with the same (^{230}Th/^{232}Th) activity ratio, (^{230}Th/^{232}Th)$_i$, from a homogeneous magma, but with different U/Th ratios (hence, (^{238}U/^{232}Th) ratios). With time, ^{238}U-^{230}Th returns to secular equilibrium by radioactive decay. Gray diamonds show the composition of the four minerals after 100 kyr. They align along an *isochron* line. Thus, the measurement of (^{230}Th/^{232}Th) and (^{238}U/^{232}Th) in minerals with different U/Th ratios can be potentially used to date the age of their crystallization. Another scenario is when samples, represented as gray dots, form a vertical array. In this case, this suggests that they have evolved with similar U/Th ratios but have crystallized at different times. The samples closest to the equiline are the oldest, and their age can be calculated using the equation given in the text. The solid line is termed *equiline* and represents the compositions in secular equilibrium.

and ^{238}U are treated like stable isotopes, since their half-lives are 14.1×10^9 and 4.4683×10^9 years, respectively (Bourdon *et al.*, 2003). If we analyse four minerals A, B, C and D that crystallized with different U/Th ratios from a homogeneous magma at the same time, their $(^{230}Th/^{232}Th)$ activity ratios will evolve with time to define an isochron, as shown in Figure 8.2. The slope, *a*, of this isochron can then be used to calculate the age of crystallization of these minerals, *T* (in yr):

$$T = -\frac{1}{\lambda} ln(a-1)$$

where λ is the decay constant of ^{230}Th (in yr^{-1}).

To use this approach, the following assumptions need to be made:
• Crystallization occurs in a closed system;
• The different minerals crystallized at the same time;
• Crystallization was rapid compared to the half-life of the system (e.g. 75 kyr for ^{238}U-^{230}Th, i.e. the half-life of the daughter nuclide);
• The minerals crystallized from magma that had a homogeneous $(^{230}Th/^{232}Th)$ activity ratio.

The ^{238}U-^{230}Th system can be used to determine crystallization ages in the range 10 to 300 kyr, and results from a number of studies listed in Table 8.1 will be discussed below.

For the ^{230}Th-^{226}Ra system, the approach is different because there is no long-lived or stable isotope of radium (Ra) that can be used to normalize ^{226}Ra and ^{230}Th activities, in the manner that ^{232}Th is used for the ^{238}U-^{230}Th system. Barium (Ba) is widely accepted to be the closest chemical analog of Ra and early studies employed an isochron diagram for ^{230}Th-^{226}Ra where ^{226}Ra and ^{230}Th activities were normalized to Ba concentrations. However, it is now well established that Ra and Ba partition differently (Blundy & Wood, 1994, 2003; Miller *et al.*, 2007). Thus, Cooper & Reid (2003) suggested an alternative method that uses a diagram of $(^{226}Ra)/$[Ba] (the ratio of ^{226}Ra activity over Ba concentration) vs time (Figure 8.3). In this diagram, the measured $(^{226}Ra)/$[Ba] ratios for mineral phases and the groundmass plot along the *y*-axis. Using the measured $(^{230}Th)/$[Ba] ratios, the evolution back in time

of $(^{226}Ra)/$[Ba] for each mineral and the groundmass can be calculated (see Appendix A for equations). In addition, the evolution of the $(^{226}Ra)/$[Ba] ratio for a melt in equilibrium with a given mineral can also be calculated. For instance, the dashed curve in Figure 8.3a describes the evolution of the $(^{226}Ra)/$[Ba] ratio in a plagioclase, *pl*. In the same diagram, the thick curve represents the evolution of the $(^{226}Ra)/$[Ba] ratio calculated for a melt, *melt$_{pl}$*, that would be in equilibrium with the plagioclase analysed. Thus, the crystallization age of the different mineral phases can be solved graphically by looking for the intersection between the evolution curve for the groundmass, *gm*, and the melts in equilibrium with the different mineral phases, *melt$_{px}$* and *melt$_{pl}$*. Note that the precision of this method can suffer from:
• Difficulty in estimating partitioning differences between Ra and Ba; and
• The contribution of impurities such as melt and mineral inclusions (Rubin & Zellmer, 2009). For this reason, in a study of Icelandic volcanic rocks, Zellmer *et al.* (2008) preferred to assume that Ra and Ba partition similarly, arguing that a correction for different behavior during crystallization would not affect significantly calculated crystal ages but would introduce larger errors.

Mineral ages have been gathered using the ^{238}U-^{230}Th, and in some instances the ^{226}Ra-^{230}Th system, from various volcanic centers around the world. A compilation is shown in Figure 8.4 where mineral ages have been corrected for eruption ages. Hence, these ages constitute pre-eruptive ages, i.e. the time elapsed since crystallization of the mineral phases before eruption. Several observations arise from these results:
• Minerals from mafic rocks have ages that can deviate significantly from their eruption ages (up to ~60 kyr; Figure 8.4a). In addition, there is a common discrepancy between the ages obtained from the ^{238}U-^{230}Th and ^{226}Ra-^{230}Th systems: the presence of ^{226}Ra-^{230}Th disequilibrium suggests crystallization ages <8,000 years (any ^{226}Ra-^{230}Th disequilibrium decays away after 8,000 years), whereas ^{238}U-^{230}Th yields ages of several tens of thousands of years. This observation, along with the study of crystal size distributions, suggests a

Table 8.1 Compilation of timescales of differentiation from U-series isotope studies

Volcanic centre	Tectonic environment	Composition range	Timescale of differentiation (kyr)	Material analysed	Reference
Cotopaxi (Ecuador)	CAB	andesite	<8	whole rock	(Garrison, et al., 2006)
Cotopaxi (Ecuador)	CAB	rhyolite	70–100	whole rock	(Garrison, et al., 2006)
Mt Shasta (USA)	CAB	andesite-dacite	8–20	minerals	(Volpe, 1992)
Mt St Helens (USA)	CAB	basalt-dacite	2–4	minerals (plag)	(Cooper and Reid, 2003)
Mt St Helens (USA)	CAB	basalt-dacite	0.15–5.7	minerals (px)	(Cooper and Reid, 2003)
Nevado del Ruiz (Colombia)	CAB	andesite-dacite	6	minerals	(Schaefer, et al., 1993)
Vesuvius (Italy)	CAB	phonotephrite	12–39	minerals	(Black, et al., 1998b)
Akutan (Aleutians)	IAB	basaltic andesite	<8	whole rock	(George, et al., 2004)
Aniakchak (Alaska)	IAB	basaltic andesite-rhyodacite	<8	whole rock	(George, et al., 2004)
Etna (Italy)	IAB	basalt-trachyte	200	minerals	(Condomines, et al., 1982)
Kos (Greece)	IAB	rhyolite	200	whole rock/minerals (zircon)	(Bachman, et al., 2007)
La Soufriere (Antilles)	IAB	andesite	30–70	whole rock	(Touboul, et al., 2007)
La Soufriere (Guadeloupe, Antilles)	IAB	andesite	34.9	minerals (px)	(Touboul, et al., 2007)
La Soufriere (Guadeloupe, Antilles)	IAB	andesite	30–70	whole rock	(Touboul, et al., 2007)
Miyakejima (Japan)	IAB	basalt	2	whole rock	(Yokoyama, et al., 2006)
Ruapehu (New Zealand)	IAB	andesite	5	whole rock	(Price, et al., 2007)
Sangeang Api (Indonesia)	IAB	trachybasalt-trachyandesite	2	whole rock/minerals	(Turner, et al., 2003a)
Santorini (Greece)	IAB	dacite	<1	minerals	(Zellmer, et al., 2000)
Seguam (Aleutians)	IAB	basalt-rhyolite	<1	minerals	(Jicha, et al., 2005)
Seguam (Aleutians)	IAB	basalt-rhyolite	130	whole rock	(Jicha, et al., 2005)
Soufriere (St. Vincent, Antilles)	IAB	andesite	60	minerals	(Heath, et al., 1998)
Taupo Volcanic Zone (New Zealand)	IAB	rhyolite	50	minerals (zircon)	(Charlier, et al., 2003)
Taupo Volcanic Zone (New Zealand)	IAB	rhyolite	250	zircon	(Brown and Fletcher, 1999)
Taupo Volcanic Zone (New Zealand)	IAB	rhyolite	27	zircon	(Charlier and Zellmer, 2000)
Toba (Indonesia)	IAB	rhyolite	150	minerals (allanite)	(Vazquez and Reid, 2004)
Alid (Eritrea)	intra-continental	rhyolite	30–50	whole rock/minerals	(Lowenstern, et al., 2006)
Ardoukoba (Djibouti)	intra-continental	basalt	0.87–1.88	whole rock	(Vigier, et al., 1999)
Bishop Tuff (USA)	intra-continental	trachybasalt-rhyolite	100	minerals (zircon)	(Reid and Coath, 2000)
Emuruangogolak (Kenya)	intra-continental	basalt-trachyte	14–40	minerals	(Black, et al., 1998a)
Erebus (Antarctica)	intra-continental	basanite-phonolite	3	minerals	(Reagan, et al., 1992)
Laacher See (Germany)	intra-continental	basanite-phonolite	10–20	minerals	(Bourdon, et al., 1994)

Location	Setting	Rock type	Value	Material	Reference
Long Valley (USA)	intra-continental	trachybasalt-rhyolite	115–230	minerals (zircon)	(Reid, et al., 1997)
Longonot (Kenya)	intra-continental	trachyte	1–2.5	whole rock	(Rogers, et al., 2004)
Oldoinyo Lengai (Tanzania)	intra-continental	carbonatite	0.007–0.018	whole rock	(Williams, et al., 1986)
Olkaria (Kenya)	intra-continental	rhyolite	22	minerals	(Heumann and Davies, 2002)
East Pacific Ridge	MORB	basalt	0.7–1.7	whole rock	(Blake and Rogers, 2005, Sims, et al., 2002)
Gorda Ridge	MORB	basalt	2–4	minerals (plag)	(Cooper, et al., 2003)
Fogo (Azores)	OIB	trachyte	0.05–4.7	glass	(Snyder, et al., 2007)
Kilauea (Hawaii)	OIB	andesite? (5% MgO)	0.55	minerals	(Cooper, et al., 2001)
Piton de la Fournaise (Reunion)	OIB	basalt-oceanite	0.025	whole rock	(Sigmarsson, et al., 2005)
Torfajökull–Veidivötn (Iceland)	OIB	basalt-rhyolite	<3.2	minerals	(Zellmer, et al., 2008)
Vestmannaeyjar (Iceland)	OIB	basalt-hawaiite-mugearite	2–2.8	whole rock	(Blake and Rogers, 2005, Sigmarsson, 1996)
Puy de Dome (France)	OIB?	trachyte	<1	minerals	(Condomines, 1997)

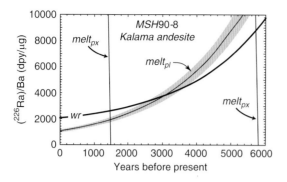

Fig. 8.3 Evolution diagram for a Mt St Helens sample (Cooper & Reid, 2003). Curves show the evolution of the $(^{226}Ra)/Ba$ (in desintegrations per year per µg; dpy/µg) back in time for the whole rock (*wr*) and as calculated for the melt in equilibrium with the plagioclase ($melt_{pl}$) and the pyroxene ($melt_{px}$) analysed in this sample. The curve $melt_{pl}$ represents the best estimate based on average of measured plagioclase compositions (An_{50}) and estimated pre-eruptive temperature of 1,000°C. The gray field encloses compositions allowing variation in the plagioclase composition (An50 ± 10) and pre-eruptive temperature (1,000–1,110°C). The crystallization age for the plagioclase is inferred from intersection between the curves *wr* and $melt_{pl}$: 3,000 years. The crystallization age for the pyroxene is inferred from intersection between the curves *wr* and $melt_{px}$. In this case, there are two possible solutions: 1,400 or 5,800 years. Modified from (Cooper & Reid, 2008). Reprinted from "*Earth and Planetary Science Letters*", **213**, Kari M. Cooper and Mary R. Reid, Re-examination of crystal ages in recent Mount St. Helens lavas: Implications for magma reservoir processes, p. 19, 2003, with permission from Elsevier.

complex crystal history with either the growth of young rims from the basaltic magma on old crystals, a mixture of old and young crystals, or a combination of the two (Charlier & Zellmer, 2000; Cooper & Reid, 2003; Turner *et al.*, 2003).
• Mineral ages from intermediate magma compositions are frequently within error of their eruption age. This suggests that the proportion of old crystals is less important in intermediate compositions than for mafic magmas. Another implication is that the complicated crystal morphologies (frequent zoning) observed in intermediate rock compositions is produced on a short timescale. These observations have been interpreted to reflect

an open system differentiation where individual andesitic magma batches crystallize successively.
• Mineral ages in felsic rock compositions range from within error of the eruption age up to several hundreds of thousands of years prior to eruption. Where short pre-eruption ages are observed, this has been interpreted as the remobilization of magmas and multi-stage ascent histories (Cottrell *et al.*, 1999). Large pre-eruption ages illustrate the time required for some large magmatic systems (e.g. Long Valley) to produce highly silicic magmas. This may reflect the time required to 'heat up' and partially melt the country rock, as crustal assimilation is likely to play a significant role in the production of silicic magmas (see next section). Before mineral ages could be inferred from U-series isotopes, Halliday *et al.* (1989) used the Rb-Sr system to infer that, at Long Valley, several magma bodies differentiated over timescales of >100,000 years. Similarly, Reagan *et al.* (2003) have shown that the generation of rhyolites at North American continental arcs may require timescales of differentiation >100,000 years. The existence of such long-lived magma bodies has been actively debated (Halliday, 1990; Mahood, 1990; Sparks *et al.* 1990). Sparks *et al.* (1990) suggested that instead rhyolites at Long Valley would be produced by rapid differentiation following re-melting of granitic plutons by mafic melts. Halliday (1990) and Mahood (1990) opposed this model, pointing out for instance that low strontium (Sr) concentrations in rhyolites preclude significant assimilation of granitic melts. Instead, Mahood (1990) proposed a model where the magma body is isolated from the country rock by a partially to completely crystallized rind that can potentially supply minerals to the magma body, in response to changes in thermal and/or physical conditions (e.g. magma recharge). In both models (Sparks *et al.* or Mahood), magma differentiation occurs as an open-system process, where take place:

  i assimilation of granitic melts in Sparks *et al.* (1990)'s models, of previously crystallized magma in Mahood (1990)'s model); and
  ii magma recharge.

Using the compilation in Figure 8.4 and Table 8.1, it is possible to investigate the differences in timescales of differentiation in various tectonic environments:

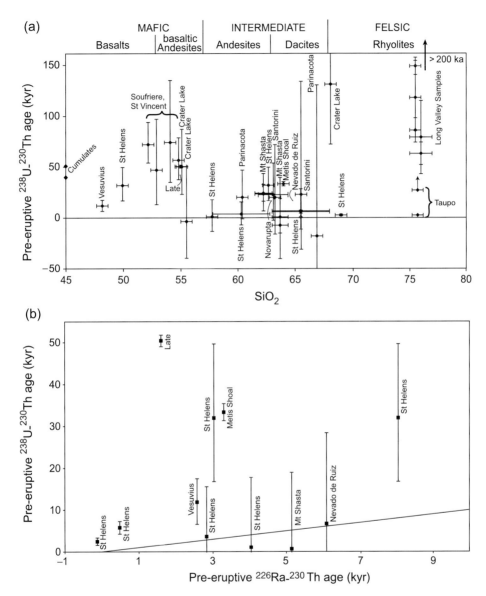

Fig. 8.4 U-series mineral ages re-calculated by Zellmer *et al.* (2005) from published data (see references in Zellmer *et al.* (2005)). (a) Pre-eruptive mineral ages from measured ^{238}U-^{230}Th disequilibrium vs SiO$_2$ content. In primitive lavas, ^{238}U-^{230}Th ages are significantly larger than eruption ages, suggesting that volcanic centers erupting primitive lavas may not differentiate in a closed system. Intermediate rocks display ^{238}U-^{230}Th ages within error of the eruption ages. Felsic rocks display ^{238}U-^{230}Th ages that range from values within error of the eruption age to several 100,000 years. (b) Comparison between mineral ages derived from ^{238}U-^{230}Th and ^{226}Ra-^{230}Th disequilibrium Significant discrepancies can be observed, such as at Mt St Helens and Metis Shoal, illustrating the complex history of minerals in a magma body, where young rims can crystallize over old mineral cores. Reprinted from "*Journal of Volcanology and Geothermal Research*", **140**, G.F. Zellmer, C. Annen, B.L.A. Charlier, R.M.M. George *et al.* Magma evolution and ascent at volcanic arcs: Constraining petrogenetic processes through rates and chronologies, p. 21, 2005, with permission from Elsevier.

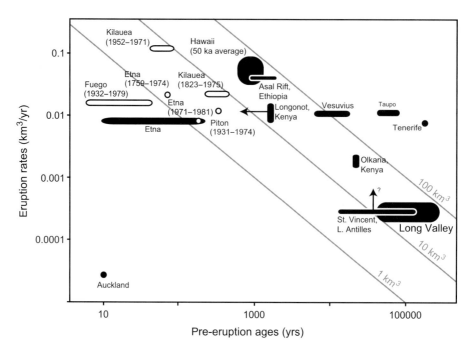

Fig. 8.5 Eruption rates vs pre-eruption ages. Pre-eruption ages are constrained from estimates of magma output rate and magma chamber volume (open symbols; Pyle, (1992)) or derived from geochronological studies of minerals and whole rocks (black symbols; Zellmer *et al.* (2005) and references therein). Note that volumes of magma erupted vary by more than three orders of magnitude between individual volcanic centres (gray contours; in km³). Reprinted from *"Journal of Volcanology and Geothermal Research"*, **140**, G.F. Zellmer, C. Annen, B.L.A. Charlier, R.M.M. George *et al.* Magma evolution and ascent at volcanic arcs: Constraining petrogenetic processes through rates and chronologies, p. 21, 2005, with permission from Elsevier.

• *Subduction zones.* Timescales of differentiation are very variable: timescales as short as 150 years have been inferred for pyroxenes from Mt St Helens (Cooper & Reid, 2003) and as long as 250,000 years for the Taupo Volcanic Zone (New Zealand; (Brown & Fletcher, 1999)). Note that the volumes of these two systems are very different and that the latter timescale is inferred from zircons that could reflect re-mobilization of old zircons into a younger magma. There is no obvious systematic difference between island and continental arcs, although rhyolitic compositions at continental arcs appear to require magma differentiation to occur over large timescales (>100,000 years; Reagan *et al.* (2003)).

• *Intra-continental magmatism.* Timescales of differentiation are often relatively short: less than several 10,000 years, except for zircon studies from large volume silicic systems (e.g. Long Valley, USA; (Reid *et al.*, 1997)).

• *Mid-ocean ridges.* Very few studies have been undertaken at mid-ocean ridges. Available results suggest short timescales, between 0.7 and 4 kyr.

• *Hotspots.* Timescales of differentiation appear relatively short: from as little as 25 years (Piton de la Fournaise, Reunion Island; (Sigmarsson *et al.*, 2005)) up to 4.7 kyr (Fogo, Azores; (Snyder *et al.*, 2007)).

At a global scale, there is a broad negative correlation between the rate of eruption at a given volcano and pre-eruption ages of crystals erupted (Figure 8.5). This shows that, on the one hand, where the eruption rate is high (e.g. Kilauea volcano in Hawaii), magmas produced in the mantle undergo little modification and crystals are produced shortly before eruption. In contrast, crystals can reside for several 100,000 years where highly silicic magmas are produced and eruption rates are low (e.g. Long Valley, USA).

TIMESCALES OF DIFFERENTIATION FROM WHOLE ROCK SAMPLES

Mineral ages, as derived from U-series isotope disequilibrium, provide insights into the dynamics of magma bodies (open vs closed system; old crystal re-mobilization, etc.), but are very often difficult to interpret. An alternative approach is to use the U-series isotope composition of whole rock samples, which span a range of compositions, in order to place time constraints on the rates of magma differentiation. Because uranium, thorium and radium are all very incompatible elements, their budget in a porphyritic volcanic rock is dominated by the groundmass, a proxy for the magma composition, rather than the phenocrysts.

Several observations can be made from U-series isotope data from whole rock samples:
• ^{238}U-^{230}Th and ^{226}Ra-^{230}Th radioactive disequilibria are largest in the more mafic rocks (i.e. $(^{238}U/^{230}Th)$ and $(^{226}Ra/^{230}Th)$ activity ratios showing the greatest deviation from 1) (Fig. 8.6). Because mafic rocks reflect magma compositions that have undergone no or little differentiation, this implies that radioactive disequilibrium in volcanic rocks is not acquired during magma differentiation in the crust, but during magma production in the mantle (see Bourdon's chapter). This also implies that magma transfer to the surface is rapid (10's to 100's of m/yr) in order to preserve these disequilibria (see Turner and Bourdon's chapter).
• $(^{238}U/^{230}Th)$ and $(^{226}Ra/^{230}Th)$ activity ratios in whole rocks are generally correlated with indexes of differentiation (Figure 8.6). Common indexes of differentiation are SiO_2 and Th contents, which increase during differentiation, or MgO content and Mg#, which decrease during differentiation. During differentiation, activity ratios typically converge towards unity, i.e. radioactive disequilibria decay away.

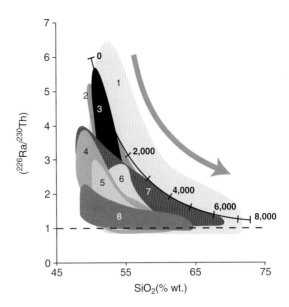

Fig. 8.6 Compilation of whole rock $(^{226}Ra/^{230}Th)$ activity ratio vs SiO_2 content in island and continental arc volcanic rocks. The general decrease in $(^{226}Ra/^{230}Th)$ during differentiation (increasing SiO_2 content; gray arrow) can be used to infer duration of differentiation and crystallization rate, assuming closed system differentiation and little fractionation between Ra and Th during crystal fractionation. For instance, the thick curve shows the compositions during differentiation of a primary magma with $(^{226}Ra/^{230}Th) = 6$ and $SiO_2 = 50$ wt. %. The curve is labeled in years of differentiation. The crystallization rate is set at 10^{-4} g/g/yr and bulk solid/melt partition coefficients for Ra, Th and SiO_2 are taken to be respectively 0.001, 0.01 and 0.5. Fields show data from different arcs: 1: Tonga-Kermadec (Turner & Hawkesworth, 1997); 2: Vanuatu (Turner *et al.*, 1999); 3: Marianas (Elliott *et al.*, 1997); 4: Sunda (Turner & Foden, 2001); 5: Nicaragua (Reagan *et al.*, 1994); 6: Lesser Antilles (Turner *et al.*, 1996; Chabaux *et al.*, 1999); 7: New Britain/Bismarck (Gill *et al.*, 1993); 8: Aleutians (George *et al.*, 1999). See Plate 8.6 for a colour version of this image.

Various mechanisms can explain the decrease in radioactive disequilibrium with increasing differentiation:
• *Time.* Radioactive disequilibrium produced during magma genesis will decay during transfer towards the surface and differentiation prior to eruption. If no other process modifies the radioactive

disequilibrium, i.e. if the magma body behaves as a closed system and differentiation (crystal fractionation) does not significantly affect the distribution of each isotope in the magma, the decrease in radioactive disequilibrium can be used to calculate a duration of differentiation (how long it takes for the magma to evolve from the most mafic to the most felsic composition) and a rate of crystallization. For instance, the abundance of a radionuclide, N_i, will vary with time as follows:

$$\frac{dN_i}{dt} = -f.(D_i - 1).N_i - \lambda_i.N_i + \lambda_{i-1}.N_{i-1}$$

where N_i and N_{i-1} are the concentrations of the radionuclide N_i and its parent N_{i-1}, respectively, f the crystallization rate (in g/g/yr), D_i the bulk solid/melt partition coefficient for N_i, and λ_{Ni} and λ_{Ni-1} the decay constants (in yr^{-1}) for nuclides N_i and N_{i-1}, respectively. In the case of ^{238}U or ^{232}Th, because these radionuclides have no parent, the last term of the right-hand side of the equation disappears. The equation can be written for any element, by simply removing the decay terms (i.e. the last two terms on the right-hand side of the equation).

• As an example, if we write this equation for ^{226}Ra, ^{230}Th and SiO_2, it is possible to solve this system of equations by setting initial abundances (the $(^{226}Ra/^{230}Th)$ activity ratio and SiO_2 concentration in the primary magma) and the crystallization rate to reproduce data for a given suite of rocks, such as those shown on Figure 8.6. In Figure 8.6, the model curve shows that it is possible to reproduce the compositions observed in the Tonga-Kermadec arc using a crystallization rate of 10^{-4} g/g/yr and a duration of differentiation up to 8,000 years. Note that this model requires knowledge of bulk solid/melt partition coefficients for Ra, Th and SiO_2. These can be calculated using experimental and theoretical mineral/melt partition coefficients (Blundy and Wood 2003) and the crystallizing mineral assemblage inferred from modeling of major element variations in the rocks. In the example in Figure 8.6, because Ra and Th are both very incompatible (which is the case in most common mineral assemblages), crystal fractionation has little effect on the $(^{226}Ra/^{230}Th)$ ratio. In fact, because Th is slightly more compatible than Ra in a gabbroic

assemblage, the activity ratio should increase slightly during differentiation from 6 to 6.02.

• *Crystal fractionation.* As seen in Section 2.1, $(^{238}U/^{230}Th)$ and $(^{226}Ra/^{230}Th)$ activity ratios in the magma can be modified during differentiation, if U, Th and Ra enter the structure of the minerals crystallizing. For instance, if we use the example shown on Figure 8.6, where a magma differentiates with a starting composition: $(^{226}Ra/^{230}Th) = 6$ and $SiO_2 = 50$ wt. %, the curve in Fig. 8.6 is calculated assuming little or no fractionation between Ra and Th, such that the decrease in $(^{226}Ra/^{230}Th)$ with increasing SiO_2 is solely accounted for by radioactive decay (i.e. *time*). The alternative scenario is where all of the variation in $(^{226}Ra/^{230}Th)$ is attributed to crystal fractionation. In this case, differentiation has to be very rapid relative to the ^{226}Ra half-life (1,622 years), i.e. a crystallization rate $>10^{-2}$ g/g/yr. In order to account for observed trends in Figure 8.6 by crystallization only, the bulk partition coefficient of Ra would need to be as high as 2, which is practically impossible (cf. Blundy & Wood, 2003). Considering a more likely value for the Ra bulk partition coefficient (e.g. 0.5), the predicted decrease in $(^{226}Ra/^{230}Th)$ with increasing SiO_2 is less than that observed in most volcanic arc volcanoes. To reproduce the decrease that is observed, the model requires a crystallization rate of 2×10^{-4} g/g/yr and a duration of differentiation of ~4,000 years. This highlights that crystal fractionation alone cannot account for all of the variation in $(^{226}Ra/^{230}Th)$, and even in the extreme scenario where Ra and Th fractionate during crystallization, a duration of differentiation of several 1,000 years is required.

• *Magma mixing.* The inverse trend between $(^{226}Ra/^{230}Th)$ and SiO_2 content (Figure 8.6) suggests that differentiation occurs over a timescale of a few 1,000 years. On this timescale, $(^{238}U/^{230}Th)$ activity ratios should remain constant for a suite of co-genetic rocks because radioactive decay over timescales of a few 1,000 years will not affect significantly the $(^{238}U/^{230}Th)$ ratio. However, data from single volcanoes often show significant variations in $(^{238}U/^{230}Th)$. In addition, at some volcanic arcs, a decrease in $^{238}U-^{230}Th$ disequilibrium with increasing indexes of differentiation is also observed, suggesting timescales for differentiation $>10,000$ years. Hence, it appears that $^{238}U-^{230}Th$

data suggest timescales of differentiation >10,000 years, whilst [226]Ra-[230]Th data suggests timescales <10,000 years (Figure 8.4b; Zellmer *et al.* (2005)). One possible explanation for these apparently contradictory observations is that the range of composition observed in co-genetic volcanic rocks result from mixing between evolved melts that resided in the magma chambers for timescales >300,000 years (hence in secular equilibrium for [238]U-[230]Th-[226]Ra) and primary melts that carry the radioactive disequilibrium generated during magma genesis. This model is similar to the 'hot zone model' proposed by Annen & Sparks (2002): in their model, silica magmas are produced as a result of repetitive intrusion of basaltic magmas in the crust. During this process, silicic compositions result from (i) residual melt from crystal fractionation and (ii) crust melting. However, it takes several 100,000 years before significant amounts of silicic melts can be produced, during which time they would have returned to secular equilibrium for [238]U-[230]Th-[226]Ra. Thus, mixing between old silica melts and primary basaltic magmas could be a viable process to account for the decrease in U-series disequilibria with increasing silica observed in volcanic rocks. Indeed, there has been a long-standing debate whether co-variations of geochemical parameters (e.g. major element concentrations) with indexes of differentiation (e.g. SiO_2 content) reflect fractional crystallization or magma mixing. This is because when these trends (termed *liquid lines of descent*) are linear, they can be potentially explained by either process. However, linear trends are not always observed and inflections in the liquid lines of descent are common. These inflections cannot be explained by two end-member mixing but more likely to result from a change in the mineral phases crystallized during differentiation.

Although magma mixing is unlikely to explain the range in U-series isotope composition observed in whole rock samples, it is important to remember that magmas may not behave as closed systems fractionating crystals. In an open system magma body, the two main sources of 'contamination' are:

1 melts derived from partial melting of the country rock (*crustal assimilation*); and

2 primary, undifferentiated magma (*magma recharge*).

In Figure 8.7, the effect of different models of differentiation on U-series isotope composition is explored. First, in the case of closed system differentiation, ([226]Ra/[230]Th) reaches secular equilibrium before ([230]Th/[238]U) becomes significantly modified. This is inconsistent with the combined [226]Ra-[230]Th and [238]U-[230]Th observations in volcanic rocks. Next we consider different models of open system differentiation. When a fraction of the magma chamber volume is frequently replenished with primary magmas (i.e. 5% of the chamber's mass every 100 years), the effect is to buffer the radioactive disequilibria such that even after several 1,000 years significant [226]Ra-[230]Th can be maintained (Fig. 8.8; Hughes & Hawkesworth (1999)). Thus, frequent magma recharge could explain how [226]Ra-[230]Th disequilibria can be preserved for timescales >8,000 years (the time after which any [226]Ra-[230]Th disequilibrium has decayed away, assuming closed system behavior). However, because basaltic magma is continuously injected, it is difficult to produce silicic compositions (e.g. dacites, rhyolites). For instance, using the parameters from Figure 8.7, if the primary magma has an SiO_2 content of 50 wt.% and assuming a bulk solid/melt partition coefficient of 0.5 for SiO_2, after 8,000 years and crystallization of 80% of the initial magma body, SiO_2 has only increased to 52.4 wt.%. Although residual magmas from crystallization represent as much as 90% of the mixture, during each recharge event, not enough time (100 years) has elapsed to produce silicic residual magmas. Alternatively, allowing longer periods of time between each recharge event can yield silicic magmas, but the consequence is that any [226]Ra-[230]Th disequilibrium will have returned to secular equilibrium.

An alternative model is one in which there is no magma recharge, but assimilation of crust-derived melts occurs (i.e. partial melting of the country rock into the magma). Berlo *et al.* (2004) have shown that there is likely to be little or no radioactive disequilibrium produced during partial melting of the crust. In Figure 8.7, we can see that assimilation of such melts rapidly drives compositions towards secular equilibrium,

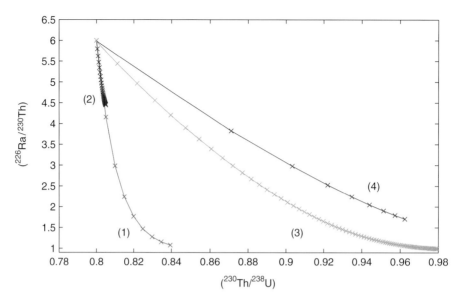

Fig. 8.7 Different models of magma differentiation are considered to explain the negative relationship commonly observed between $(^{226}Ra/^{230}Th)$ and $(^{230}Th/^{238}U)$ in volcanic rocks. Model (1): closed-system differentiation. Crystallization rate = 10^{-4} g/g/yr. Tick marks are every 1,000 years. Models (2)–(4): open-system differentiation. Model (2): discrete magma recharge. Every 100 years, 10% of the mass of the magma chamber is replenished with primary magma and an equivalent volume is erupted (in order to keep the size of the magma chamber constant). Crystallization rate = 10^{-4} g/g/yr. Tick marks are every 100 years. Model (3): crustal assimilation. Crystallization rate = 10^{-4} g/g/yr. The rate of crustal assimilation is set to be 30% of the crystallization rate ($r = 0.3$; DePaolo (1981)), i.e. 3×10^{-5} g/g/yr. Concentrations in the crustal melt are: $[U]_c = 2$ ppm, $[Th]_c = 5$ ppm. Tick marks are every 100 years. Curves (1)–(3) end at a duration of differentiation of 8,000 years. Model (4): discrete magma recharge + assimilation. This model is a combination of Models 2 and 3. Crystallization rate = 10^{-3} g/g/yr. $r = 0.3$, i.e. an assimilation rate of 3×10^{-4} g/g/yr. Same concentrations in the crustal melt as in Model (3). Tick marks are every 100 years. The curve ends at a duration of differentiation of 800 years. Common parameters to all four models: total fraction crystallized = 80%. $D_U = 0.05$, $D_{Th} = 0.01$, $D_{Ra} = 0.25$. Concentrations in the primary magma are: $[U]_m = 0.1$ ppm, $[Th]_m = 0.2$ ppm. Activity ratios in the primary magma are: $(^{230}Th/^{238}U) = 0.8$, $(^{226}Ra/^{230}Th) = 6$.

without requiring extensive amounts of time. Dosseto *et al.* (2008) found that, in a model involving assimilation, it only takes 1,500 years to produce dacitic compositions at Mt St Helens, compared with 2,000 to 2,500 years in a closed system model. This model explains simultaneously the observed decreasing $(^{226}Ra/^{230}Th)$ and $(^{230}Th/^{238}U)$ disequilibria during differentiation and also allows for the production of silicic compositions. However, this model predicts that silicic magmas will be near or at secular equilibrium, which contrasts with significant radioactive

disequilibria observed in most andesitic and dacitic volcanic rocks.

A more complex model is one in which both crustal assimilation and discrete magma recharge occur together (Figure 8.7, curve (4); equations are given in Appendix B). In this case, it is possible not only to reproduce the simultaneous decrease in $(^{226}Ra/^{230}Th)$ and $(^{230}Th/^{238}U)$ disequilibria during differentiation, the silicic compositions, but also to obtain silicic magmas with significant U-series disequilibrium. Note that this model requires fast crystallization rates (10^{-3} g/g/yr) and

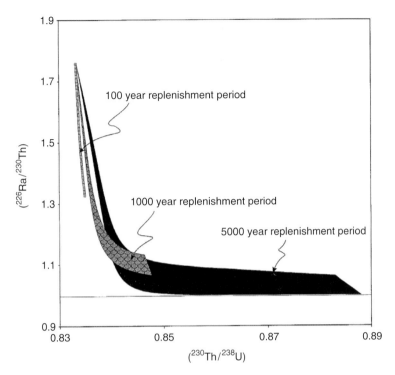

Fig. 8.8 Combined U-Th-Ra systematics demonstrating the effects of different replenishment timescales. The difference in half-life between ^{226}Ra and ^{230}Th is sufficiently large that if the timescale is short enough to produce significant effects on $(^{226}$Ra$/^{230}$Th$)$, then $(^{230}$Th$/^{238}$U$)$ reaches a steady-state close to that of the parental magma. The model conditions used were: $(^{230}$Th$/^{232}$Th$)_{init}$ = 0.75, $(^{238}$U$/^{232}$Th$)$ = 0.9, $[$Th$]_{init}$ = 2 ppm, $[^{226}$Ra$]_{init}$ = 300 fg g^{-1}, $[$Ba$]_{init}$ = 200 ppm, $D_U = D_{Th}$ = 0.05, $D_{Ra} = D_{Ba}$ = 0.25, Chamber volume = 50 km^3, Eruption volume = 5 km^3, Crystallization = 10% per iteration. Reprinted from *"Geochimica et Cosmochimica Acta"*, [63(23/24)], R. D. Hughes and C. J. Hawkesworth. The effects of magma replenishment processes on 238U–230Th disequilibrium, [4101–4110], 1999, with permission from Elsevier.

dacitic compositions are obtained after only 800 years. This has important implications for the dynamics of magma differentiation and this is discussed below.

To summarise, a number of mechanisms can affect the U-series isotope composition of magmas during differentiation: time, crystal fractionation, crustal assimilation and primary magma recharge. Thus, the ability of different models (closed-system crystallization, open-system magma recharge, crustal assimilation and crystallization) to explain whole rock U-series isotope data can be tested using the timescale constraints. These models need to be able to reproduce:

• The common observation of a simultaneous decrease of $(^{226}$Ra$/^{230}$Th$)$ and $(^{230}$Th$/^{238}$U$)$ disequilibria during differentiation; and
• The occurrence of significant $(^{226}$Ra$/^{230}$Th$)$ and $(^{230}$Th$/^{238}$U$)$ disequilibria in silicic magmas (andesitic or dacitic).

Open system differentiation seems to be the most adequate model. It probably involves frequent recharge of a magma body undergoing both fractional crystallization and assimilation of the country rock. Inferred timescales of differentiation are very rapid and it takes less than a few tens of thousands years to produce dacitic compositions from a basaltic magma.

PHYSICAL IMPLICATIONS FOR MAGMA DIFFERENTIATION

In the previous section, we have seen that in order to explain U-series isotope data in co-genetic suites of volcanic rocks:

• Differentiation is often required to occur over very short timescales (less than a few 1,000 years); and

• Crustal assimilation, probably combined with magma recharge, is likely to play a significant role during differentiation.

Note that the role of crustal assimilation has not always been readily identified in the past, since common approaches (e.g. Sr isotopes) required the country rock to be isotopically distinct from the primary magmas (i.e. at least 100 million years old). This is not required when using U-series isotopes. Thus, on the basis of the whole rock U-series isotope data, we suggest that crustal assimilation may be more common than previously believed (Handley *et al.*, 2008).

Annen & Sparks (2002) have shown that repetitive magmatic intrusions over several 100,000 years are required in order to 'heat up' the country rock to the point where it can produce significant amounts of crustal melts that can become assimilated into a differentiating magma. This suggests that below volcanic centers erupting silicic magmas, there has been repetitive magmatic intrusions for several 100,000 years. Nevertheless, once this 'incubation time' is achieved, the differentiation from primary basaltic to silicic compositions only takes less than a few 1,000 years, as suggested by U-series data. This latter timescale can be used to infer the depth where differentiation occurs. The geometry of the magma body is a major control on the rate of magma cooling (hence, differentiation) and needs to be considered in order to constrain the depth of magma differentiation. Spherical magma bodies (aspect ratio, AR = 1) are slower than sheet-like magma bodies (AR < 1). Thus, in the example of Mt St Helens, for a 10 km^3 spherical magma body to produce dacitic compositions in ~2,000 years (as inferred from whole rock U-series isotope data modeling), it should be located at a depth of ~8 to 10 km, which is consistent with geophysical imaging (Dosseto *et al.*, 2008). Sheet-like magma bodies (i.e. *sills*) can cool and differentiate much faster, hence could be located deeper in the crust. Nevertheless, timescales of differentiation as short as a few 100 years (see Figure 8.7, curve (4)) may be incompatible with differentiation in the lower crust, where hot country rock inhibits fast cooling.

CONCLUDING REMARKS

Studies of the U-series isotope composition of minerals suggest a wide range of ages: from a few tens of years to several hundred thousand years. In general, short timescales are inferred at mid-ocean ridges and hotspots, longer timescales at subduction zones and for intra-continental volcanism. However, in many cases, long timescales inferred from minerals may reflect the re-mobilization of old crystals, produced during previous cycles of crystallization, into a young magma. Nevertheless, old crystal ages suggest that some volcanic centers have undergone magma replenishment and crystallization over several hundreds of thousands of years. These long timescales are generally associated with larger volume volcanoes erupting silicic compositions (dacites, rhyolites, etc.; see Bachmann, 2010, this volume). They may reflect the time required to produce such magma compositions, because large periods of time are required to heat up the country rock and produce significant amounts of crustal melt.

Studies of the U-series isotope composition of whole rocks can be used to constrain the dynamics of magma bodies. It is likely that magma differentiation is an open-system process in most cases, where recharge of the magma body is combined with fractional crystallization and assimilation of crustal melts also plays a significant role. Modeling of existing data suggest that the range of magma compositions erupted require fast differentiation (a few hundred years to a few thousand years) with frequent magma recharge and crustal assimilation. This generally favors a model where magmas differentiate in the upper crust (upper 10 km), in a cool country rock, unless it can be otherwise demonstrated that fast differentiation is possible in sills emplaced in the lower crust.

APPENDIX A – CALCULATION OF CRYSTALLIZATION AGES USING ^{226}RA-^{230}TH

As explained in the text, using ^{226}Ra-^{230}Th in minerals to calculate crystallization ages requires accounting for differences in chemical behavior of Ra and Ba. Cooper & Reid (2003) suggested a method where the evolution of the composition of the melts in equilibrium with the minerals analysed is calculated and compared to the calculated evolution of the composition of the groundmass (see text for details). The crystallization age can be inferred graphically, as shown on Fig. 8.3b, or calculated analytically. Here, we give the equations to calculate the evolution of the composition in the minerals, the groundmass, and the melts in equilibrium with the minerals (Cooper & Reid, 2003).

Note that the second equation can be used to calculate the evolution of the groundmass composition.

The model crystallization age for phase A is the time at which the melt in equilibrium with phase A had the same Ra/Ba as the groundmass:

$$\left(\frac{D_{Ba}}{D_{Ra}}\right)_A \left\{ \left[\frac{(^{226}Ra)_{meas, A}}{[Ba]_{meas, A}}\right](e^{\lambda 226^t}) - \left[\frac{(^{230}Th)_{meas, A}}{[Ba]_{meas, A}}\right](e^{\lambda 226^t} - 1) \right\}$$
$$= \left[\frac{(^{226}Ra)_{meas, GM}}{[Ba]_{meas, GM}}\right](e^{\lambda 226^t}) - \left[\frac{(^{230}Th)_{meas, GM}}{[Ba]_{meas, GM}}\right](e^{\lambda 226^t} - 1)$$

Solving for t yields the following equation to calculate model crystallization ages:

$$t_{GM-A} = \frac{1}{\lambda_{226}} \ln \left(\frac{\left[\frac{(^{230}Th)_{meas, GM}}{[Ba]_{meas, GM}}\right] - \left(\frac{D_{Ba}}{D_{Ra}}\right)_A \left[\frac{(^{230}Th)_{meas, A}}{[Ba]_{meas, A}}\right]}{\left(\frac{D_{Ba}}{D_{Ra}}\right)_A \left\{ \left[\frac{(^{226}Ra)_{meas, A}}{[Ba]_{meas, A}}\right] - \left[\frac{(^{230}Th)_{meas, A}}{[Ba]_{meas, A}}\right] \right\} - \left[\frac{(^{226}Ra)_{meas, GM}}{[Ba]_{meas, GM}}\right] + \left[\frac{(^{230}Th)_{meas, GM}}{[Ba]_{meas, GM}}\right]} \right)$$

The $(^{226}Ra)/[Ba]$ ratio of a melt in equilibrium with a mineral A at the time of crystallization (t) is given by:

$$\left[\frac{(^{226}Ra)}{[Ba]}\right]_{melt, t} = \frac{\left[(^{226}Ra)_A \Big/ D_{Ra}^A\right]}{\left[[Ba]_A \Big/ D_{Ba}^A\right]} = \left[\frac{(^{226}Ra)}{[Ba]}\right]_{A, t}\left[\frac{D_{Ba}}{D_{Ra}}\right]_A$$

where $(^{226}Ra)/Ba$ in mineral A at time t (years from present; ybp), is calculated from the measured ratio at present (subscript '*meas*') by the standard decay equation:

$$\left[\frac{(^{226}Ra)}{[Ba]}\right]_{A, t} = \left[\frac{(^{226}Ra)_{meas, A}}{[Ba]_{meas, A}}\right](e^{\lambda 226^t})$$
$$- \left[\frac{(^{230}Th)_{meas, A}}{[Ba]_{meas, A}}\right](e^{\lambda 226^t} - 1)$$

APPENDIX B – DISCRETE MAGMA RECHARGE, CONTINUOUS CRUSTAL ASSIMILATION AND CRYSTALLIZATION

In this model, the evolution of a nuclide abundance in the differentiating magma body is subjected to the effect of crystallization and crustal assimilation, and is described by the following equations:

For ^{238}U and ^{232}Th (which have no parent nuclide):

$$\frac{dN_{i,m}}{dt} = r.f.N_{i,a} - f(r + D_i - 1) - \lambda_i.N_{i,m}$$

where $N_{i,m}$ is the abundance of the nuclide i in the magma, r is the ratio assimilation rate over crystallization rate, f is the crystallization rate (in g/g/yr), D_i the bulk solid/melt partition coefficient and λ_i, the decay constant (in yr^{-1}) for the nuclide i.

For ^{230}Th and ^{226}Ra:

$$\frac{dN_{i,m}}{dt} = r.f.N_{i,a} - f(r + D_i - 1) - \lambda_i.N_{i,m} + \lambda_{i-1}.N_{i-1,m}$$

where λ_{i-1} is the decay constant (in yr^{-1}) and $N_{i-1,m}$ the abundance of the parent nuclide, $i-1$.

During the first increment of differentiation, a nuclide i has an initial abundance in the magma $N_{i,prim}$ and its abundance evolves following the equations given above. At a given time, $T_{recharge}$, which is the period of recharge, the magma that has undergone differentiation (crystallization + crustal assimilation) is mixed with a primitive magma. The abundance of a nuclide i in the mixture is given by the following equation:

$$N_{i,mix} = X_{prim}.N_{i,prim} + (1 - X_{prim}).N_{i,m}$$

where X_{prim} is the ratio of the mass of primitive magma to the mass of the system 'residual magma + primitive magma'. $N_{i,m}$ is the abundance of nuclide i in the residual magma after differentiation for a duration $T_{recharge}$.

The composition of the mixture, $N_{i,mix}$, then evolves for a duration $T_{recharge}$ following the first two equations above, until a new batch of primitive magma is added and the composition of the new mixture is calculated using the equation above. The number of recharge events is calculated as follows:

$$n_{recharge} = \frac{F}{f.T_{recharge}}$$

where F is the total fraction of magma that has crystallized.

REFERENCES

Allègre CJ. 1968. ^{230}Th dating of volcanic rocks: a comment. *Earth and Planetary Science Letters* **5**: 209–210.

Annen C, Sparks RSJ. 2002. Effects of repetitive emplacement of basaltic intrusions on thermal evolution and melt generation in the crust. *Earth and Planetary Science Letters* **203**(3–4): 937–955.

Berlo K, Turner, S, et al. 2004. The extent of U-series disequilibria produced during partial melting of the lower crust with implications for the formation of the Mt St Helens dacites. *Contributions to Mineralogy and Petrology* **148**(1): 122–130.

Blundy J, Wood B. 1994. Prediction of crystal-melt partition coefficients from elastic moduli. *Nature* **372**(6505): 452–454.

Blundy J, Wood B. 2003. Mineral-melt partition of Uranium, Thorium and their daughters. In: *Uranium-series Geochemistry*, Bourdon, B, Henderson GM, Lundstrom CC, and Turner SP (eds), Geochemical Society – Mineralogical Society of America, Washington DC **52**: 59–123.

Bourdon B, Elliott T. 2010. Melt production in the mantle: constraints from U-series. In: *Timescales of Magmatic Processes: From Core to Atmosphere*, Dosseto, A, Turner S, Van Orman JA (eds), Wiley-Blackwell, Oxford.

Bourdon B, Henderson, GM, et al. 2003. *Uranium-series Geochemistry*. Washington, Geochemical Society – Mineralogical Society of America, Washington DC.

Brown SJA, Fletcher IR. 1999. SHRIMP U-Pb dating of the pre-eruption growth history of zircons from the 340 ka Whakamaru Ignimbrite, New Zealand: Evidence for >250 ky magma residence times. *Geology* **27**(11): 1035–1038.

Cashman K, Blundy J. 2000. Degassing and crystallization of ascending andesite and dacite. *Philosophical Transactions: Mathematical, Physical and Engineering Sciences* **358**(1770): 1487–1513.

Chabaux F, Hémond C, et al. 1999. ^{238}U-^{230}Th-^{226}Ra disequilibria in the Lesser Antilles arc: Implications for mantle metasomatism. *Chemical Geology* **153**: 171–185.

Charlier B, Zellmer G. 2000. Some remarks on U-Th mineral ages from igneous rocks with prolonged crystallisation histories. *Earth and Planetary Science Letters* **183**(3–4): 457–469.

Condomines M. 1997. Dating recent volcanic rocks through ^{230}Th-^{238}U disequilibrium in accessory minerals: Example of the Puy de Dôme (French Massif Central). *Geology* **25**(4): 375–378.

Cooper KM, Reid MR. 2003. Re-examination of crystal ages in recent Mt St Helens lavas: Implications for magma reservoir processes. *Earth and Planetary Science Letters* **213**(1–2): 149–167.

Cooper KM, Reid MR. 2008. Uranium-series Crystal Ages. *Reviews in Mineralogy and Geochemistry* **69**(1): 479–544.

Cottrell E, Gardner JE, *et al.* 1999. Petrologic and experimental evidence for the movement and heating of the pre-eruptive Minoan rhyodacite (Santorini, Greece). *Contributions to Mineralogy and Petrology* **135**(4): 315–331.

DePaolo DJ. 1981. Trace element and isotopic effects of combined wall-rock assimilation and fractional crystallization. *Earth and Planetary Science Letters* **53**: 189–202.

Dosseto A, Turner SP, *et al.* 2008. Uranium-series isotope and thermal constraints on the rate and depth of silicic magma genesis. Dynamics of crustal magma transfer, storage and differentiation. Annen C, Zellmer G (eds), *London, Geological Society*. **304**: 169–181.

Elliott T, Plank, T, *et al.* 1997. Element transport from subducted slab to juvenile crust at the Mariana arc. *Journal of Geophysical Research* **102**: 14, 991–15, 019.

Fukuoka T. 1974. Ionium dating of acidic volcanic rocks. *Geochemistry Journal* (Nagoya) **8**(3): 109–116.

George RM, Turner S, *et al.* 1999. Along-arc U-Th-Ra systematics in the Aleutians. AGU Fall meeting, San Francisco, American Geophysical Union.

Gill JB, Morris JD, *et al.* 1993. Timescale for producing the geochemical signature of island arc magmas: U-Th-Po and Be-B systematics in recent Papua New Guinea lavas. *Geochimica et Cosmochimica Acta* **57**: 4269–4283.

Halliday AN. 1990. Reply to comment of R.S.J. Sparks, H.E. Huppert, and C.J.N. Wilson on 'Evidence for long residence times of rhyolitic magma in the Long Valley magmatic system: The isotopic record in pre-caldera lavas of Glass Mountain.' *Earth and Planetary Science Letters* **99**(4): 390–394.

Halliday AN, Mahood GA, *et al.* 1989. Evidence for long residence times of rhyolitic magma in the Long Valley magmatic system: The isotopic record in pre-caldera lavas of Glass Mountain. *Earth and Planetary Science Letters* **94**(3–4): 274–290.

Handley HK, Turner SP, Smith IEM, Stewart RB, Cronin SJ. 2008. Rapid timescales of differentiation and evidence for crustal contamination at intra-oceanic arcs: Geochemical and U-Th-Ra-Sr-Nd isotopic constraints from Lopevi Volcano, Vanuatu, SW Pacific. *Earth and Planetary Science Letters* **273**(1–2): 184–194.

Hughes RD, Hawkesworth CJ. 1999. The effects of magma replenishment processes on ^{238}U-^{230}Th dise-

quilibrium. *Geochimica et Cosmochimica Acta* **63**(23/24): 4101–4110.

Kigoshi K. 1967. Ionium dating of igneous rocks. *Science* **156**(3777): 932–934.

Mahood GA. 1990. Second reply to comment of Sparks RSJ, Huppert HE, Wilson CJN on 'Evidence for long residence times of rhyolitic magma in the Long Valley magmatic system: the isotopic record in the precaldera lavas of Glass Mountain.' *Earth and Planetary Science Letters* **99**(4): 395–399.

McBirney AR, Noyes RM. 1979. Crystallization and layering of the Skaergaard Intrusion. *Journal of Petrology* **20**(3): 487–554.

Miller SA, Burnett, DS, *et al.* 2007. Experimental study of radium partitioning between anorthite and melt at 1 atm. *American Mineralogist* **92**(8–9): 1535–1538.

Peate DW, Chen JH, *et al.* 1996. ^{238}U-^{230}Th Dating of a Geomagnetic Excursion in Quaternary Basalts of the Albuquerque Volcanoes Field, New Mexico (USA). *Geophysical Research Letters* **23**: 2271– 2274.

Pyle DM. 1992. The volume and residence time of magma beneath active volcanoes determined by decay-series disequilibria methods. *Earth and Planetary Science Letters* **112**(1–4): 61–73.

Reagan MK, Morris JD, *et al.* 1994. Uranium series and beryllium isotope evidence for an extended history of subduction modification of the mantle below Nicaragua. *Geochimica et Cosmochimica Acta* **58**(19): 4199–4212.

Reagan MK, Sims, KWW, *et al.* (2003). Timescales of differentiation from mafic parents to rhyolite in North American continental arcs. *Journal of Petrology* **44**(9): 1703–1726.

Reid MR, Coath CD, *et al.* 1997. Prolonged residence times for the youngest rhyolites associated with Long Valley Caldera: 230Th–238U ion microprobe dating of young zircons. *Earth and Planetary Science Letters* **150**(1–2): 27–39.

Rubin KH, Zellmer GF. 2009. Reply to comment on 'On the recent bimodal magmatic processes and their rates in the Torfajökull-Veidivötn area, Iceland,' Cooper KM. *Earth and Planetary Science Letters* **281**(1–2): 115–123.

Sigmarsson O, Condomines M, *et al.* (2005). Magma residence time beneath the Piton de la Fournaise Volcano, Reunion Island, from U-series disequilibria. *Earth and Planetary Science Letters* **234**(1–2): 223–234.

Snyder DC, Widom E, *et al.* 2007. Timescales of formation of zoned magma chambers: U-series disequilibria in the Fogo A and 1563 AD trachyte deposits, São Miguel, Azores. *Chemical Geology* **239**(1–2): 138–155.

Sparks RSJ, Huppert HE, *et al.* 1990. Comment on 'Evidence for long residence times of rhyolitic magma in the Long Valley magmatic system: the isotopic record in precaldera lavas of Glass Mountain,' by Halliday, AN, Mahood, GA, Holden, P, Metz, JM, Dempster TJ, Davidson JP. *Earth and Planetary Science Letters* **99**(4): 387–389.

Spera F. 1980. Thermal evolution of plutons: a parameterized approach. *Science* **207**(4428): 299–301.

Taddeucci A, Broecker WS, *et al.* (1967).^{230}Th dating of volcanic rocks. *Earth and Planetary Science Letters* **3**: 338–342.

Turner S, Bourdon B. 2010. Melt transport from the mantle to the crust – U-series isotopes. In: *Timescales of Magmatic Processes: from Core to Atmosphere*. Dosseto A, Turner S, Van Orman JA (eds), Wiley-Blackwell, Oxford.

Turner S, Foden J. 2001. U, Th and Ra disequilibria, Sr, Nd and Pb isotope and trace element variations in Sunda Arc lavas; predominance of a subducted sediment component. *Contributions to Mineralogy and Petrology* **142**(1): 43–57.

Turner S, George R, *et al.* 2003. Case studies of plagioclase growth and residence times in island arc lavas from Tonga and the Lesser Antilles, and a model to reconcile discordant age information. *Earth and Planetary Science Letters* **214**(1–2): 279–294.

Turner S, Hawkesworth C. 1997. Constraints on flux rates and mantle dynamics beneath island arcs from Tonga-Kermadec lava geochemistry. *Nature* **389**: 568–573.

Turner S, Hawkesworth, C, *et al.* (1996). U-series isotopes and destructive plate margin magma genesis in the Lesser Antilles. *Earth and Planetary Science Letters* **142**: 191–207.

Turner S, Peate DW, *et al.* (1999). Two mantle domains and the time scales of fluid transfer beneath the Vanuatu arc. *Geology* **27**(11): 963–966.

Vigier N., Bourdon B, *et al.* 1999. U-decay series and trace element systematics in the 1978 eruption of Ardoukoba, Asal rift: timescale of magma crystallization. *Earth and Planetary Science Letters* **174**: 81–97.

Winter JD. 2001. An *Introduction to Igneous and Metamorphic Petrology*. Prentice Hall, Upper Saddle River, NJ.

Zellmer GF, Annen C, *et al.* 2005. Magma evolution and ascent at volcanic arcs; constraining petrogenetic processes through rates and chronologies. *Journal of Volcanology and Geothermal Research* **140**(1–3): 171–191.

Zellmer GF, Rubin KH, *et al.* 2008. On the recent bimodal magmatic processes and their rates in the Torfajökull-Veidivötn area, Iceland. *Earth and Planetary Science Letters* **269**(3–4): 388–398.

9 Defining Geochemical Signatures and Timescales of Melting Processes in the Crust: An Experimental Tale of Melt Segregation, Migration and Emplacement

TRACY RUSHMER[1] AND KURT KNESEL[2]

[1]GEMOC ARC National Key Centre, Department of Earth and Planetary Sciences, Macquarie University, Sydney, NSW, Australia
[2]Department of Earth Sciences, University of Queensland, Brisbane, Australia

SUMMARY

Both the mechanism of melt segregation and the nature of the short-range transport paths developed during partial melting can exert significant control over melt geochemistry because these variables influence the degree of equilibrium achieved between the melt and the solid restite. We suggest that the physical process of melt segregation will therefore leave a chemical signature that may be detected in the anatectic melts and the rocks which melted to form them. The experimental studies described below suggest that there are distinct differences in composition between melts derived statically and distributed in crack networks from those formed in permeable shear zones. These differences in composition can be exploited by studies of natural anatectic granites to estimate rates of melting, extraction and crystallization. As these melts rise into the upper crust, assimilation can also become an important contributor to final magma composition. We describe the use of radiogenic isotopes because they are a powerful means of assessing crustal melting and contamination models. However, the evidence presented here for disequilibrium suggests that the isotopic composition of crustal melts may not necessarily be the same as their source. We have seen that the use of trace elements may be problematic as well (e.g. REE in our pelite melting experiments) in that melt compositions may deviate from those predicted using established phase equilibrium and mineral-melt partitioning data. If melting in the crust really is a disequilibrium process, would it not make constraining source and contaminant signatures a less tractable problem? We propose that in accepting that this can happen, we are now in a position to make real progress. We propose that the isotopic and trace element composition of crustal melt is a function not only of the composition of its precursor, but also of the processes by which and rates at which it is formed and extracted. Therefore, if we can establish a general framework for modelling the compositions of crustal

Timescales of Magmatic Processes: From Core to Atmosphere, 1st edition. Edited by Anthony Dosseto, Simon P. Turner and James A. Van Orman.

melts, we then have a powerful tool to constrain the mechanisms and timescales over which melts are extracted or assimilated. The results are consistent with a growing body of theoretical and field evidence indicating that processes, ranging from melting in the lower crust to assimilation in the upper crust, are rapid, efficient and can occur over timescales of decades or less.

INTRODUCTION

A key to understanding the chemical differentiation of continental landmasses, as well as the great diversity of chemical compositions of magmatic rocks, concerns the rates and mechanisms for the formation of melt and its transport and emplacement on or near Earth's surface. Our knowledge of the physical processes controlling melt generation, extraction and accumulation has evolved rapidly over the past decade (Brown, 2004; Brown, 2007; Brown & Solar, 1999; Clemens & Mawer, 1992; Jackson et al., 2003; Vigneresse et al., 2008). Progress has come from collective consideration of insights gained through:
• Studies of structures, textures and compositions of melt source regions (e.g. migmatite terrains), complimented by parallel investigations focusing on zones of magma accumulation (e.g. plutons and volcanoes), each underpinned by continuing advances in analytical geochemistry, geochronology and magnetic fabric measurements;
• Laboratory experimentation of melting behavior and the rheology of partially molten rocks; and
• New approaches to mathematical formulation and numerical simulation of melt generation, extraction and transport processes (Annen et al., 2006, Beaumont et al., 2006; Jackson et al., 2003; Petford et al., 2000; Petford & Koenders, 1998). The emerging picture is one in which accumulation of bodies of magma at shallow crustal levels takes place by a stepwise process governed by rapid and discontinuous extraction and ascent of melt generated at deeper crustal levels. The rate-limiting step in the growth and evolution of shallow magma bodies, which feed volcanic eruptions or solidify to form plutons and batholiths, thus appears to reside at deep structural levels in the crust. This

realization leads us to examine the dynamic nature of the mid- to deep-crustal melting environment, where the generation of intermediate to silicic magma is principally driven by of influx hot mantle-derived magma (Bergantz & Barboza, 2006).

Given that melting is generally thermally driven (although fluid influx can also be important), it is tempting to attribute the apparent pulsed nature of melt extraction to periodic basalt invasion in the lower crust. Although numerical simulations provide insight into the degree and timescales of melting associated with repeated influx of basalt into the mid- to lower crust (Petford & Gallagher, 2001; Annen & Sparks, 2002; Dufek & Bergantz, 2005), poor constraints on emplacement periodicity in nature prevent a direct assessment at present. Nevertheless, an important finding of recent numerical modeling is that, in many cases, residual melt from the incomplete crystallization of the mafic magma may dominate over that produced by partial melting of the crust. One exception is for basalt emplaced into fertile pelitic crust. Annen et al. (2006) have shown, through thermal models combined with experimental data, that injected sills of basaltic composition can produce substantial crustal melt from pelitic sources; such assemblages are also a major control in orogenic settings with no recognized mantle input.

Melting may alternatively be viewed as a more or less continuous and pervasive process, such that piecemeal accumulation of magma reflects discontinuous extraction from the source. This model finds support in the observation that the thermal anomalies associated with repeated basalt influx in the crust take several millions of years to decay through diffusion. It is also consistent with the piecemeal nature of magmatism in orogenic settings lacking direct input from the mantle (e.g. Himalayan leucogranites). Therefore, the timing and tempo of melt segregation and ascent may be ultimately governed by the episodic nature of hydrous phase reactions and the non-linear rheology of partially molten rocks.

We address this problem through laboratory experimentation. Our goal is to examine the complex links between melting environment (e.g. depth, mineralogy, reaction-controlled permeability, presence or absence of deformation) and melt chemistry, and the constraints these place on the rates and timescales of melt extraction. However, in keeping with the multi-faceted approach that has led us to our current understanding of melt generation and migration in the crust, we strive to evaluate these experimental constraints in light of insight gained from multiple means of investigation and by considering the crustal magmatic column from the bottom up.

MELT GENERATION AND MOBILITY IN THE MID- TO DEEP CRUST: INSIGHT THROUGH EXPERIMENTATION

Experimental constraints on the importance of reaction mechanisms and deformation

Partial melting experiments on crustal rocks have helped to advance our understanding of melt segregation rates and melting processes active in the crust. Experiments on natural rocks have clearly provided an important approach determining the variables that control melting, melt segregation and the rates at which these processes can occur. Through experiments we can establish the starting conditions and record through analyses of run products the crucial melting reactions and the melt compositions. The availability of new technology, such as *in-situ* analytical techniques, has greatly enhanced our ability to measure important trace-element signatures that can then be used to interpret rocks that represent continental crustal magmas.

This section reviews experimental results performed on crustal rock cores under hydrostatic conditions, where there is no applied stress to the sample, and under applied stress (deformation), where samples are deformed during partial melting. We also present new work on muscovite and biotite-bearing crustal assemblages. The early experiments have shown that the dihedral angle model, which uses equilibrium partial melt distribution in a solid-melt system to assess melt segregation potential, cannot be directly applied to

crustal rocks (Laporte & Watson, 1995). Results show that the interfacial energy controlling melt distribution is anisotropic and, therefore, the textures deviate significantly from that predicted for ideal systems. As a result, this approach has not been widely applied to the crust, and cannot predict melt fractions at which melt segregation should occur (Laporte, 1994; Rushmer, 1996). In addition, the 'critical melt fraction' (CMF) model, which requires granitic melt to remain in the source until melt fractions reach >25 vol.%, has also been shown to be an unreliable model for segregation of granitic melts (Rutter & Newmann, 1995; Rosenberg & Handy, 2005). The results of deformation experiments suggest that melt connectivity and segregation is possible at very low melt fractions, much lower than considered possible by the CMF model, and is controlled by a variety of factors. Through textural and rheological observations, experimental studies in the past decade have shown that melt segregation is controlled by several variables, including depth of melting, permeability development, type of reaction and volume change associated with reactions (some that can greatly enhance permeability) and the importance of deformation. Therefore, in addition to the pressure and temperature of the potential melting reactions, melt migration can occur at a variety of different melt fractions, depending on the tectonic environment (e.g. whether or not deformation is present). These variables need to be considered to assess the development of permeable pathways for melt segregation at different levels in the crust. Once the nature of melt segregation is known we can then investigate its implications for the geochemistry of the magmas themselves. Ultimately, the compositions of the magmas might in turn be used to establish conditions under which melt migrated from its source.

Figure 9.1 shows a framework in which we can consider melt segregation processes at different permeabilities (Oliver, 1996). Under conditions where deformation is at a minimum, with very little strain and homogeneous distribution of permeability, reaction-controlled permeability can dominate. In the mid-crust, the common partial melting reactions involve the mineral muscovite and these reactions, in themselves, can produce

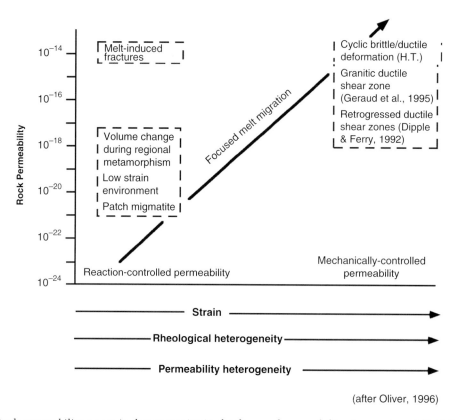

Fig. 9.1 Rock permeability vs strain, heterogeneity in rheology and permeability. In a review of fluid flow in the crust, Oliver (1996) suggests that these important physical parameters that control whether or not fluid flow is pervasive or channelized depends on grain-scale mechanisms. Shown here includes melt migration processes. Reaction-controlled permeability will likely be more pervasive and dominant in rocks with low-strain. In high-strain environments, deformation-produced permeability will cause melt flow to be more focused, such as along shear zones (after Oliver, 1996).

permeable networks that allow for rapid segregation of melt (Connolly, 1997; Connolly et al., 1997; Rushmer, 2001). Regional metamorphism, where dehydration reactions dominate, has also been shown to generate permeability (Connolly, 1997). In the lower crust, permeable networks can be set up by focused deformation (shear zones) producing pathways for melt extraction (Brown & Rushmer, 1997). With increasing strain and strain rates, rheological heterogeneity and permeability heterogeneity, the development of melt pathways in the lower crust becomes more mechanically controlled (Oliver, 1996; Manning & Ingbretson, 1999; Dipple & Ferry, 1992; Brown & Solar, 1999). This

framework provides a solid base from which we can understand melt segregation and transport mechanisms. For example, the geochemistry and emplacement of low-melt-fraction granites (i.e. Himalaya leucogranites), originally considered too viscous to segregate efficiently, are now much better understood (Ayers & Harris, 1997; Harris et al., 1995, 2000).

Review of static experimental results

Hydrostatic experiments on rock cores of various crustal rocks are specifically designed to allow us to characterize the distribution of granitoid melt at different levels of the crust (e.g. along grain boundaries or in cracks). The key to trigger melt segregation at the partial melting sites is inter-connectivity between pockets of melt. The melt distribution data provide information on the fraction of melt that is necessary before interconnectivity can be achieved. Dihedral angle studies on crustal assemblages have shown moderate to low dihedral angles, which should allow melt to easily interconnect. However, the studies also found anisotropy in melt distribution, so that planar solid-melt interfaces are developed in addition to the triple-junction melt pockets (Laporte *et al.*, 1997). In general, the presence of anisotropy in crustal mineral assemblages in the lower crust appears to enhance melt interconnectivity (Brown, 2004), at least for silicic melt systems, but we cannot use the textures to determine a definitive melt fraction under which silicic melt will segregate. In addition, these studies do not consider the generation of melt and how the melting reaction itself may modify textures or, more importantly, produce microcracks that can allow melt to migrate at the grain scale – i.e. become interconnected simply through reaction. Textures from static experiments on natural rock cores of crustal composition have been used to assess whether microcracking by reaction is a common process in the mid- to lower crust (Rushmer, 2001). Results show that muscovite fluid-absent melting (melting without free H_2O fluid) is the most common reaction that produces microcracks, although textures produced in experiments on mafic lowermost crustal rocks suggest that hornblende + clinozoisite breakdown will also produce cracking (Klepeis *et al.*, 2003). Otherwise, most common lower crustal melting reactions do not induce microcracking. These observations imply that melt that escapes at low melt volumes from the mid-crust is likely to have segregated rapidly and bear a muscovite-melting geochemical signature, resulting in leucogranitic bodies such as the High Himalayan examples (Viskupic *et al.*, 2005). Melting in the lower portions of the crust, where biotite fluid-absent reactions or hornblende reactions, without the presence of clinozoisite, may result with melt being retained

along grain boundaries and remaining distributed at the grain scale until higher volumes of melt are generated. Numerical modeling of orogenic belts has suggested that their rheology is controlled by the prolonged presence of a weak crustal layer (Hodges, 2006). Lower crustal melt retained along grain boundaries may establish weakened zones and keep melt in the system. However, deformation plays an important role in melt segregation. The interplay between deformation and melting reactions therefore needs to be addressed.

Review of dynamic experimental results

Deformation associated with external tectonic processes can induce melt segregation in partially molten source rocks in which melt, under static conditions, would not be able to become interconnected unless high melt fractions are reached (Petford *et al.*, 2000). Deformation experiments, in which samples are placed under applied stress while undergoing partial melting, therefore better simulate active tectonic environments and the generation of mechanically produced permeability. High melt pore pressures produced by the applied stress develop in the partially molten rocks in the laboratory, under both middle and lower crustal conditions (Rutter & Newmann, 1995; Dell'Angelo & Tullis, 1988). High internal melt pressure produces an extensive fracture network (cataclasis), even at pressure and temperature conditions under which rocks should deform by ductile flow. Figure 9.2a shows melt fraction against rock strength in log viscosity (Pa s). The melt fraction at the transition from grain-supported rheology to suspension is the theoretical CMF (and used in early studies to suggest the point where crustal melt segregates). This transition is shown with two experimental studies, one on amphibolite at 1.8 GPa and one on granite at 0.25 GPa. These data show that weakening occurs in the natural rocks at different points, and at melt fractions at or below 25 vol.% melt. The figure also shows a cataclastic texture, formed by high internal pore pressure in the amphibolite sample. This texture is associated with weakening of the sample. In an

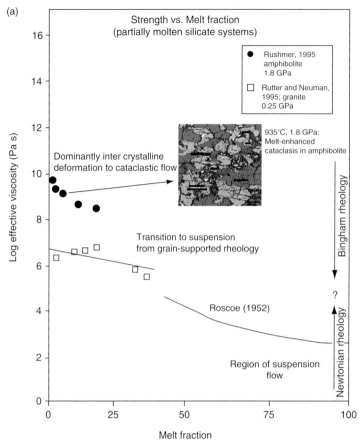

(a)

Strength vs. Melt fraction
(partially molten silicate systems)

● Rushmer, 1995
   amphibolite
   1.8 GPa

□ Rutter and Neuman,
   1995; granite
   0.25 GPa

935°C, 1.8 GPa:
Melt-enhanced
cataclasis in amphibolite

Dominantly inter crystalline
deformation to cataclastic flow

Transition to suspension
from grain-supported rheology

Roscoe (1952)

Region of suspension
flow

Bingham rheology

Newtonian rheology

?

Log effective viscosity (Pa s)

Melt fraction

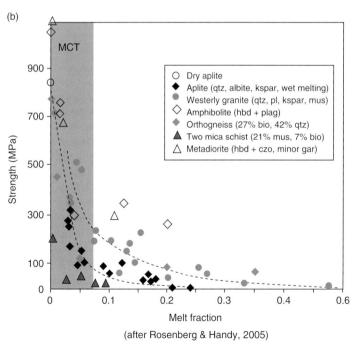

(b)

MCT

○ Dry aplite
◆ Aplite (qtz, albite, kspar, wet melting)
● Westerly granite (qtz, pl, kspar, mus)
◇ Amphibolite (hbd + plag)
◆ Orthogneiss (27% bio, 42% qtz)
▲ Two mica schist (21% mus, 7% bio)
△ Metadiorite (hbd + czo, minor gar)

Strength (MPa)

Melt fraction

(after Rosenberg & Handy, 2005)

Fig. 9.2 (a) Log effective viscosity (shear stress/strain rate) vs melt fraction for experiments on partially molten granite (Rutter & Neumann, 1995) and amphibolite (Rushmer, 1995). Grain-supported rheology dominates between 0 and approximately 30 vol.% melt (there are variations depending on the type of melting reactions) up to the transition zone to suspension, where crystals can no longer interact. This transition zone is the 'rheological critical melt fraction'. Also included is the theoretical relationship for the viscosity of suspended spheres (Roscoe, 1952). (b) Strength (MPa) vs melt fraction for experiments on crustal rock types. This is a modified compilation of strength data from Rosenberg & Handy (2005). Included are data from the Delegate Aplite (van der Molen & Paterson, 1979), Westerly granite (Rutter & Neumann, 1995), amphibolite (Rushmer, 1995) and orthogneiss (Holyoke & Rushmer, 2002). Rosenberg & Handy included the two dashed lines that are least squares exponential fits to the Delegate Aplite and Westerly granite data. Rosenberg & Handy also show the percentage of melt-bearing grain boundaries from static melting experiment of van der Molen & Paterson (1979) that suggest the MCT (melt connectivity threshold) is below 10 vol.% melt. In addition, the 2-mica pelite data are shown (Holyoke & Rushmer, 2002) along with metadiorite data (unpublished). Both the metadiorite and the 2-mica pelite show dramatic weakening as melting begins in contrast to other crustal rock types. These samples are also associated with positive ΔV of reaction.

extensive review by Rosenberg & Handy (2005), crustal rock types that have been used in deformation studies were compiled and plotted on a sample strength (MPa) vs melt fraction diagram. As shown in Figure 9.2a, the usual approach was to plot in logarithmic strength vs melt-fraction diagrams, but they found this blurred the large strength drops found at low melt fractions. From this review, Rosenberg & Handy (2005) suggested that crustal-scale localization of deformation (and therefore possible melt segregation) occurs at the onset of melting, and that melt escapes before ever reaching the high melt fractions that would coincide with the CMF.

This graph is shown in Figure 9.2b, modified from Rosenberg & Handy (2005). We also include new strength data for a 2-mica schist and a metadiorite. Crustal rock deformation data are plotted together and show that weakening occurs below 10 vol% melt. The graph also shows that different crustal rock types have different strength responses to deformation. The weak nature of the 2-mica schist is noteworthy, and will be discussed in detail below. This schist is dominated by muscovite melting, and the high volume change associated with melting reaction may attribute to its rapid weakening. We continue to review the role of deformation and melt segregation in the section below, but focus on pelitic compositions.

Muscovite and biotite-bearing crustal rocks

Muscovite and biotite-bearing assemblages comprise a significant portion of the continental crust and play an important role in the geochemical and rheological evolution of many orogenic belts. Pelitic mineral assemblages are responsible for increasing the radiogenic heat budget, and creating thermal anomalies, and their relatively low solidus temperatures (due to quartz and high abundance of feldspar) allow for extensive partial melting. Orogenic belts containing significant amounts of mica-rich metapelite are, as a result, more susceptible to widespread, melt-induced weakening (Teyssier & Vanderhaege, 2001). Fluid-absent melting reactions of muscovite and biotite

consume mainly mica, quartz and feldspar and therefore partial melt compositions will be granitic in composition.

In view of their significant role in generating partial melts in the crust, we focus now on an experimental study of a 2-mica schist and the chemistry of the experimentally derived melts. We consider geochemical and textural data from a series of experiments performed under both hydrostatic and dynamic conditions that represent different physical environments in which melt segregation is possible. We have performed hydrostatic partial melting experiments on solid cores of a muscovite + biotite-bearing pelite (2M pelite), in which reaction-induced microcracking, produced by muscovite-dehydration melting, forms a permeable, interconnected crack network (Rushmer, 2001). Complimentary dynamic experiments (partial melting under applied stress) were performed on the same starting material to simulate deformation-assisted melt segregation (Holyoke & Rushmer, 2002). The contrasts in textural development between the different experimental designs provide insights into the roles of melting and extraction processes in different tectonic environments. Therefore, *in-situ* analyses of the glass (melt) were carried out for both the static and deformation experiments. Data from the starting micas were also collected, to establish whether there are distinct geochemical signatures associated with the different experimental conditions. Early geochemical data from mineral separates are a combination of mica plus accessory phases, which occur as common inclusions in the grains, and cannot be used for our purposes (Yang & Rivers, 2000). We find that *in-situ* LA-ICPMS (laser ablation – inductively coupled mass spectrometry) and SIMS (secondary ion mass spectrometry) techniques alleviate this problem.

Constraints on melting of pelitic crust from the laboratory: A tale of two experiments

For the partial melting experiments, we used a high-pressure gas vessel to perform the static experiments and a solid-media deformation

Table 9.1   Summary of the static and deformation experimental results

Experiment Name	Pressure (GPa)	Temperature (C)	Melt volume Volume %	Strain (%)	Duration (Hours)	Comments
Static Exps.						
2MS32	0.7	740	<5%	Static – 0%	12	Muscovite reaction; melt cracks in quartz grains observed.
2M5	0.7	850	15–20%	Static – 0%	504	Extensive melting of muscovite; melt cracks/biotite reaction to melt, spinel, opx, and oxides.
2M2	0.7	900	15–20%	Static – 0%	16	Same observations as above; melt cracks reach 25–50 microns in width.
Def. Exps.						
2M-D19	0.7	760	5%	20%	12	Two zones of cataclasis, one of ~50% grain-size reduction; and one zone of ~90% grain size reduction.
2M-D16	0.7	800	40%	25%	12	High melt fraction, relic grains of quartz and feldspar; strain partitioned completely into melt. Areas of muscovite-derived melt (clear glass) and areas of biotite-derived melt (dark glass).
2M-D3	0.7	900	50%	20%	12	Same features as above. Both muscovite-derived melt (clear glass) and biotite-derived melt (dark glass) observed.

apparatus ('Griggs Rig') to perform dynamic experiments. All experiments were performed at pressures of 0.7 to 1.0 GPa, and between 740 and 950°C. The details of the experimental approaches are described in Rushmer (2001) and Holyoke & Rushmer (2002). The pelitic starting material is a greenschist- to amphibolite-grade muscovite-biotite metapelite, composed of 21% muscovite, 7% biotite, 34% quartz, 25% plagioclase, 13% alkali-feldspar and accessory zircon, apatite and monazite. The pelite represents a common crustal rock and exhibits the typical melting reactions that occur in the mid- and lower crust. The fluid-absent reaction begins at approximately 740°C under lower crustal conditions (~21–30 km depth). Fluid-absent melting reactions observed in the pelite experiments, close to the first point of melting (solidus) are best described by balanced reaction (based on oxygen): 1.0 Musc + 0.32 Plag + 0.36 Qtz = 1.14 Melt + 0.22 Kfs + 0.22 Sil + 0.009 Bt (Patino-Douce & Harris, 1998) and, at higher temperatures, by: 1.0 Bt + 0.54 Qtz + 0.053 Plag = 0.73 Opx + 0.002 Spl + 0.59 Kfs + 0.28 Melt (Rushmer, 2001). Table 9.1 shows a summary of the experimental study.

In these solid-core experiments, textures associated with the development of permeability are apparent, either though cracking, resulting from the volume change associated with the melting reaction, or in melt-enhanced deformation, which produces zones composed of melt (glass) and fractured grains that have undergone grain-size reduction (cataclasis). These textures were used to establish melt segregation mechanisms. Figures 9.3a, b and c show examples of the run results and textures associated with static conditions (Figure 9.3a) and under deformation (Figures 9.3b and c).

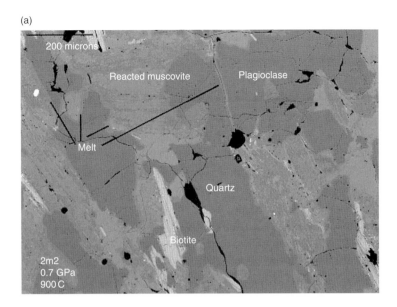

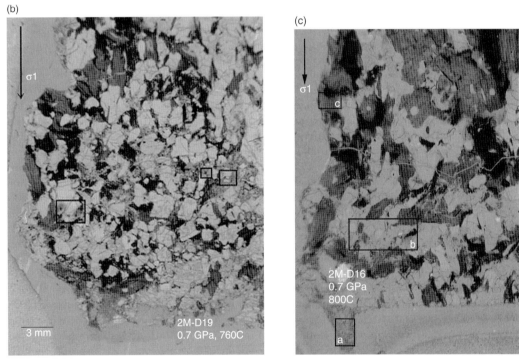

Fig. 9.3 (a) Static core experiment 2M-2 (0.7 GPa and 900°C). Backscatter image shows melt-bearing cracks connecting reacted muscovite grains. LA-ICPMS data were collected on the melt phase in rims and in cracks (labeled 'melt'); (b) Macro-photomicrograph providing view of 2M-19 (0.7 GPa and 760°C) in plane polarized light. Sample is strained 20% and cataclasis fracturing plagioclase and quartz grains (light-colored phases) is observed in the high strain zones. Grain size reduction is up to 90% in the matrix (for additional images, see Holyoke & Rushmer, 2002). Biotite grains are dark and show cleavage. LA-ICPMS sites are shown in black boxes. Arrow shows direction of shortening (σ_1); c) In 2M-D16, the two distinct glasses are especially well developed. Clear glass has accumulated in low pressure sites at the bottom of the charge (box a) and is not in contact with dark glass (boxes b and c) Texturally, development of the dark glass is clearly associated with the melting of biotite. As above, LA-ICPMS sites are also shown in the boxes.

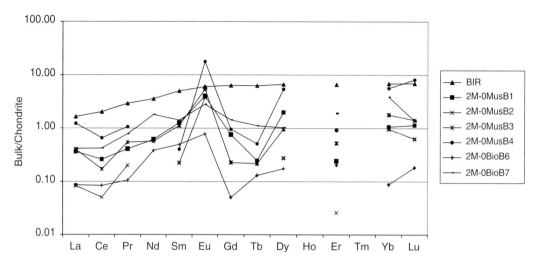

Fig. 9.4 Rare Earth Element (REE) concentrations in the micas present in the starting material, 2-mica schist. The REE data for the muscovite and biotite in 2M were collected by LA-ICPMS (as with the glasses). BIR is the standard and Zr, La, Ce and Y abundances above the detection limit were attributed to the presence of micro-inclusions (micron to <micron in size) of accessory phases in the grains. These data are not shown. This method has also been used as criteria by Yang & Rivers (2000). Data reduction was processed using Lasyboy (J. Sparks, Boston University) with Ti values from external microprobe analysis of TiO_2 as the internal standard. Time slice intervals ranging from 3.15 to 8+ seconds were selected based on the lowest Zr, Ce, La and Y peaks. This procedure allows for minimum zircon, xenotime and monazite micro-inclusion contamination. The resulting abundances are first compared to other 2M mica analyses for consistency and reproducibility.

Textural observations of the static and dynamic experiments also indicate that the melting reactions were dominated by muscovite fluid-absent melting at lower temperatures followed by biotite fluid-absent melting at higher temperatures as expected.

Analyses have been obtained for the glass (quenched melt) in the cracks and in highly permeable glass-bearing cataclastic zones in the products of the deformation experiments. In addition to major-element data, trace-element analyses of the glasses and starting micas were collected. For trace elements, *in-situ* SIMS and LA-ICP-MS techniques were used. Laser-ablation data for the products of static experiments were collected in the large melt cracks that usually connect two reacted muscovite grains (Figure 9.3a). In general, biotite is not extensively melted in the static core experiments.

Figures 9.3b and c shows the melting textures associated with the dynamic experiments. Melt is distributed in highly permeable cataclastic zones (area b in Figure 9.3c) or along the edges of the capsule (areas a and c in Figure 9.3c), rather than in melt-filled cracks. Biotite fluid-absent melting is more extensive in the dynamic experiments and we observe higher volumes of melt under similar P-T conditions (Table 9.1). We observe both clear (colorless) glass, derived by muscovite fluid-absent melting and darker colored glass, derived through biotite fluid-absent melting (Figure 9.3c, areas b, c). In the cataclastic zones, melt derived from both biotite and muscovite reactions are observed (Figure 9.3c, area b). Laser-ablation sites are also shown in the area boxes in Figures 9.3 b and c.

To interpret the geochemical data from the experiments, the starting mica compositions had to be determined. Figure 9.4 shows the rare earth element abundances (REE) of the micas normalized to CI chondrite. Both muscovite and biotite

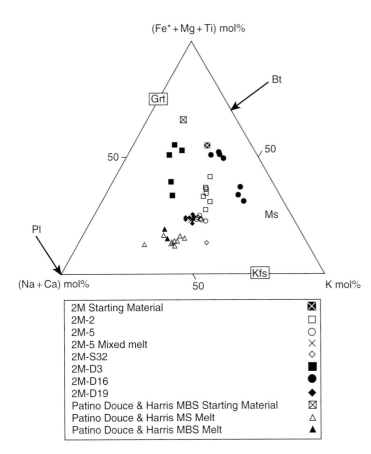

Fig. 9.5 Major element glass data plotted in (Na + Ca) – K – (Fe + Mg + Ti) compositional space. Data are shown for static 2M experiments (2MS32, 5 and 2. 2M-5 mixed melt includes a biotite component) and dynamic 2M experiments (2M-19, 16 and 3). Initial partial melts from the muscovite-bearing schist experiments (MS = muscovite schist partial melts, MBS = muscovite-biotite partial melts) from Patino-Douce & Harris (1998) are shown for reference as are starting material bulk compositions for Patino-Douce & Harris (1998) and the 2M pelite. Mineral phases are also plotted for reference; Bt = biotite, Kfs = K-feldspar, Ms = muscovite, Pl = plagioclase, Grt = garnet.

are low in REE, overall. However, there is a distinct positive Eu anomaly in the muscovites (2M-MusB1-4), which is not present in the biotites (2M-BioB6,7). This signature is inherited by the melt formed by muscovite breakdown.

Rb, Sr and Ba are the most abundant trace elements in the micas, with Sc, V, Cr and Co also present in variable but appreciable quantities. The 2-mica biotite Rb, Sr and Ba values range from 509 to 573 ppm, 1 to 2 ppm and 647 to 837 ppm, respectively. Li, Sc and Co values range from 70 to 80 ppm, 7 to 10 ppm and 43 to 45 ppm. Two-mica muscovite trace element concentrations show that Rb (250–268 ppm) is lower in muscovite than biotite, higher in Sr (24–27 ppm), and higher in Ba (1032–1422 ppm).

Static experiments: glass compositions

Figure 9.5 shows the major-element compositions of glasses and starting compositions plotted in a (Na + Ca) – K – (Fe + Mg + Ti) ternary diagram. Glass data from the static experiments are similar to the experimental melt compositions derived from piston-cylinder experiments performed by Patino-Douce & Harris (1998) on a similar pelitic assemblage.

Rare Earth Element (REE) glass data for the hydrostatic experiments are presented in Figure 9.6. Muscovite-derived melts produced at the lowest temperatures (2M-5 and S32) have consistently low REE abundances, ranging between 0.01 and 1.0×CI (chondrite normalized), and are

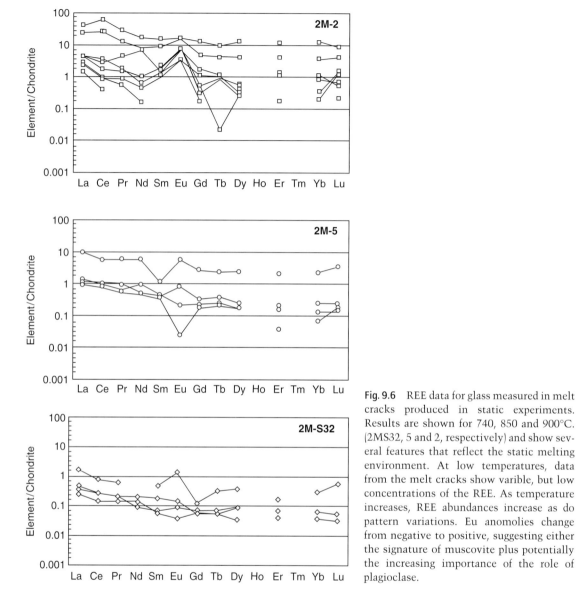

Fig. 9.6 REE data for glass measured in melt cracks produced in static experiments. Results are shown for 740, 850 and 900°C. (2MS32, 5 and 2, respectively) and show several features that reflect the static melting environment. At low temperatures, data from the melt cracks show varible, but low concentrations of the REE. As temperature increases, REE abundances increase as do pattern variations. Eu anomolies change from negative to positive, suggesting either the signature of muscovite plus potentially the increasing importance of the role of plagioclase.

characterized by gently sloping REE patterns with LREE-enrichments relative to HREE and with a positive Eu-anomaly (although not in all cases). At higher temperature, (900°C; 2M-2), the glasses show increases in REE concentrations and become distinctly heterogeneous in their REE abundances, which range between 1.0 and 100.0 × CI. However, the patterns are similar to those obtained in the lower temperature experiments and we see the positive Eu-anomaly.

Dynamic experiments: glass compositions

Major-element compositional data from glass in the permeable cataclastic zones are also plotted in Figure 9.5. Melts produced at 760°C are clear glasses with homogeneous major element compositions that overlap those produced in the static experiments at the same temperature. They also have similar compositions to low temperature experiments from Patino-Douce & Harris (1998). As temperature increases, two compositionally distinct glasses develop in the samples. One is dark in color and is enriched in Mg, Fe and Ti (Figure 9.5) and the second is clear and has a composition that is comparable to the clear glass produced in the lower temperature experiments. The two distinct glasses are especially well-developed in experiment 2M-D16, where clear glass has accumulated at the sides of the charges and is not in contact with dark glass. Texturally, development of the dark glass is clearly associated with the melting of biotite. The presence of two distinct glasses is also evident to some extent for the highest temperature experiment, 2M-D3 but more mixing of the two melts is observed.

REE data for the dynamic experiments are presented in Figure 9.7 and show that the melts are closer to equilibrium than the static experiments. In the lowest temperature experiment (2M-D19) with limited melting, the glass has low REE abundances (>0.1 to <10 × CI), but shows a slight enrichment relative to the glasses in the static experiments but noticeably no positive Eu-anomaly. The shapes of the REE patterns from different analytical sites are very similar and glass in the dynamic experiments appears to be more homogeneous. REE abundances increase systematically with temperature and increasing degree of melting, an effect that is most noticeable in the dark glasses. For example, in 2M-D16 dark glass is highly enriched in REE, whereas the clear glass has similar REE patterns to those in low melt volume samples in the static experiments, including the positive Eu-anomaly. At higher temperatures, high melt volume samples (2M-D3) are enriched in LREE and have positive Eu anomalies.

Partial melting of muscovite and biotite-bearing assemblages: Models for the lower to mid crust from the view of trace elements

REE concentrations are a common geochemical tool and they have been used extensively to assess melting processes in the crust and mantle. There are few studies of the trace-element geochemistry of micaceous phases in the literature (Yang & Rivers, 2000 and references therein). The majority of these studies, with the exception of Yang & Rivers (2000), employed bulk analysis of mineral separates to obtain trace-element data. Our analyses of the starting biotite and muscovite show that they have low trace-element contents and are essentially devoid of REE (but with positive Eu-anomalies in the muscovite). The analyses show that Rb and Ba are the principle trace elements concentrated in the micas and have the potential to be contributed to the melt during the onset of partial melting. The data indicate that Sr strongly partitions into muscovite rather than biotite and consequently will contribute to the melt at the muscovite fluid-absent solidus, although this will depend on the presence of plagioclase too.

REE are present in detectable concentrations all in hydrostatic and dynamic melts produced during partial melting at all temperatures. At the lowest temperatures, the concentrations are similar to the starting micas. We interpret the higher abundances observed at higher temperatures to be due to the dissolution of accessory phases (plus those included in the biotite) during melting. LREE may also be from feldspar, when it is involved in the melting reactions. REE data collected on feldspars from amphibolite and granulite grade metasediments from the Ivrea Zone (Bea & Montero, 1999) show that plagioclase and potassium feldspar, when normalized against chondrite, have moderate LREE (10,100 × CI), contains a Eu positive anomaly (100 × CI) and is devoid of HREE (<1 × CI). In addition to Sr, feldspar involved in the higher temperature muscovite fluid-absent melting reactions will contribute some amount of LREE and Eu (along with the muscovite phase) to the melt, but negligible MREE and HREE. Monazite REE data show that abundances are greatly concentrated in this accessory phase (Bea & Montero, 1999). LREE are between 1.0^3 and $1.0^4 \times$ CI and abundances steadily increase to 1.0^5, to almost $1.0^6 \times$ CI for HREE. As discussed in Watson *et al.* (1996) and in Zeng *et al.* (2005), monazite, in addition to zircon, controls the REE element budget for many crustal rock types.

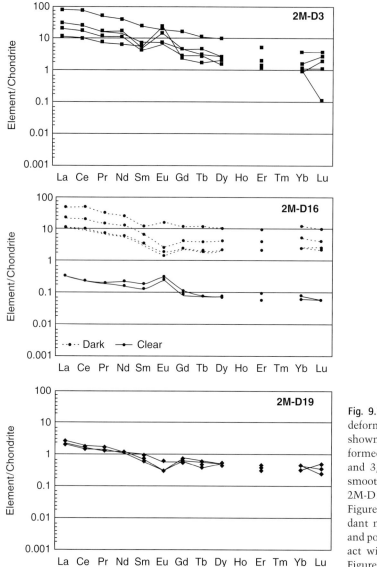

Fig. 9.7 REE data for glass measured in deformation experiments. Results are shown for deformation experiments performed at 760, 800 and 900°C (2M-D19, 16 and 3, respectively). Patterns are notably smoother than in the static experiments. In 2M-D16, note that the clear melt (area a in Figure 9.3c) represents the low REE abundant muscovite-derived glass that escaped and pooled in this edge before it could interact with biotite (dark) melt (areas b, c in Figure 9.3c).

From these experiments, we interpret the REE abundance to be dependant on at least two major factors, in addition to the melting reaction itself: i) temperature; and ii) the bulk composition of the melt. As temperature increases, the degree of fluid-absent melting of the micas increases releasing additional, larger 'phase armored' accessory phases into the melt (Watt *et al.*, 1996). These phases, predominantly zircon and monazite, will undergo progressively more dissolution into the melt with increasing temperature. The effect of temperature on melting reactions and major element geochemistry is also demonstrated in Figure 9.5. The location of the 900°C 2M-2 melts plot closer to biotite relative to 2M-5 and 2M-S32 lower temperature melts. Once experimental temperatures increase to where biotite begins to be involved in the melting process, production of biotite-derived melt increases. Melt also begins to migrate, perhaps using some of the melt network paths produced by the fluid-absent muscovite reaction. The resulting hydrostatic melt REE patterns show an overall increase in abundance with temperature but with heterogenous distribution, dependant on the site of analysis.

Bulk composition of the melt is also critical in determining REE abundance. At a constant temperature, each phase has a specific set of parameters that controls the amount of dissolution that will occur in to the melt. For example, the solubility of zircon is proportional to the magnitude of the cation ratio $M = (Na + K + 2Ca)/(AlSi)$, as shown by Watson & Harrison (1983). The concentration of the LREEs in the melt and the H_2O content are the two parameters that control the solubility of monazite (Rapp & Watson, 1986; Zeng *et al.*, 2005). The melt produced by biotite and muscovite breakdown will incorporate the inclusions of zircon and monazite released into the melt until saturation is achieved, if given enough time. At constant temperature, the biotite derived melts (dark glasses) demonstrate elevated REE abundances when compared to muscovite-derived melts (colorless glasses). This observation can be attributed to the slightly elevated M values of the biotite-derived melts and the higher solubility of LREEs in a melt with higher concentrations of ferromagnesian components.

Whereas the abundance of trace-elements in the melt phase are attributed to factors described above, trace-element distribution and REE patterns of melts appear to be controlled mainly by the nature of the melt network paths that occur in the experiment, and therefore can then be used to help assess rates of melt migration. Melt pathways in hydrostatic experiments produce permeable, interconnected crack networks formed by the reaction-induced microcracking associated with the volume change of the muscovite fluid-absent reaction. In the dynamic experiments, melt-enhanced deformation produces permeability, but deformation in the presence of melt reduces grain size by cataclasis, and melt is more

evenly distributed, more mixed and intimately associated with solid mineral phases than in the crack network system. The hydrostatic network therefore limits chemical communication relative to the dynamic produced pathways, and melts derived from muscovite reactions will be very low in REE and essentially be in disequilibrium. This is certainly the case on the scales of length and time in the experiments and may happen in nature, as discussed by Sawyer (1988), Nabelek & Glascock (1995) and Watt *et al.* (1996, 2000). Watt *et al.* (1996) examined a migmatite suite (Moine Kirtomy Migmatitie Suite in Scotland) and the leucosomes (representing melt that had migrated from its source) with similar chemistry as the experimental melts derived from muscovite breakdown, including the positive Eu anomaly. The positive Eu anomaly may arise due to disequilibrium melting of feldspar combined with LREE- and HREE-bearing accessory phases being left behind in the solid restite. This is also likely in the experimental study. This field study inferred that small batches of melt escaped rapidly and was enhanced by deformation. In the dynamic experiments reaction, kinetics are enhanced and local variability in melt composition is reduced but there are still sites of melting that have not equilibrated. This may, in part, explain the variable REE abundances observed in the melts and why some glasses are similar in composition to the static experiment glasses. The variable glasses reflect localized differences in melt composition, even at the highest run temperatures. In a later study, Watt *et al.* (2000) found evidence for microfracturing in residual quartz from three granulite facies migmatites. Crack densities were higher than observed in the static experiments described above and approach theoretical interconnectivity thresholds for percolation, as also shown by Connolly *et al.* (1997). Watt *et al.* (2000) suggest that increased permeability arising from reaction-induced fractures may enhance melt extraction during partial melting.

Deformation is an important additional variable affecting the major- and trace- element melt geochemistry in the dynamic experiments. This is due to the difference in melt distribution and enhanced reaction kinetics. The major-element melt compositions in the dynamic and static experiments are initially very similar to those in the lowest-temperature experiments. However, at higher temperatures, there is a significant departure in melt composition. This is best demonstrated by 2M-D19 conducted at 760°C, which has melt compositions that plot in close to 2M-5 clear melts formed at 850°C (Figure 9.5). Whilst both melt compositions show low REE concentrations overall, 2M-D19 is slightly elevated and may have more involvement of biotite (lack of a strong Eu anomaly) relative to 2M-5 and is distinctly smoother in pattern from 2M-S32, the static experiment conducted at the same temperature. We interpret this difference as being a consequence of enhanced reaction kinetics in the dynamic experiments because melt-enhanced cataclasis reduces the grain-size and brings reacting phases into contact more readily during deformation.

A spread in the major-element data is observed in the dynamic experiments as temperature increases. 2M-D16 consists of distinct populations of clear and dark melts (Figure 9.3c). Textural and compositional evidence suggests that the clear, colourless melts are early quartz plus muscovite +/– plagioclase melts, which segregated into local dilational sites in the initial stages of deformation. The REE abundances of 2M-D16 clear melts are comparable to those of melts in the lower temperature static experiments. In contrast, 2M-D16 dark melts plot closer to biotite in Figure 9.5 and have REE patterns reflecting the higher ferromagnesian content and the possible incorporation of the larger grains of accessory phases as biotite reacts. Deformation has also enhanced the communication between the solid and melt phases in each melt population, resulting in overall smoother REE patterns. There appears to be very limited communication between the dark and clear melts outside of the cataclastic zone where extensive mixing does occur.

2M-D3, like 2M-D16, also has two melt populations. The higher temperature (900°C) of this deformation experiment has promoted more extensive melting and much more extensive

compositional equilibration between distinct melt compositions. As a consequence, the distinct, low REE signature of the muscovite-dominated melt has been largely lost. At this temperature, extensive involvement of biotite and plagioclase, in addition to dissolution of larger accessory phase grains, is reflected in the more ferromagnesian melt chemistries and elevated REE abundances.

There have been observations of these types of signatures in studies on leucogranites. Muscovite dehydration melting is thought to be the source for many of the leucogranites in the High Himalaya (Harris et al., 1995, 2000). Isotopic and trace element studies from these natural settings have helped with timescales of melting and melt segregation.

Timescales of crustal melting: insights from Himalayan leucogranites trace element geochemistry

Harris et al. (1995, 2000) performed quantitative modeling and geothermometry to relate the leucogranites of the High Himalaya with potential pelitic sources and to help establish extraction and crystallization timescales. In order to determine the relationship between melting, extraction and crystallization and their associated timescales, REE and trace elements Rb, Sr and Ba were used to assess equilibrium, along with monazite and zircon thermometry.

Disequilibrium due to rapid segregation of muscovite-derived melts, was inferred to have occurred by the observation of differences in temperatures from monazite and zircon thermometry. In addition, some of the studied leucogranites were inferred to have been composed of melts that were undersaturated in REE (most notably LREE), which implied that the melts were extracted in <10 ka. Experimental studies on a Himalayan pelite (Patino-Douce & Harris, 1998) suggested that some Himalayan melts were also undersaturated in Zr and therefore segregation may have occurred within 100 years. Short timescales, such as these, have been discussed in Clemens & Mawer, (1992) and Petford et al. (1993,

2000) for the emplacement of granite through a dyke network. The transport distances of Himalayan granitic melts have been suggested by Harris *et al.* (2000) to be up to 10 km and may be achieved by dykes in about 1 day. At these rates even the largest leucogranite could be emplaced in around 10 years. Crystallization of Himalayan melts involved longer timescales from the modeling of Harris *et al.* (2000). They calculated that if the magmas were emplaced as thin sheets (~100 m wide; see also Le Fort *et al.*, 1987 & Cruden, 2006) a timescale of >500 years would be required. Therefore it is likely that, for sheet complexes such has the leucogranites in the High Himalaya, it is the tectonics of space creation plus perhaps the crystallization processes rather than melt segregation and transport that is the rate-determining step on the establishment of a solid intrusive body.

From these combined field and experimental studies, the overall timescales of melt segregation and emplacement for many granites related to mountain building are rapid and may be <1 ka. This suggests that deformation must also be present and rapid, because melts would have solidified in a few hundred years. However, the timescale required for prograde heating of the magma sources in the crust is much >1 Ma (Connolly & Thompson, 1989; Thompson & Connolly, 1995). As the melt production rate is determined by heat flow into the source, it is likely the rates of magma production in the crust is therefore determined by the heating processes (underplating, intrusion of mafic magmas) and the thermal diffusivities of crustal rocks, not the ability of the melt to segregate.

During partial melting in the mid- to lower crust, both the mechanism of melt segregation and the nature of the short-range transport paths can exert significant control over melt geochemistry, because these variables influence the degree of equilibrium achieved between the melt and the solid restite (Sawyer, 1991; Watt *et al.*, 1996). It is therefore likely that the physical process of melt segregation will leave a chemical signature that may be detected in the anatectic melts and the rocks that melted to form them. The above studies suggest that there are distinct differences

in composition between melts derived statically and distributed in crack networks from those formed in permeable shear zones. These differences in composition can be exploited by studies of natural anatectic granites to estimate rates of melting, extraction and crystallization. As these melts rise into the upper crust, assimilation can also become an important contributor to final magma composition.

MELTING AND ASSIMILATION IN THE UPPER CRUST

The picture that we have constructed thus far is one in which segregation, ascent and accumulation of silicic magma generated in the mid- to lower crust are rapid, potentially discontinuous, processes. Much of the chemical diversity of magma compositions appears to be acquired at depth, where the type of melting reaction and the interplay between deformation and melting can play critical roles in the composition of individual melt batches. However, some volcanic and plutonic rocks still require assimilation of upper-crustal material to fully explain their chemical and isotopic characteristics (Verplank *et al.*, 1995; Tepley & Davidson, 2003; Ramos *et al.*, 2005; Knesel & Duffield, 2007).

Accordingly, in this section we turn our attention to the processes and rates of melting in the upper crust. Here we expand the scope to include mafic, as well as silicic, systems. We begin with a summary of evidence from some natural examples of xenolith and wall-rock melting and disaggregation. These studies demonstrate that melting is rapid, and attendant assimilation of partial melts, dispersion of refractory minerals (xenocrysts or antecrysts) and associated crystallization reactions (Beard *et al.*, 2004, 2005) may operate over timescales that are comparable to or shorter than the typical eruption repose periods of many volcanoes at convergent plate margins, as recently highlighted by Dungan (2004). Such evidence for rapid melting and assimilation in nature provides the rationale for re-examining what laboratory experiments can tell us. In this context, we

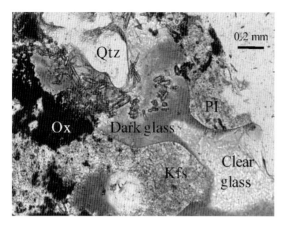

Fig. 9.8 Photomicrograph of partially melted granitoid, showing development of dark and clear melt (quenched to glass) during near-surface melting along the margin of basaltic dyke in the Sierra Nevada Batholith, California. *Kfs* alkali feldspar, *Ox* is Fe-Ti oxide, *Pl* plagioclase, *Qtz* quartz. See Plate 9.8 for a colour version of this image.

examine the results on melting experiments of solid granite blocks conducted at surface conditions, and compare these results to those from melting at deeper structural levels. Here again we find that experimental investigations provide insight into the factors governing the geochemical characteristics of melt formation in the crust.

Review of melting processes and timescales in natural systems

Petrographic observations from partially melted wall rocks and xenoliths provide a natural starting point for examination of the physiochemical nature and timescales of melting and assimilation processes in the shallow crust. Figure 9.8 shows an example of a partially melted granodiorite, where melting was induced by intrusion of basalt on the eastern edge of the Sierra Nevada batholith, California, USA. First reported by A. Knopf in 1918, the partially melted rock contains both clear and dark glasses (quenched melt), similar to those in the dynamic experiments presented earlier. In this case, formation of the clear glass reflects disequilibrium melting along quartz and feldspar grain boundaries, while the dark glass appears to have resulted from contamination of

this melt by Fe from oxide grains after biotite (Knopf, 1918, 1938. These observations reiterate the importance of textural (grain-boundary) controls on the formation and chemistry of melts (Mehnert *et al.*, 1973; Büsch *et al.*, 1974; Wolf & Wyllie, 1991, 1995; Kaczor *et al.*, 1988; Green, 1994). Whilst we will investigate some geochemical implications for such disequilibrium processes below, it is essential to first evaluate some of the available constraints on the timescales over which melting occurs in the upper crust.

Evidence for rapid melting and assimilation in nature

It is well established that detailed study of partially melted and variably digested xenoliths brought to the surface in volcanic rocks provide insights into the mechanisms of upper-crustal assimilation (Sigurdsson, 1968; Le Maitre, 1974; Maury & Bizouard, 1974; Harris & Bell, 1982; Grapes, 1986; Grove *et al.*, 1988; Green, 1994). However, most investigations generally fail to identify rates of melting and assimilation (where it has occurred), due to lack of tightly constrained xenolith residence times. The challenge is not only to identify the sources of xenoliths, but also when they were introduced into the host magma. Several geologically well-constrained cases arising at two well-known arc volcanoes are worth examining here. Lavas and pyroclastic flows of the 1980–1986 Mt St Helens dacite eruption, for instance, contain abundant partially melted gabbroic xenoliths scavenged from the plutonic residue of pre-1980 eruptive periods (Heliker, 1995). The xenoliths also include lesser quantities of quartz diorite, hornfelsic basalt, dacite, andesite and vein quartz, and are notably less abundant in the initial cataclysmic eruptive phase. It therefore appears that the violence of the May 18, 1980 eruption shattered the magma reservoir and conduit wall rock, and thereby introduced the inclusions carried to the surface in the subsequent dacite lava extrusion and pyroclastic flows. If so, the short residence times demonstrate that melting and disaggregation of xenoliths can be strikingly rapid, on the order of a few years or less.

Mt Mazama, another Cascade arc volcano well-known for erupting partially melted plutonic rocks, provides additional insight into the timescales of melting in the upper crust. Blocks of variably melted (0–50 vol.%) granodiorite, along with minor diabase, quartz diorite, granite, aplite and granophyre, were ejected in the climatic eruption that formed Crater Lake caldera at Mt Mazama (Bacon, 1992). Because the majority of the ejected blocks lack adhering juvenile magma, it appears that they represent fragments of the composite pluton associated with growth of the system that eventually produced the climatic caldera-forming eruption at 7.7 ka. In this case, however, melting has been inferred to occur prior to entrainment (Bacon, 1992). Diffusion considerations of Zr gradients in glass (La Tourrette *et al.*, 1991), homogeneity of titanomagnetite and preservation of oscillatory zoning in plagioclase (Bacon, 1992), and O-isotope re-equilibration among quartz and plagioclase (Bacon *et al.*, 1994) in partially melted rocks constrain the duration of melting to as little as 10 years (oxygen isotopes) to perhaps as much as 10^4 years (plagioclase zoning). Although it is not possible to more tightly constrain the timescale at present, an important point worth making here is that, if the duration of melting did approach 10^4 years at Crater Lake, it was not sufficient to achieve chemical equilibration between melt and residual minerals for some trace elements (Bacon, 1992) and Sr isotopes (Bacon *et al.*, 1994) in some of the partially melted blocks.

Crater Lake also provides constraints on the timescale for assimilation. On the basis of trace-element (Bacon & Druitt, 1988) and isotopic evidence (Bacon *et al.*, 1994), it has been suggested that the climatic rhyodacite incorporated up to 25% partial melt and restitic minerals from the shallow plutonic portions of the system. These rocks contain the accessory mineral zircon. However, because it lacks zircon and because it was not zircon saturated, the climatic rhyodacite seems to have dissolved any zircon liberated from the digested plutonic blocks. Therefore the mineralogical evidence of assimilation was erased within a few thousand years or less (Bacon & Lowenstern, 2005). In contrast, despite

also being undersaturated with respect to zircon, precaldera rhyodacite erupted at ~27 ka does contain resorbed zircon similar in size, shape and zoning to those in the granodiorite blocks.[238]U-[230]Th ages determined by secondary ion mass spectrometry (SIMS) suggest that the zircons, along with 80% of the plagioclase, in the rhyodacite were recycled from the plutonic rocks (Bacon & Lowenstern, 2005). When coupled with models for zircon dissolution (Watson, 1996), these results constrain the timescale for melting, disaggregation and dispersal of granodiorite into the pre-caldera rhyodacite to a few 10s of years before eruption.

Such short timescales may seem impractical given some conventional views of protracted accumulation and differentiation of silicic magma at long-lived volcanic centers. They are nonetheless consistent with:
• Kinetic models predicting a characteristic timescale for assimilation on the order of weeks to years (Edwards & Russell, 1998);
• Theoretical analysis of xenolith assimilation complimented by laboratory analog experiments yielding melting rates on the order of 1 mm/hr (McLeod & Sparks, 1998); and
• Independent estimates for assimilation timescales on the order of decades or less, arising at a variety of magmatic systems ranging from relatively small-volume basalts (Costa & Dungan, 2005) to some of the largest rhyolitic eruptions known on Earth (Gardner *et al.*, 2002).
Thus, there is now a growing body of theoretical and field evidence that melting and assimilation in the upper crust can be strikingly rapid and efficient processes, regardless of the composition or size of the system. With this insight in hand, we now return to the geochemical characteristics of our Sierran example of contact melting.

Geochemical consequences of rapid melting and assimilation

A result of the apparent rapid nature of melting and assimilation in the upper-crustal environment is that contaminant compositions may deviate from those expected from equilibrium melting models (Maury & Bizouard, 1974; Kaczor

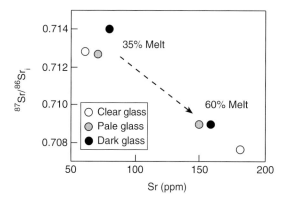

Fig. 9.9 Plot of initial Sr-isotope ratio against concentration for clear, pale and dark glass from two wall-rock samples representing moderate (~35%) to high degrees (~60%) of partial melting of granitic crust near Earth's surface.

et al., 1988; Philpotts & Asher, 1993). Examination of the Sr-isotopic compositions of glasses from our partially melted Sierran granitoid (Figure 9.8) serves to underline this notion. Initial $^{87}Sr/^{86}Sr$ ratios of separates of dark to clear glass range from 0.7140 to 0.7076 (Figure 9.9) and reflect rapid disequilibrium melting with compositions controlled by local mineralogy. Of course, evidence for isotopic disequilibrium in nature and the concurrent ramifications for interpretation of isotopic data from igneous rocks are not new; such disequilibrium was observed over 30 years ago by Pushkar & Stoesser (1975) in an elegant study of a partially melted xenolith from the San Francisco volcanic field and has since been well documented in other xenolith and wall-rock settings (Bacon *et al.*, 1994; Tommasini & Davies, 1997; Knesel & Davidson, 1999; Davis & Tommasini, 2000). Yet, as provocative as these observations are, poor constraints on the conditions and duration of melting, and even initial rock compositions in some cases, plague attempts to establish a general framework for understanding the isotopic characteristics of anatectic melts from field analogs. A different approach is needed, and here again is where experiments can help.

Constraints on melting of granitic crust from the laboratory: A tale of two experiments

In view of the foregoing evidence for disequilibrium in the upper-crustal melting environment where granitoid rocks are prevalent, we consider here two experimental approaches designed to study the isotopic compositions of melts formed from mica-bearing granitic starting materials. The first employs large (in experimental terms), ~27 cm^3 (40–50 g) granite blocks. The use of solid granite, rather than a powder, allows us to study the formation of melt along grain boundaries as would occur in natural systems. Due to the large sizes of the granite blocks, these experiments were conducted in a muffle furnace, at atmospheric pressure between 1,050 and 1,250°C. The second set of experiments was performed using 0.3 g of fine sand (75–100 μm grain size) derived from the same granite. The smaller sample size affords us the opportunity to study the composition of melts formed under confining pressure. The experiments on the crushed granite were performed in a piston-cylinder apparatus at 600 MPa over temperatures between 850 and 1,000°C.

The starting material is a coarse-grained Proterozoic granite from the Marble Mountains, Mojave desert, California. Although the rock is mineralogically variable on the hand-sample scale, it consists primarily of subequal proportions of alkali feldspar (up to 2 cm in length), plagioclase, and quartz along with 10 to 15% biotite, minor magnetite and secondary muscovite. While somewhat similar to the pelite examined earlier, in that both starting materials contain two micas, the muscovite in the granite is subordinate to biotite and it represents a hydrothermal alteration product primarily present as fine sericite and small flakes (up to several hundreds of microns) within variably altered feldspar grains. The granite represents a common upper-crustal rock and exhibits the typical melting reactions observed for partially melted xenoliths (solid-granite experiments) and in conventional experimental studies of mica-bearing rock powders conducted under hydro- or lithostatic pressure (crushed-granite

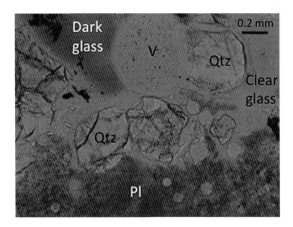

Fig. 9.10 Photomicrograph of partially melted granite showing development of dark and clear melt generated by melting of solid granite blocks at 1 atm in the laboratory. *Pl* plagioclase, *Qtz* quartz, *V* vesicle. See Plate 9.10 for a colour version of this image.

experiments). The details and full results of both experimental approaches are described in Knesel & Davidson (1996, 1999) and Knesel & Davidson (2002), respectively.

Isotopic compositions of melts generated at Earth's surface

In the solid-granite experiments, the lack of confining pressure limited melting at 1,050°C to trace quantities (<1%) of colorless glass. The melt was localized along quartz-feldspar contacts, both at the margins of course grains and where quartz occurred as small inclusions within feldspars. The total melt fraction rose dramatically to 56 wt.% at 1,100°C. Melt at this temperature consisted of clear and dark varieties (Figure 9.10). The textural and geochemical characteristics of both glass types indicate they are best described by disequilibrium reactions involving plagioclase + alkali feldspar (± quartz) to produce clear melt and biotite + plagioclase (± alkali feldspar and quartz) to produce dark melt and residual iron oxides. Following this large initial melting step, melt fraction increased gradually with increasing temperature to 77 wt.% at 1,250°C, mainly through dissolution of remaining plagioclase with only minor contributions from quartz.

Sr-isotopic analysis of glass separates for both melt types demonstrates striking isotopic disequilibrium at the onset of melting. In general, dark glass is more radiogenic than the bulk granite due to the significant contribution of biotite-derived Sr, and clear glass is less radiogenic because its Sr is mainly derived from the two feldspars (Figure 9.11a). With increasing temperature, the $^{87}Sr/^{86}Sr$ ratios for both glasses evolve towards the whole-rock composition, whilst Sr concentrations increase, as the restite evolves towards a Sr-free quartz-rich residue. Electron microprobe and micro-drill sampling (for major-element and Sr-isotope analyses, respectively) demonstrate that the extreme compositions of the two melt types reflects incomplete mixing over the short duration of the experiments (Figure 2 of Knesel & Davidson, 1996). Given time to mix completely through diffusion, which would take in the order of a few years or less for the conditions employed here (Figure 3 of Knesel & Davidson, 1996), bulk-melt $^{87}Sr/^{86}Sr$ ratios start *high* and evolve (*decrease*) towards the composition of the source rock with the advance of melting (Figure 9.11b). This behavior is similar to that observed in our natural Sierran analog (Figure 9.9), as well as in other studies of partially melted rocks (Pushkar & Stoesser, 1975; Bacon *et al.*, 1994; Tommasini & Davies, 1997) and in 1-atm experiments involving phlogopite and plagioclase (Hammouda *et al.*, 1996). Collectively these studies underscore the significance of the instability of hydrous minerals on melt composition under the low-pressure conditions prevailing during melting in the upper crust. A question remains, however, as to how this generalization extends to melting deeper in the crust. We address this problem below.

Isotopic compositions of melts generated at depth

In agreement with previous experimental studies (Gardien *et al.*, 2002, and reference therein), melting in the piston-cylinder experiments conducted at 600 MPa began by H_2O released initially through

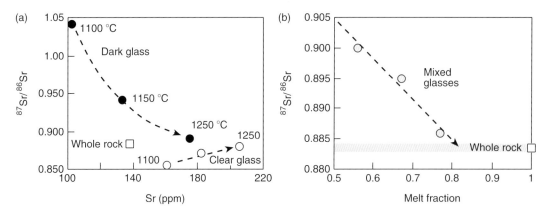

Fig. 9.11 (a) Plot of Sr-isotope ratio against concentration for clear and dark glass, showing glass evolution with increasing temperature during progressive melting of solid granite blocks at 1 atm in the laboratory. The high-^{87}Sr/^{86}Sr ratio of first dark glass reflects the large contribution of radiogenic Sr derived from widespread biotite breakdown; the melt ^{87}Sr/^{86}Sr then decreases with increasing temperature, primarily due to addition of unradiogenic Sr released by progressive dissolution of the remaining plagioclase. In contrast, the low-^{87}Sr/^{86}Sr ratio of initial clear glass reflects melting mainly along plagioclase + alkali feldspar ± quartz grain boundaries, but the gradual increase in ^{87}Sr/^{86}Sr is the result of more extensive diffusive mixing with radiogenic dark melt with increasing temperature. (b) Plot of Sr-isotope ratio against melt fraction for fully mixed glass for 1-atm experiments. Initially high-^{87}Sr/^{86}Sr (bulk) melt systematically decrease towards the granite whole-rock value (square) with increasing melt fraction, similar to that seen for the natural example of partially melted granite in Figure 9.9.

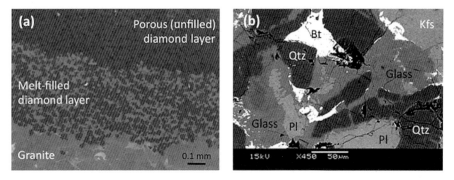

Fig. 9.12 (a) SEM backscatter image of 600 MPa piston-cylinder experiment, showing migration of melt (gray phase in middle portion of image) from granite (bottom) into the overlying porous diamond layer (dark-gray aggregates in top and middle part of image). (b) SEM backscatter image showing glass pools and films located between and along grain boundaries in crushed granite (850°C, 6 days). *Bt* biotite, *Kfs* alkali felspar, *Pl* plagioclase, *Qtz* quartz.

breakdown of muscovite and progressed to biotite-dehydration melting with increasing experiment temperature. When brought to run conditions, some of the partial melt migrated by porous flow along grain boundaries, into a layer of diamonds lying above the granite (Figure 9.12). On termination of each experiment, the melt was quenched to glass and could be isolated from the restite in the lower portion of the charge (for isotopic analysis) by sectioning with a diamond wafering saw.

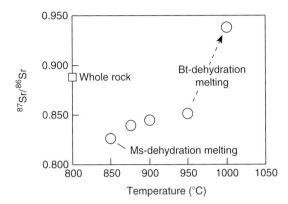

Fig. 9.13 Plot of Sr-isotope ratio against temperature for 600 MPa granite melting experiments, showing control of melting relations on melt composition. Initial melt $^{87}Sr/^{86}Sr$ ratios determined by regression of time-series isotopic data for 850, 875, 900, 950 and 1,000°C experiments. Muscovite-dehydration melting produces liquid with an $^{87}Sr/^{86}Sr$ ratio below that of the starting granite material. However, melt $^{87}Sr/^{86}Sr$ ratios increase gradually with increasing temperature because of dissolution of alkali feldspar, until finally incongruent breakdown of biotite above 950°C yields melt with an $^{87}Sr/^{86}Sr$ ratio greater than the source material.

Figure 9.13 shows the variation of $^{87}Sr/^{86}Sr$ ratios as a function of temperature in the experimental melts extrapolated to the onset of melting from time-series data ranging from 2 to 28 days. The striking result is that, in contrast to the 1-atm experiments and numerous examples of low-pressure xenolith and wall-rock melting in natural systems, the Sr-isotopic composition of the 600-MPa melts start *below* the whole-rock value and *increase* with the progress of melting. In this case, the low-$^{87}Sr/^{86}Sr$ ratios of the lower-temperature melts are a result of the unradiogenic Sr in the plagioclase consumed in the initial muscovite-dehydration melting reaction. With increasing temperature, the melt $^{87}Sr/^{86}Sr$ ratios increase gradually through dissolution of alkali feldspar and ultimately by widespread incongruent breakdown of biotite, which at temperatures above 950°C produced melt with an $^{87}Sr/^{86}Sr$ ratio greater than the starting granite.

Partial melting (and assimilation) of muscovite- and biotite-bearing assemblages revisited: Models for the mid to upper crust from the view of Sr isotopes

Radiogenic isotopes are a powerful means of assessing crustal melting and contamination models. However, the evidence presented here for disequilibrium suggests that the isotopic composition of crustal melts may not necessarily be the same as their source. We have seen that the use of trace elements may be problematic as well (e.g. REE in our pelite melting experiments) in that melt compositions may deviate from those predicted using established phase equilibrium and mineral-melt partitioning data. The question that needs to be addressed is whether this recognition represents a step forward or backward. If melting in the crust really is a disequilibrium process, would it not make constraining source and contaminant signatures a less tractable problem? We propose that in accepting that this is what happens (at least at some times and in some places), we are in a position to make real progress. The claim here is that the isotopic (and trace-element) composition of a crustal melt is a function not only of the composition of its precursor, but also of the processes by which and rates of which it is formed and extracted. The corollary to this is that, if we can establish a general framework for modeling the isotopic compositions of crustal melts, we then have a powerful tool to constrain the mechanisms and timescales over which melts are extracted or assimilated.

In addition to knowledge of the source mineralogy, the depth and temperature of melting, because of their control on mineral stability and their proportions involved in melting reactions, are essential parameters for constructing any melting model. Temperature may vary between or even during different melting events, and must be assessed on a case-by-case basis. However, observations drawn from our experiments provide a framework for evaluating melting behavior and resultant melt isotope compositions as a function of depth.

From our experiments at 600 MPa, we conclude that the Sr-isotopic compositions of melts generated at mid-crustal depths conform to those expected from traditional phase equilibria. Near the solidus, fluid-absent melting yields melts with $^{87}Sr/^{86}Sr$ ratios below those of the source, whilst melting at temperatures sufficient for biotite incongruent melting, produces high-$^{87}Sr/^{86}Sr$ melt (Figure 9.13). This agreement gives confidence in modeling the isotopic compositions of partial melts in natural systems using experimentally determined melting reactions (Knesel & Davidson, 2002; Zheng et al., 2005). In turn, through comparison of such models with the products of melting in nature (e.g. granites), we can evaluate the types of melting reactions, the mechanisms driving melting, and the timescales and factors governing melt extraction and ascent. Application of this approach, for instance, has shed light on the controversial role of fluids and the influence of deformation on the generation and extraction of crustally derived leucogranites of the Himalayan orogen (Knesel & Davidson, 2002). Isotopic disequilibrium during the production of some of these leucogranites is consistent with independent estimates for extraction times of 10 ka or less (Harris et al., 2000).

In contrast, melting at shallow crustal levels appears to take place far from equilibrium. Under these conditions, melt compositions are controlled by the texture mineralogy of the source and cannot simply be modeled by phase relations derived from experimental studies of rock powders in the laboratory. Nevertheless, our 1-atm experiments (Figure 9.11b), along with those of Hammouda et al. (1996), suggest that the instability of biotite and, to a lesser extent, amphibole at low pressure may lead to production of partial melts with $^{87}Sr/^{86}Sr$ ratios greater than those of the source rocks. Whilst this scheme is clearly supported by observations from the case studies discussed above (Figures 9.8 and 9.9), the consequences for crustal contamination processes are less clear.

If melting is rapid and runs nearly to completion, as might be expected during entrainment of granitoid xenoliths in mafic magma, the isotopic composition of the assimilated melt is likely to approach that of the inclusion as a whole. The resulting contamination may therefore approach that produced by bulk assimilation. In this case, the physical record of contamination, apart from

scattered xenocrysts (primarily of quartz), may be cryptic. A word of caution is warranted here, however, as this does not mean that the contaminated hybrid can be assumed a priori to be a simple, uniform mixture of contaminant and host magma. If mixing is not complete at the time of eruption, considerable isotopic variability can develop, even within a single lava flow (Laughlin et al., 1972).

Whether upper-crustal xenoliths entrained in silicic magma should meet the same fate is less certain. The lower temperatures of dacitic to rhyolitic magmas should limit the extent of partial melting and therefore encourage production of high-$^{87}Sr/^{86}Sr$ melt through initial breakdown of biotite and/or amphibole. Assimilation of this melt would impart a very different signature than that of bulk contamination. However, insofar as shear stresses related to flow of viscous magma may promote rapid xenolith disaggregation (Green, 1994; Dungan, 2004; Beard et al., 2005), we might envisage an alternative scenario. Since disaggregation disperses both melt and restitic fragments into the host magma, contamination may again approximate a bulk process. Conversely, sedimentation of refractory xenocrysts might still push the net effect towards one of selective contamination. There is thus a need for a better understanding of two- or even three-phase flow that may attend melting and assimilation of crustal materials. Nevertheless, in any of these scenarios, the xenocrysts (or antecrysts) are likely to be significantly more abundant and diverse than in our previous example of assimilation by mafic magma. Plagioclase, alkali feldspar, pyroxene, Fe-Ti oxides and common accessory minerals, such as zircon, monazite and apatite, along with any unreacted hydrous phases may be released into the host magma. Geochemical characterization of these disaggregated crystals provides a useful tool to clarify the nature of the mechanisms and timescales involved in shallow-level assimilation and differentiation. As we have seen, it is the efforts in this area that have begun to show us how rapid such processes can be. Finally, it is also worth pointing out that other features, such as wispy patches, stringers and other indistinct concentrations of minerals that are common in plutonic rocks at thin-section to outcrop scales but are commonly avoided in geochemical investigations, may be remnants of incompletely assimilated material. The study of these features may ultimately help to further constrain the assimilation process, as recently highlighted by Beard *et al.* (2005).

SUMMARY

We see in convergent zones and in some intraplate settings the expression of many of the aspects of continental crust melting, segregation and assimilation processes discussed above, on both the outcrop and grain-scale. We see not only the development of compositional and isotopic variations in the crystallized magmas, but the rheological and structural response to melt-enhanced weak zones. Experiments on natural crustal rocks have provided an important approach determining the variables that control melting, melt segregation and the rates at which these processes can occur. Using state-of-the-art analytical equipment, we can determine trace-element abundances and distribution in the experimental melts to establish melting reactions, and the influence of deformation on melt composition. REE patterns of melts appear to be mainly controlled by the nature of the melt network paths occurring in the experiment and therefore can then be used to help assess rates of melt migration. Melt pathways in hydrostatic experiments are permeable, interconnected crack networks produced by the reaction-induced microcracking associated with positive volume changes. However, in the dynamic experiments, melt-enhanced deformation produces permeability, and deformation in the presence of melt reduces grain size by cataclasis. Melt is therefore more evenly distributed and intimately associated with mineral phases than in the crack-network system. The result is that the melt is more homogenous in trace-element and isotope composition. Physically, the presence of melt along and within grains can strongly influence rheology, producing weakened zones in the crust.

On the large scale, partial melting and subsequent migration and emplacement of that melt are the fundamental processes by which the crust chemically differentiates. Therefore, magma compositions themselves provide a resource through which we can further explore crustal chemical and physical evolution. We find that the isotopic composition of a crustal melt is a function not only of the composition of its precursor, but also of the processes and rates by which it formed and is extracted. We suggest that the trace-element and isotopic compositions of crustal melts provide tools to constrain the mechanisms and timescales over which melts are generated and extracted or assimilated. From combined field and experimental studies, the overall timescales of melt segregation to emplacement for many granites generated during mountain building are <1 ka. These rapid rates are also found in a variety of studies on melting and melt-transport processes in the continental crust.

Figure 9.14 shows a summary of different segregation, transport and emplacement rates taken from this study and the literature. These rates represent a fracture-dominated system and are most applicable to areas of very active deformation, such as the Himalaya. Melting is driven thermally, not by fluid influx, so the rates are for processes occurring under dominantly fluid-absent conditions. Rates for melt segregation in the lower crust come from Sawyer (1991) and Watt *et al.* (1996) and are based on disequilibrium in leucosomes. These studies used trace elements to estimate the degree to which equilibrium was obtained, particularly Zr, in the leucosome. Melt segregation is driven by pressure gradients developed during deformation (plus positive volume changes as noted by Watt *et al.*, 1996, 2000) and occurs at a rate of about 1 to 2 cm/year. As shown here, experimentally, we observe melt accumulation at the edge of the capsules at the end of the deformation experiments. This demonstrates that permeability was achieved during the run time and, as the pressure gradients develop, migration can occur during this time frame. A minimum rate of ~0.1 cm/hour is estimated for these experiments. The high strain rate applied to the samples in the deforma-

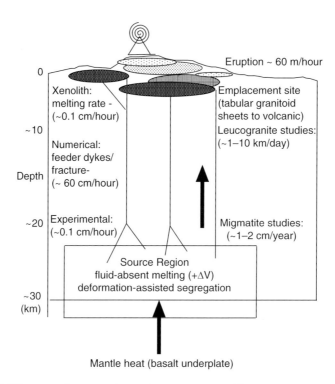

Eruption ~ 60 m/hour

Xenolith:
melting rate -
(~0.1 cm/hour)

Emplacement site
(tabular granitoid
sheets to volcanic)
Leucogranite studies:
(~1–10 km/day)

Numerical:
feeder dykes/
fracture-
(~ 60 cm/hour)

Depth

Experimental:
(~0.1 cm/hour)

Migmatite studies:
(~1–2 cm/year)

Source Region
fluid-absent melting (+ΔV)
deformation-assisted segregation

0

~10

~20

~30
(km)

Mantle heat (basalt underplate)

Fig. 9.14 Summary of different melt segregation and transport rates. These rates represent a fracture dominated system, thus the rates shown here are likely faster than in tectonically inactive areas. Rates for melt segregation in the lower crust come from Sawyer (1991) and Watt *et al.* (1996) from their studies on disequilibrium in leucosomes. Both studies show rates have been enhanced by deformation, plus positive volume changes during melting. Note, we have considered thermally driven melting for the lower crust and not fluid-influx, which will change the melting reactions from fluid-absent to fluid-present and the likely the rate of melting. Observations from the experimental results on muscovite and biotite under lower crustal conditions described here show that melt moves along and through grains to the edge of the capsule during the experimental run time, providing us with a rough estimate of migration rate. Clemens & Mawer (1992) consider fracture propagation, triggered by positive volume reactions, for granitic melt transport through the crust. In their numerical treatment of the problem, they show that granitic magmas are able to travel rapidly through fractures (dykes), to high crustal levels, without freezing. Their results imply filling times (emplacement) can build batholiths <1 ka, which agrees with Harris *et al.* (2000). Eruption rates are extremely rapid. The estimate of 60 m/hour comes from two experimental studies by Rutherford & Hill (1993) and Castro & Dingwell (2009). Finally, a melting rate is given for the upper crust. The rate is from a combined theoretical and experimental study on xenoliths by McLeod & Sparks (1998).

tion study allows for a large pressure gradient to be generated, but we also see melt accumulation on the capsule edge in some static experiments. Therefore, melt migrates at this rate without the aid of deformation (Connolly *et al.*, 1997).

Clemens & Mawer (1992) modeled fracture propagation and granite magma transport by dykes from the source to emplacement. They

considered that fracturing developed during positive-volume melting reactions. Their results show that granitic magmas are able to travel rapidly through fractures to high crustal levels, without reaching thermal 'death' by freezing. Their transport rates are ~60 cm/hour. This suggests that magmas can travel 20 km in about 8 months. Thus, filling times (emplacement) for even large batholithic complexes can be achieved in <1 ka, which agrees with Harris *et al.* (2000).

Figure 9.14 also includes results from two experimental studies by Rutherford & Hill (1993) and Castro & Dingwell (2009), who both found eruption rates on the order of ~60 m/hour. Rutherford & Hill (1993) adopted an experimental approach to amphibole breakdown in lavas from Mt St Helens to estimate that magma ascent from 8 km depth to the surface occurred at between 15 and 66 m/hr or 131 to 578 km/yr. Castro & Dingwell (2009) used a combination of petrological, experimental and rheological considerations to evaluate magma transport at the Chaiten volcano in Chile, which recently erupted rhyolitic magma with very little warning. Their results suggested that rhyolite transport from the magma chamber (~5 km) to the near surface occurred in about 4 hours, or also about 60 m/hr. The estimated eruption rates from these two studies are in good agreement.

We also include a melting rate from McLeod & Sparks (1998). Their approach combined theoretical and experimental study of xenoliths assimilation. They find that melting rates are controlled by several variables, including magmatic temperature, initial xenolith temperature, melting temperature and melt viscosity. Melting rates are on the order of 1 to 2 mm/hr and are most rapid when the xenolith temperature is close to the host magma temperature. These results are consistent with a growing body of theoretical and field evidence, indicating that melting and assimilation in the upper crust can be rapid and efficient processes, occurring over timescales of decades or less.

CONCLUDING REMARKS

For future progress in this field, there is a need to incorporate more numerical modelling. Current models need to address fluid fluxing as a melting agent, incorporate the influence of melt extraction on melting and melt-source evolution, and be thermodynamically constrained (calculated) so that melt compositions change for each time step (Jackson *et al.*, 2003). They also need the ability to assess the 'true' shorter repose intervals to melting

and melt extraction. In terms of experimentation, we still need detailed experimental (static and deformation) investigations of melt reactions, connectivity, migration and chemistry, but with special attention to accessory minerals and spatial variations in composition. The development of techniques that allow *in-situ* measurements has been, and will continue to be, a major factor in our improved understanding of melting and segregation on the grain-scale.

This is GEMOC publication #674.

REFERENCES

Annen C. Sparks RSJ. 2002. Effects of repetitive emplacement of basaltic intrusions on thermal evolution and melt generation in the crust. *Earth and Planetary Science Letters* **203**: 937–955.

Annen C, Blundy JD, Sparks RSJ. 2006. The genesis of intermediate and silicic magmas in deep crustal hot zones. *Journal of Petrology* **47**: 505–539.

Ayers M, Harris N. 1997. REE fractionation and Nd-isotope disequilibrium during crustal anatexis: constraints from Himalayan leucogranites. *Chemical Geology* **139**: 247–267.

Bacon CR. 1992. Partially melted granodiorite and related rocks ejected from Crater Lake caldera, Oregon. *Transactions of the Royal Society of Edinburgh* **83**: 27–47.

Bacon CR, Druitt TH. 1988. Compositional evolution of the zoned calcalkaline magma chamber of Mt Mazama, Crater Lake, Oregon. *Contributions to Mineralogy and Petrology* **98**: 224–256.

Bacon CR, Gunn SH, Lanphere MA, Wooden JL. 1994. Multiple isotopic components in Quaternary volcanic rocks of the Cascade Arc near Crater Lake, Oregon. *Journal of Petrology* **35**: 1521–1556.

Bacon CR, Lowenstern JB. 2005. Late Pleistocene granodiorite source for recycled zircon and phenocrysts in rhyodacite lava at Crater Lake, Oregon. *Earth and Planetary Science Letters* **233**: 277–293.

Bea F, Montero P. 1999. Behavior of accessory phases and redistribution of Zr, REE, Y, Th, and U during metamorphism and partial melting of metapelites in the lower crust: An example from the Kinzigite Formation of Ivrea-Verbano, NW Italy. *Geochimica et Cosmochimica Acta* **63**: 1133–1153.

Beard SJ, Ragland PC, Rushmer T. 2004. Petrogenetic implications of reactions between anhydrous minerals and hydrous melt to yield amphibole and biotite (hydration crystallization). *Journal of Geology* **112**: 617–621.

Beard JS, Ragland PC, Crawford ML. 2005. Reactive bulk assimilation: A model for crust-magma mixing in silicic magmas. *Geology* **33**: 681–684.

Beaumont C, Ngugen MH, Jamieson RA, Ellis S. 2006. Crustal flow modes in large hot orogens. In: *Channel Flow, Ductile Extrusion and Exhumation in Continental Collision Zones*, Law RD, Searle MP, Godin L (eds), **2**: 91–145. Geological Society Special Publication, London.

Bergantz GW, Barboza SA. 2006. Elements of a modeling approach to the physical controls on crustal differentiation. In: *Evolution and Differentiation of the Continental Crust*, Brown M, Rushmer T (eds), Cambridge University Press, Cambridge, 520–549.

Brown M. 2004. Melt extraction from lower continental crust. *Transactions of the Royal Society of Edinburgh: Earth Sciences* **95**: 35–48.

Brown M. 2007. Crustal melting and melt extraction, ascent and emplacement in orogens: Mechanisms and consequences. *Jounal of the Geological Scoiety, London* **164**:709–730.

Brown, M. & Rushmer, T. 1997. The role of deformation in the movement of granite melt: Views from the laboratory and the field. In *Deformation-enhanced Fluid Transport in the Earth's Crust and Mantle* (ed. M. B. Holness), pp. 111–144. The Mineralogical Society Series: 8. Chapman and Hall, London.

Brown M, Solar GS. 1999. The mechanism of ascent and emplacement of granite magma during transpression: a syntectonic granite paradigm. *Tectonophysics* **312**: 1–33.

Büsch W, Schneider G, Mehnert KR. (1974). Initial melting at grain boundaries. Part II: Melting in rocks of granodioritic, quartzdioritic and tonalitic composition. *Neues Jahrb Mineral Monatsch* **8**: 345–370.

Castro JM., Dingwell DB. 2009. Rapid ascent of rhyolitic magma at Chaite'n volcano, Chile. *Nature* **461**: 780–784.

Clemens JD, Mawer CK. 1992. Granitic magma transport by fracture propagation. *Tectonophysics* **204**: 339–360.

Connolly JAD. 1997. Devolatilization-generated fluid pressure and deformation propagated fluid flow during prograde regional metamorphism. *Journal of Geophysical Research* **102**(18): 149–73.

Connolly JAD, Thompson AB. 1989. Fluid and enthalpy production during regional metamorphism. *Contributions to Mineralogy and Petrology* **102**: 347–366.

Connolly JAD, Holness MB, Rubie DC, Rushmer T. 1997. Reaction-induced microcracking: An experimental investigation of a mechanism for enhancing anatectic melt extraction. *Geology* **25**: 591–594.

Costa F, Dungan M. 2005. Short timescales of magmatic assimilation from diffusion modeling of multiple elements in olivine. *Geology* **33**: 837–840.

Cruden AR. 2006. Emplacement and growth of plutons: implications for rates of melting and mass transfer in continental crust. In: *Evolution and Differentiation of the Continental Crust*, Brown M, Rushmer T (eds), Cambridge University Press, Cambridge, 455–519.

Davies GR, Tommasini S. 2000. Isotopic disequilibrium during rapid crustal anatexis: implications for petrogenetic studies of magmatic processes. *Chemical Geology* **162**: 169–191.

Dell'Angelo LN, Tullis J. 1988. Experimental deformation of partially melted granitic aggregates. *Journal of Metamorphic Geology* **6**: 495–515.

Dipple GM, Ferry JM. 1992. Metasomatism and fluid flow in ductile fault zones. *Contributions to Mineralogy and Petrology* **112**: 149–164.

Dufek J, Bergantz GW. 2005. Lower crustal magma genesis and preservation: a stochastic framework for the evaluation of basalt-crust interaction. *Journal of Petrology* **46**: 2167–2195.

Dungan MA. 2004. Partial melting at the Earth's surface: implications for assimilation rates and mechanisms in subvolcanic intrusions. *Earth and Planetary Science Letters* **140**: 193–203.

Edwards BR, Russell JK. 1998. Timescales of magmatic processes: new insights from dynamic models for magmatic assimilation. *Geology* **26**: 1103–1106.

Gardien V, Thompson AB, Grujic D, Ulmer P. 2002. Experimental melting of biotite + plagioclase + quartz ±; muscovite assemblages and implications for crustal melting. *Journal of Geophysical Research* **100**: 15,581–15,591.

Gardner JE, Layer PW, Rutherford MJ. 2002. Phenocrysts versus xenocrysts in the youngest Toba Tuff: implications for the petrogenesis of $2,800 \text{ km}^3$ of magma. *Geology* **30**: 347–350.

Grapes RH. 1986. Melting and thermal reconstitution of pelitic xenoliths, Wehr Volcano, East Eifel, West Germany. *Journal of Petrology* **27**: 343–396.

Green NL. 1994. Mechanism for middle to upper crustal contamination: Evidence from continental-margin magmas. *Geology* **22**: 231–234.

Grove TL, Kinzler RJ, Baker MB, Donnelly-Nolan JM, Lesher CE. 1988. Assimilation of granite by basaltic magma at Burnt Lava flow, Medicine Lake volcano, northern California: Decoupling of heat and mass

transfer. *Contributions to Mineralogy and Petrology*
99: 320–343.

Hammouda T, Pichavant M, Chaussidon M. 1996.
Isotopic equilibration during partial melting: An
experimental test of the behaviour of Sr. *Earth and
Planetary Science Letters* **144**: 109–121.

Harris C, Bell JD. 1982. Natural partial melting of syen-
ite blocks from Ascension Island. *Contributions to
Mineralogy and Petrology* **79**: 107–113.

Harris NBW, Ayres M, Massey J. 1995. Geochemistry of
granitic melts produced during the incongruent melting of muscovite: Implications for the extraction of
Himalayan leucogranite magmas. *Journal of
Geophysical Re*search **100**: 15,767–15,778.

Harris NBW, Vance D, Ayres MW. 2000. From sediment
to granite: timescales of anatexis in the upper crust.
Chemical Geology **162**: 155–167.

Heliker C. 1995. Inclusions in Mt St. Helens dacite
erupted from 1980 through 1983. *Journal of Volcanology
and Geothermal Research* **66**: 115–135.

Hodges KV. 2006. Evolution of Ideas on channel flow
and ductile extrusion in the Himalaya-Tibetan
Plateau System: A synthesis of the Channel Flow-
Extrusion hypothesis as developed for the Himalayan-
Tibetan orogenic system. *Geological Society, London,
Special Publications* **268**: 71–90. doi: 10.1144/GSL.
SP.2006.268.01.04.

Holyoke C, Rushmer T. 2002. An experimental study
of grain-scale melt segregation mechanisms in
crustal rocks. *Journal of Metamorphic Geology* **20**:
493–512.

Jackson MD, Cheadle MJ, Atherton MP. 2003.
Quantitative modeling of granitic melt generation
and segregation in the continental crust. *Journal of
Geophysical Research* **108**: Article Number 2332,
doi:10.1029/2001JB001050.

Kaczor SM, Hanson GN, Peterman ZE. 1988.
Disequilibrium melting of granite at the contact with
a basic plug: A geochemical and petrographic study.
Journal of Geology **96**: 61–78.

Klepeis KA, Clarke, GL, Rushmer T. 2003. Magma
transport and coupling between deformation and
magmatism in the continental lithosphere. *GSA
Today* **13**(1): 4–11.

Knesel KM, Davidson JP. 1996. Isotopic disequilibrium
during melting of granite and implications for crustal
contamination of magmas. *Geology* **24**: 243–246.

Knesel KM, Davidson JP. 1999. Sr isotope systematics
during melt generation by intrusion of basalt into
continental crust. *Contributions ot Mineralogy and
Petrology* **136**: 285–295.

Knesel KM, Davidson JP. 2002. Insight into collisional
magmatism from isotopic fingerprints of melting
reactions. *Science* **296**: 2206–2208.

Knesel, KM, Duffield WA. 2007. Gradient in silicic
eruptions caused by rapid inputs from above and
below rather than protracted chamber differentiation.
Journal of Volcanology and Geothermal Research
167: 181–197.

Knopf A. 1918. A geologic reconnaissance of the Inyo
Range and the eastern slope of the southern Sierra
Nevada, California. US Geol Survey Prof. Paper 110.

Knopf A. 1938. Partial fusion of granodiorite by intrusive basalt, Owens Valley, California. *American
Journal of Science* **235–236**: 373–376.

Laughlin AW, Brookins DG, Carden JR. 1972. Variations
in the initial strontium ratios of a single basalt flow.
Earth and Planetary Science Letters **14**: 79–82.

Laporte D. 1994. Wetting behavior of partial melts during crustal anatexis: the distribution of hydrous silicic melts in polycrystalline aggregates of quartz.
Contributions to Mineralogy and Petrology **116**: 486–
499. doi: 10.1007/BF00310914.

Laporte D, Watson EB. 1995. Experimental and theoretical constraints on melt distribution in crustal
sources – the effect of crystalline anisotropy on melt
interconnectivity. *Chemical Geology* **124**: 161–84.

Laporte D, Rapaille C, Provost A. 1997. Wetting angles,
equilibrium melt geometry, and the permeability
threshold of partially molten crustal protoliths. In:
*Granite: From Segregation of Melt to Emplacement
Fabrics*, Bouchez JL, Hutton DHW, Stephens WE (eds),
Kluwer Academic Publishers, Dordrecht, 31–54.

La Tourrette TZ, Burnett DS, Bacon CR. 1991, Uranium
and minor-element partitioning in Fe-Ti oxides and
zircon from partially melted granodiorite, Crater
Lake, Oregon. *Geochimica et Cosmochimica Acta*
55: 457–469.

Le Fort P, Cuney M, Deniel C, *et al.* 1987. Crustal generation of Himalayan leucogranite. *Tectonophysics*
134: 39–57.

Le Maitre RW. 1974. Partially fused granite blocks from
Mt Elephant, Victoria, Australia. *Journal of Petrology*
15: 403–412.

Manning CE, Ingebritsen SE. 1999. Permeability of the
continental crust: implications of geothermal data
and metamorphic systems. *Reviews of Geophysics*
37: 127–50.

Maury RC, Bizouard H. 1974. Melting of acid xenoliths
into a basanite: An approach to possible mechanisms
of crustal contamination. *Contributions to
Mineralogy and Petrology* **48**: 275–286.

McLeod P, Sparks RSJ. 1998. The dynamics of xenolith assimilation. *Contributions to Mineralogy and Petrology* **132**: 21–33.

Mehnert KR, Büsch W, Schneider G. 1973. Initial melting at grain boundaries of quartz and feldspar in gneisses and granulites. *Neues Jahrb Mineral Monatsch* **4**: 165–183.

van der Molen I, Paterson MS. 1979. Experimental deformation of partially-melted granite. *Contributions to Mineralogy and Petrology* **70**: 299–318.

Nabelek PI, Glascock MD. 1995. REE-depleted leucogranites, Black Hills, South Dakota: a consequence of disequilibrium melting of monazite-bearing schists. *Journal of Petrology* **36**: 1055–2322.

Oliver NHS. 1996. Review and classification of structural controls of fluid flow during regional metamorphism. *Journal of Metamorphic Geology* **14**: 477–492.

Patino-Douce AE, Harris NBW. 1998. Experimental constraints on Himalayan anatexis. *Journal of Petrology* **39**: 689–710.

Petford N, Kerr RC, Lister JR. 1993. Dike transport of granitoid magmas. *Geology* **21**: 845–848.

Petford N, Koenders MAC. 1998. Self-organization and fracture connectivity in rapidly heated continental crust. *Journal of Structural Geology* **20**: 1425–1434.

Petford N, Cruden AR, McCaffery KJW, Vigneresse J-L. 2000. Granite magma formation, transport and emplacement in the Earth's crust. *Nature* **408**: 669–673.

Petford N, Gallagher K. 2001. Partial melting of mafic (amphibolitic) lower crust by periodic influx of basaltic magma. *Earth and Planetary Science Letters* **193**: 483–499.

Philpotts AR, Asher PM. 1993. Wallrock melting and reaction effects along the Higganum diabase dike in Connecticut: Contamination of a continental flood basalt feeder. *Journal of Petrology* **34**: 1029–1058.

Pushkar P, Stoeser DB. 1975. ^{87}Sr/^{86}Sr ratios in some volcanic rocks and some semifused inclusions of the San Francisco volcanic field. *Geology* **3**: 669–671.

Ramos FC, Wolff JA, Tollstrup DL. 2005 Sr isotopic disequilibrium in Columbia River flood basalts: Evidence for rapid shallow-level open-system processes. *Geology* **33**: 457–460.

Rapp RP, Watson EB. 1986. Monazite solubility and dissolution kinetics: implications for the thorium and light rare earth chemistry of felsic magmas. *Contributions to Mineralogy and Petrology* **94**: 304–316 doi: 10.1007/BF00371439.

Roscoe R. 1952. The viscosity of suspensions of rigid spheres. *British Journal of Applied Physics* **3**: 267–269.

Rosenberg CL, Handy MR. 2005. Experimental deformation of partially melted granite revisited: Implications for the continental crust. *Journal of Metamorphic Geology* **23**: 19–28.

Rushmer T. 1995. An experimental deformation study of partially molten amphibolite: Applications to low-fraction melt segregation. In: *Journal Segregation of Melts from Crustal Protoliths: Mechanisms and Consequences*, Brown M, Rushmer, T, Saywer EW (eds), *Journal of Geophysical Research Special Section* **100**(B8): 15,681–15,696.

Rushmer T. 1996. Melt segregation in the lower crust: how have experiments helped us? In: *The Origin of Granites and Related rocks – Third Hutton Symposium*. Brown M, Candela P. (eds), *Transactions of the Royal Society of Edinburgh: Earth Sciences* **87**: 73–83.

Rushmer T. 2001. Volume change during partial melting reactions: Implications for melt extraction, melt geochemistry and crustal rheology. *Tectonophysics* **342**: 389–405.

Rutherford MJ, Hill, PM. 1993. Magma ascent rates from amphibole breakdown: an experimental study applied to the 1980–1986 Mt St Helens eruption. *Journal of Geophysical Research* **98**: 19,667–19,686.

Rutter E, Newmann D. 1995. Experimental deformation of partially molten Westerly granite under fluid-absent conditions, with implications for the extraction of granitic magmas. *Journal of Geophysical Research* **100**: 15,697–15,715.

Sawyer EW. 1988. Formation and evolution of granite magmas during crustal reworking, the significance of diatexites. *Journal of Petrology* **39**: 1147–1167.

Sawyer EW. 1991. Disequilibrium melting and the rate of melt-residuum separation during migmatization of mafic rocks from the Grenville Front, Quebec. *Journal of Petrology* **32**: 701–738.

Sigurdsson H. 1968. Petrology of acid xenoliths from Surtsey. *Geology Magazine* **105**: 440–453.

Tepley FJ, Davidson, JP. 2003. Mineral-scale Sr-isotope constraints on magma evolution and chamber dynamics in the Rum layered intrusion, Scotland. *Contributions to Mineralogy and Petrology* **145**: 628–641.

Teyssier C, Vanderhaege O. 2001. Partial melting and the flow of orogens. *Tectonophysics* **342**: 451–472.

Thompson AB, Connolly JAD. 1995. Melting of the continental crust: Some thermal and petrological constraints on anatexis in continental collision zones and other tectonic settings. *Journal of Geophysical Research* **100**: 15,565–15,579.

Tommasini S, Davies GR. 1997. Isotopic disequilibrium during anatexis: a case study of contact melting,

Sierra Nevada, California. *Earth and Planetary Science Letters* **148**: 273–285.

Verplanck PL, Farmer GL, McCurry M, Mertzman S, Snee LW. 1995. Isotopic evidence on the origin of compositional layering in an epizonal magma body. *Earth and Planetary Science Letters* **136**: 31–41.

Vigneresse J-L, Burg J-P, Moyen J-F. 2008. Instabilities development in partially molten rocks. *Boll. Soc. Geol. It.* **127**: 235–242.

Viskupic K, Hodges KV, Bowring SA. 2005. Timescales of melt generation and the thermal evolution of the Himalayan metamorphic core, Everest region, eastern Nepal. *Contributions to Mineralogy and Petrology* **149**: 1–21 doi: 10.1007/s00410-004-0628-5.

Watson EB, Harrison TM. 1983. Zircon saturation revisited: temperature and compositional effects in a variety of crustal magma types. *Earth and Planetary Science Letters* **64**: 295–304.

Watson EB. 1996. Dissolution, growth and survival of zircons during crustal fusion: Kinetic principles, geological models and implications for isotopic inheritance. *Transactions of the Royal Society of Edinburgh: Earth Science* **87**: 43–56.

Watson EB, Vicenzi EP, Rapp RP. 1996. Inclusion/host relations involving accessory minerals in high grade metamorphic and anatectic terrains. *Contributions to Mineralogy and Petrology* **101**: 220–231.

Watt GR, Burns IM, Graham GA. 1996. Chemical characteristics of migmatites: Accessory phase distribution and evidence for fast melt segregation rates. *Contributions ot Mineralogy and Petrology* **125**: 100–111.

Watt GR, Oliver NHS, Griffen BJ. 2000. Evidence for reaction-induced microfracturing in granulite facies migmatites. *Geology* **28**(4): 327–330.

Wolf MB, Wyllie PJ. 1991. Dehydration-melting of solid amphibolite at 1o Kbar: textural development, liquid interconnectivity and applications to the segregation of magmas. *Mineralogy and Petrology* **44**: 151–179.

Wolf MB, Wyllie PJ. 1995. Liquid segregation parameters from amphibolite dehydration melting experiments. *Journal of Geophysical Research* **100**: 15,611–15,621.

Yang P, Rivers T. 2000. Trace element portioning between coexisting biotite and muscovite from metamorphic rocks, Western Labrador: Structural, compositional and thermal controls. *Geochimica et Cosmochimica Acta* **64**: 1451–1472.

Zeng L, Asimow P, Saleeby J. 2005. Coupling of anatectic reactions and dissolution of accessory phases and the Sr and Nd isotope systematics of anatectic melts from a metasedimentary source. *Geochimica et Cosmochimica Acta* **69**(14): 3671–3682.

10 Timescales Associated with Large Silicic Magma Bodies

OLIVIER BACHMANN

Department of Earth and Space Sciences, University of Washington, Seattle, WA, USA

SUMMARY

While emptying >100–1,000 km³ of silicic magma from upper crustal reservoirs takes a few weeks at most, the accumulation of these gigantic magma bodies, and therefore the recurrence intervals between supervolcanic eruptions, is up to 10^7 times longer; high-precision geochronology providing eruption ages (^{39}Ar-^{40}Ar) and magma residence times (U-Pb and ^{238}U-^{230}Th) indicate that the growth and maturation of these bodies in the upper crust generally takes between 10^5 and 10^6 years. During that extended period of time, we now know that many processes, such as recharge from deeper sources, convective overturn, crystal growth and dissolution, crystal-liquid separation, can repeatedly occur. While some processes (e.g. cooling and crystallization) will tend to trap magmas in the crust, others (e.g. recharge, partial melting, convective overturn) will favor eruption. Better constraining the relative importance and rates of these competing processes involved in the formation and evolution of large silicic magma bodies will lead us to more accurate predictions on when and where the next cataclysmic event will take place.

Timescales of Magmatic Processes: From Core to Atmosphere, 1st edition. Edited by Anthony Dosseto, Simon P. Turner and James A. Van Orman.

INTRODUCTION

Discovering the existence of large, eruptible silicic magma bodies (with SiO_2 contents >~60 wt.%) in the Earth's crust is a recent fundamental advance in volcanology. About 50 years ago, with the advent of modern geoanalytical methods, scientists studying volcanic products in long-lived magmatic provinces started realizing that some eruptions involved gigantic volumes of magma; up to several thousands of cubic kilometers (km³) evacuated from shallow reservoirs in just a few days to weeks. These are now commonly called 'supereruptions' (see the February 2008 issue of *Elements*) and are equivalent to erupting, in a geological instant, up to 10 to 30 times the total volumes of large, long-lived stratovolcanoes (i.e. Mt Rainier in the Cascades Range in the western USA; (Figure 10.1). Apart from the eruption that destroyed part of the island of Santorini (Greece) around ~1,600 BC, such events have never been recorded in any detail by human society, but their deposits and associated craters (collapse structures named 'calderas') provide unambiguous evidence for the presence of kilometer-sized dominantly liquid silicic magma bodies in the upper crust (Smith, 1960; Lipman, 1984).

Realizing that volcanic eruptions can span an enormous range of erupted volumes, volcanologists have designed scales to rank explosive magnitudes of eruptions. The most commonly used scale, proposed in 1982 by Newhall & Self

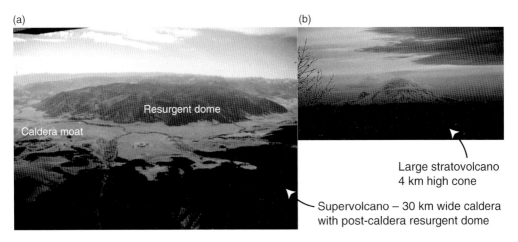

Fig. 10.1 (a) The 26.87 Ma Creede caldera in the Southern Rocky Mountain volcanic field, CO, source of the Snowshoe Mountain Tuff. Snowshoe Mountain in the resurgent dome in the center of the picture, which formed by intrusion-related uplift after the caldera collapsed. The Creede eruption emptied a magma chamber about 2 to 5 times the volume of Mt Rainier (b), taken from Seattle, WA. Some super-eruptions can evacuate magma bodies 10 times the size of these large stratovolcanoes in just a few days.

Table 10.1 Range and characteristics of volcanic eruptions occurring on Earth (based on Newhall & Self, 1982)

VEI	Description	Plume height	Volume	Example	Number of occurences over last 10 ky*
0	non-explosive	<100 m	1,000s m³	Kilauea	>10,000
1	gentle	100–1000 m	10,000s m³	Stromboli	>10,000
2	explosive	1–5 km	0.001 km³	Galeras, 1992	>3,400
3	severe	3–15 km	0.01 km³	Ruiz, 1985	>850
4	cataclysmic	10–25 km	0.1 km³	Galunggung, 1982	>270
5	paroxysmal	>25 km	1 km³	St. Helens, 1981	>80
6	colossal	>25 km	10s km³	Krakatau, 1883	>35
7	super-colossal	>25 km	100 km³	Tambora, 1815	>5
8	mega-colossal	>25 km	1,000s km³	Yellowstone, 2 Ma	0 (last at 26 ky)

*Number of occurrences is approximate and based on accounts from the Global Volcanism Program of the Smithsonian Institution (http://www.volcano.si.edu/). Plume height refers to as the maximum altitude reached by the ash column ("plume") during eruption.

(Table 10.1 and Figure 10.2), is called the Volcanic Explosivity Index (VEI), and is based on the volume of erupted material. This scale grades from VEI 1, characterizing 'gentle' small volume lava eruptions occurring daily at volcanoes such as Kilauea (Hawaiian Islands, USA) to VEI 8 'mega-colossal' eruptions that occur very infrequently

(approximately every 50,000–100,000 years, see following discussion) and lead to gigantic outpourings of rock fragments into the atmosphere. The Magnitude scale (Pyle, 2000) is a similar method to rank volcanic eruptions (although it is based on the mass of a given eruption, using erupted volume divided by estimated average

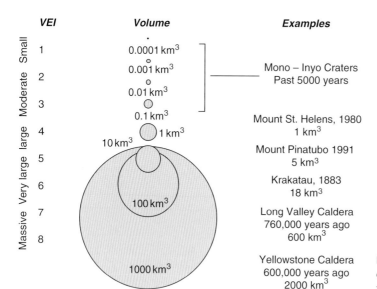

Fig. 10.2 Comparing size of different eruptions within the VEI scale framework (from Newhall & Self, 1982).

densities of deposits), and is calculated by the following equation:

$$M = \log_{10}(m)-7.0$$

where m is the mass of the eruption in kilograms. Estimating erupted volumes can be challenging, particularly for old eruptions, and/or in areas covered by sea or forests. The VEI and magnitude scale classifications are only as good as the volume estimates are.

A better understanding of explosive volcanic eruptions becomes a more urgent priority every day as the human population expands (see the Naples area for a striking example of highly populated and volcanically highly active region; Figure 10.3). This is particularly true for large explosive eruptions (VEI ≥6, with volumes ≥10 km³) that are known to dramatically impact life on Earth (both directly because of eruption related hazards and indirectly because of resulting catastrophic climate repercussions; Robock, 2000). Infamous examples include the Minoan eruption of Santorini (VEI 6, ~30 km³), which may have partly triggered the decline of the Minoan civilization

(LaMoreaux, 1995), and the eruption of the Young Toba Tuff (VEI 8, ~2,500 km³), which is thought to have nearly brought the human species to extinction around 70,000 years ago (Rampino & Self, 1992; Ambrose, 1998; Rampino & Ambrose, 2000). The 20th century was fortuitously largely devoid of volcanic events on this scale, as the only eruption with over 10 km³ of erupted material, the Valley of Ten Thousand Smokes (VTTS) ignimbrite, occurred early in 1912 in a remote area (Alaska Peninsula, Hildreth & Fierstein, 2000). The most recent event that created a significant environmental impact on the planet was the 1991 eruption of Mt Pinatubo (Luzon, Philippines; 5–7 km³ of erupted S-rich material), which ejected ~20 Mt of SO_2 into the atmosphere (Wallace & Gerlach, 1994) and induced significant global cooling (Soden et al., 2002).

The accumulation and rapid evacuation of 100s to 1,000s km³ of silicic magmas (VEI 7–8) does not happen anywhere on Earth, but is favored by a number of geodynamical parameters that restrict their occurrence in some specific areas (Figure 10.4). First, the largest and most silicic supervolcanic eruptions are only found within

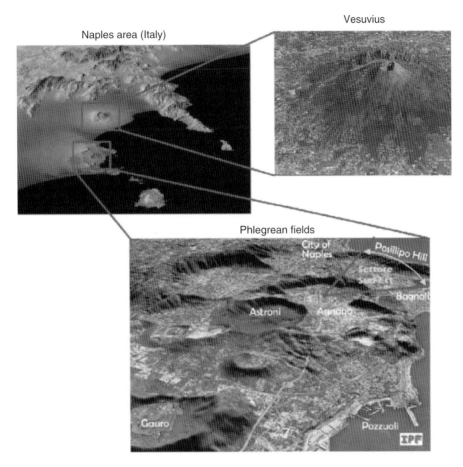

Fig. 10.3 Pictures of the two active volcanic complexes in the area of Naples (Italy), showing densely populated zones in and around the volcanoes (pictures courtesy of Dr A. Neri, INGV Pisa).

continental landmasses, where low-density and fusible pre-existing wall rocks increase the likelihood of forming and accumulating silicic magmas (Hughes & Mahood, 2008; although see Tamura *et al.*, 2009 for rhyolitic units in an oceanic setting). They are also particularly common in subduction zones, where volatile-charged mantle-derived basalts fuel magmatism, creating a particularly explosive mix by the time silicic magmas are produced (Figure 10.4). Within volcanic arcs, these eruptions mostly take place in areas of high convergence rate (Hughes & Mahood, 2008); high basaltic magma generation in the mantle driven by fast convergence fosters a very

productive magmatic distillation column. On the other hand, back-arc spreading appears to impede supereruptions, while duration of magmatism and obliquity of subduction have no apparent effect on where such events occur (Hughes & Mahood, 2008).

An eruption with a volume of $>10\,km^3$, which will have appreciable consequences on a worldwide scale, is statistically likely to occur within the next few decades (Table 10.1). As summarized by a recent report on large volcanic eruptions (Sparks *et al.*, 2005), it is not a question of if, but rather a question of when. The world's population should be aware that such global threats exist,

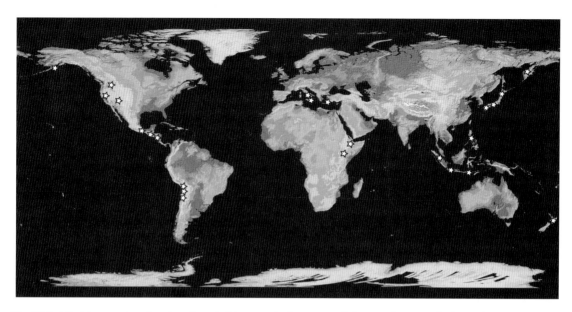

Fig. 10.4 World map showing locations (white stars) of some well-known large explosive volcanic eruptions (^{3}100 km^3) over the last 1 Ma (based on Self 2006 and Sparks *et al.* 2005). Note the abundance of supereruptions around the Pacific Ring of Fire. See Hughes and Mahood 2008 for a more exhaustive database. Base map from http://geology.com/nasa/world-topographic-map/. See Plate 10.4 for a colour version of this image.

and volcanologists are working towards constraining as accurately as possible the time frames related to these phenomena. As many natural hazards, these supervolcanic eruptions are generally perceived as less dangerous than other, manmade threats (i.e. global warming and nuclear wastes), although they can be at least as destructive. As a matter of fact, VEI 8 events are the ultimate geological hazards, even compared to large meteorite impacts, which are five to ten times less likely to occur in the future (Woo, 1999). Better constraining the timescales involved in the construction and eruption of these large silicic magma bodies is obviously of utmost importance in terms of hazards mitigation and our understanding of the processes involved is our only tool to assess the risks and reduce damage.

Constraining timescales of magmatic phenomena is a daunting task, as they span about 14 orders of magnitude. Processes in volcanic conduits during explosive eruptions can occur within seconds (e.g. magma fragmentation), whilst the

lifespan of a large volcanic province commonly exceeds ten millions years, much in excess of what humans can directly observe. This chapter will focus mainly on the longer timescales related to the formation and lifespan of large silicic magma bodies, relying on indirect methods to infer timescales. The more volcanological aspects related to eruption mechanisms are discussed in other chapters of this book and related publications (see suggested reading at the end of this chapter).

Silicic magmas do not necessarily always erupt; vast granitic plutons, such as those found in a nearly continuous chain from Alaska to Southern Chile, also form large silicic magma bodies, although they reach complete solidification before they can escape their reservoirs. What do they tell us about the timescales of silicic magma generation, and how can we link them to the volcanic world? A short subsection on the plutonic perspective will briefly develop some of these topics that remain an active area of research.

HOW TO ESTIMATE TIMESCALES IN LARGE SILICIC MAGMA CHAMBERS?

There are several ways to estimate timescales related to large silicic magmas. The most important ones are:

1 radiometric dating;
2 modeling diffusive re-equilibration of chemical elements in minerals and glasses;
3 modeling rates of physical processes (e.g. how long does melting, cooling, crystal sedimentation, gas percolation, etc. take in the conditions relevant for magmatic systems).

All three methods are discussed below. However, since some aspects (particularly diffusion modeling (Costa & Morgan, this volume) and U-series disequilibrium dating (Dosseto & Turner, this volume)) are treated in-depth in other chapters of this book, they will not be discussed in excessive details in the following pages.

RADIOMETRIC DATING

Due to the time-dependent decay of radioactive nuclides (referred to as *half-life* or *decay rate*), a number of possible techniques to measure timescales in geological material have been developed over the last century (most common are U-Pb, U-Th, K-Ar, Ar-Ar, ^{14}C). Technically, all pairs of radioactive-radiogenic elements (*radioactive* being the unstable parent, and *radiogenic* the stable daughter nuclide) could be used to obtain an age as long as the half-life of the parent is commensurate with the age of the rock of interest, and that the basic tenets of radiometric dating hold:

• The system has to be chemically closed with respect to the elements of interest;
• The number of daughter atoms since the time of closure can be accurately measured (by mass spectrometry); and
• The half-life of the radioactive element is known with precision and accuracy.

In this chapter, we will concentrate on two of the most commonly used methods that are particularly useful for dating large silicic magma bodies:

1 ^{39}Ar-^{40}Ar (related to ^{40}K-^{40}Ar dating method) and
2 U-Pb (and ^{238}U-^{230}Th).

^{14}C can also be used in volcanic areas when the emplacement of lava and pyroclastic flows kills vegetation, but is limited to very young deposits (<~60,000 years). More details on K-Ar, Ar-Ar and U-Th-Pb dating methods can be consulted in McDougall & Harrison, 1999 and Dickins, 1997.

Before going further into geochronology, the concept of closure temperature needs to be introduced. This notion is derived from the fact that chemical elements can diffuse through any medium and the diffusion rate is highly temperature-dependant. In short, at high temperature, radiogenic elements produced by the decay of their parents will exchange quickly with the outside world, and will never accumulate to any extent within their mineral host. In contrast, at low temperatures, daughter products cannot leave the site of decay, and therefore accumulate where they are produced. Of course, depending on the element and the structure of the host, the temperature at which the system starts accumulating the radiogenic elements ('when the clock starts ticking') varies widely. This temperature at which the system becomes closed to diffusive re-equilibration is referred to as the closure temperature.

Dodson (1973) provided a commonly used formulation to estimate the closure temperature:

$$\frac{E}{RT_c} = \ln\left(-\frac{AD_0 RT_c^2}{E\frac{dT}{dt}a^2}\right)$$

where T_c is the closure temperature; E is the activation energy for diffusion of the daughter isotope; R is the gas constant; D_0 is the pre-exponential factor of the diffusion; A is a geometric factor; dT/dt is the cooling rate; and a is the diffusing distance (Dodson, 1973). This equation is valid for the following assumptions:

• A constant cooling rate;
• T_c is a *mean closure temperature* (ignoring compositional variability within the crystal of interest);

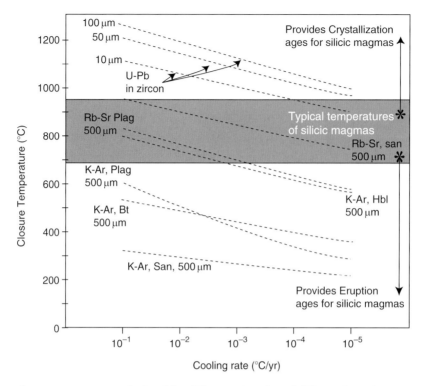

Fig. 10.5 Mean closure temperatures calculated for different minerals and different isotopic systems based on the Dodson (1973) equation. Modified from Costa (2008). Calculations were done using a spherical geometry of 500 microns, except for zircons, for which spheres of 10, 50 and 100 microns were used. Mineral abbreviations: Plag = plagioclase, Bt = biotite, Hbl = hornblende, San = sanidine. See Costa (2008) for source of diffusion coefficients.

- It considers a finite crystal within an infinitely large reservoir characterized by infinitely fast diffusion; and
- Sufficient diffusion has occurred that the whole volume of the crystal has been chemically modified.

K-Ar, and its derivative, Ar-Ar, involves the noble gas argon as the radiogenic product of potassium decay. Since Ar does not fit well and is loosely bonded to any mineral or melt structure in which it is released, its closure temperature is generally low (<650–700°C but generally below 400°C for most realistic cooling rates and common hosts). Therefore, K-bearing minerals or melts remain open to diffusive exchange with surroundings above these temperatures (Figure 10.5). As silicic magmas reach complete solidification

(defined as the *solidus* of the rock) around ~650 to 700°C, the K-Ar and Ar-Ar method can only cross its closure temperature during rapid cooling due to eruption. This method therefore records the eruption age of the unit of interest (Figure 10.5).

U-Pb dating behaves in an entirely different way than K-Ar and Ar-Ar. U and Pb atoms diffuse very slowly in most hosts, including in the mineral zircon, which is by far the most commonly used host for obtaining U-Pb ages in magmatic rocks. Closure temperatures for the U-Pb in zircon are in excess of 900°C for most geologically reasonable conditions (Lee *et al.*, 1997). Therefore, ages obtained by this method provide the crystallization age of the analysed material, whatever happens to it afterwards (e.g. eruption, reheating, exhumation; Figure 10.5). This is the reason why zircon yields the oldest ages found on Earth (~ 4,400 Ma; Wilde *et al.*, 2001).

Most volcanologists tend to be drawn towards active magmatic provinces, for obvious reasons. Therefore, many zircons extracted from silicic volcanic rocks are so young that the amount of radiogenic Pb produced since crystallization is below detection limit (because of the long half-life of ^{238}U). Fortunately, there is a second way to obtain a crystallization age. By exploiting the fact that ^{238}U and ^{206}Pb are separated by a series of intermediate daughter products (often referred to as U-series nuclides, see several chapters in this book; Bourdon & Elliott, this volume; Turner & Bourdon, this volume; Dosseto & Turner, this volume), we can use one of the longest-lived U-series daughter, ^{230}Th, to obtain a 'disequilibrium' age. As zircon crystallizes, it preferentially incorporates U over Th, which causes an initial disequilibrium between these two nuclides. The systems moves back to a stable decay chain ('secular equilibrium') in about 5 half-lives of ^{230}Th (~350,000 years). By knowing the relative abundance of ^{238}U and ^{230}Th corrected for their decay rate, an age can be obtained. Therefore, U-Pb and ^{238}U-^{230}Th dating of zircons can provide crystallization ages from a few thousands of years to the age of the Earth.

Recurrence intervals between supervolcanic eruptions

The first and most obvious way of estimating the recurrence interval between large volcanic eruptions is to determine their eruption ages using K-Ar or Ar-Ar dating. Over the last half century, a large number of volcanic units have been dated, allowing a statistical estimate of the time between eruptions of a given volume (or mass). The VEI scale has this recurrence interval built into its definition (daily for VEI 1–2 and >10,000 years for VEI 8, Table 10.1 and Figure 10.6). Similar recurrence intervals can be obtained with the magnitude scale (see Mason *et al.*, 2004 for compilation of all of the largest known eruptions on Earth).

A first-order observation that arises from these studies is that the longer the volcano waits, the

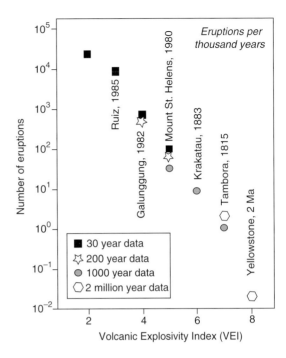

Fig. 10.6 Recurrence intervals for eruptions at a given Volcanic Explosivity Index (VEI). Data points were gathered at different intervals (the more common low VEI eruptions gathered on a shorter timescales). Modified from Simkin, 1993.

bigger the next eruption will be. This inverse correlation between size and recurrence interval, highlighted already 40 years ago (Smith, 1979), is a common feature of natural systems; earthquake magnitude and sizes of meteorite impact display the same trend. In detail, however, there are significant variations in space and time in volcanic provinces; The Taupo Volcanic zone (New Zealand) has been generating (and erupting) a lot of silicic magmas over the last few hundreds of thousands of years (often referred to as 'flare-up' mode), while others areas (e.g. Aegean Arc, Greece) are much less productive despite producing the same magma types, and being in a similar tectonic setting (both the Taupo Volcanic Zone and the Aegean Arc are subduction zones set in an highly extensional area of continental crust). This variation is exemplified by the range in

Table 10.2 Eruption ages of the main ignimbrites of the Oligocene Central San Juan caldera cluster (Southern Rocky Mountain Volcanic Field, Colorado, USA) (from Lipman & McIntosh, 2008)

	Approximate volume (km³)	Eruption Age (Ma) by Ar-Ar data
Masonic park Tuff	500	~28.6
Fish Canyon Tuff	5000	28.03
Carpenter Ridge Tuff	1000	27.55
Blue Creek Tuff	250	~27.4
Wason Park Tuff	500	27.38
Rat Creek Tuff	150	26.91
Cebolla Creek Tuff	250	26.90
Nelson Mountain Tuff	500	26.90
Snowshoe Mountain Tuff	500	26.87

rhyolite magma productivity around the world that spans 3 orders of magnitude (from 10^{-5}–10^{-2} km³/yr, with a global average for silicic magmas of $4 \pm 1.4 \times 10^{-3}$ km³/yr; White et al., 2006), most likely related to variations in major geodynamical parameters, such as mantle productivity and convergence rate (Hughes & Mahood, 2008).

Recurrence intervals calculated on the basis of dating multiple eruptions of a given magnitude worldwide has proved very useful in providing a global statistical average. But can we estimate how often a given volcano erupts in a given time period? For small eruptions on an active volcano (e.g. Hawaiian Islands, USA; Stromboli, Italy; Santiaguito, Guatemala), this can be assessed by recording the activity in the course of a human lifetime (or documented eruptive history). However, for supervolcanoes, we need to find an area that has witnessed numerous large eruptions and obtain precise eruption ages for all of them.

Several areas around the world have emitted multiple supereruptions ('multicyclic caldera complexes'). The youngest such areas are found in the North Island of New Zealand (Taupo Volcanic Zone), in Sumatra (Toba caldera), in the Central Andes (Altiplano Puna Ridge Complex) and in the western USA (Long Valley, Jemez Mountains, Yellowstone). In the course of our planet's evolution, there were many more such areas, some with nearly ten VEI 7–8 eruptions occurring in the same place over just a few millions of years. The southwestern USA (Colorado, Utah, Nevada, New Mexico, Arizona) has experienced such an extreme event in the early to mid-Tertiary (around 35–25 Ma ago). Areas such as the Southern Rocky Mountain Volcanic Field have emitted dozens of large pyroclastic units from focalized areas ('caldera clusters'). For example, the Central San Juan caldera cluster has produced at least nine deposits larger than 100 km³ in <2 Ma (Lipman & McIntosh, 2008; Table 10.2). The recurrence interval in this area only is about a 1 in every 200,000 years, including the largest supervolcanic unit recorded thus far (the 5,000 km³ Fish Canyon Tuff; Lipman, 2000; Mason et al., 2004).

Magma residence times

Another way to obtain information about the timescale related to the construction and lifespan of a magma chamber in the crust is to combine ages obtained by Ar-Ar and U-Th-Pb on the same unit. As Ar-Ar is providing an eruption age and the U-Pb the crystallization age of zircon, this comparison allows us to define the different zircon populations within a magma, and estimate how long zircon crystals have been growing within a given reservoir. This is known as the 'residence time' of the magma body close to or above its solidus.

Due to the robustness of zircon to age resetting, it has been known for a long time that 'xenocrysts' (crystals that are foreign to the

Table 10.3 Examples of silicic magma bodies and their residence times

Unit name	Volcanic Province	Volume	Eruption age	Magma residence time
Fish Canyon Tuff	Southern Rock Mountain Volcanic Field	5,000 km³	28.03 Ma	>300–700 ky
Kos Plateau Tuff	South Aegean Arc	> 60 km³	0.161 Ma	>200–300 ky
Young Toba Tuff	Sumatra	~2,800 km³	0.075 Ma	>400 ky
Huckleberry Ridge Tuff	Yellowstone	~2,500 km³	2.0 Ma	>200 ky
Whakamaru ignimbrite	Taupo	~1,000 km³	0.34 Ma	>250 ky
Oruanui Ignimbrite	Taupo	530 km³	0.026 Ma	> 300 ky
Unzen 1991	Unzen	<1 km³	0.00001 Ma	> 100 ky
La Pacana ignimbrite	Altiplano Puna Ridge Complex	2,700 km³	~4 Ma	>500–750 ky
Bishop Tuff	Long Valley	600 km³	0.76 Ma	>100–350 ky
Ammonia Tank Tuff	SW Nevada volcanic field	900 km³	11.45 Ma	>1500 ky

Residence time is estimated here using the oldest co-magmatic zircon (or zircon area for in-situ technique) dated in a given unit (oldest antecryst) and subtracting the eruption age given by Ar-Ar. When dating multiple zircons, the age distribution frequently forms peaks at certain times, indicating periods on higher zircon preservation and/or crystallization. See Costa, 2008 for list of references.

magma they are in, inherited from older crustal lithologies) are commonly found in magmatic rocks. This recycling of older crystals has proved to be problematic to accurately constrain magma residence time with bulk zircon ages (involving many different crystals) are used. However, with the advent of *in-situ* dating techniques (Secondary Ion Microprobe Spectrometry – SIMS and Laser Ablation Inductively Coupled Plasma Mass Spectrometry – LA-ICPMS), this problem has been reduced. *In-situ* techniques permit the determination of ages within parts of zircon crystals as small as 30 to 40 microns in diameter. Therefore, zircon crystals (or cores of crystals) significantly older than the period of magmatism under investigation can generally be isolated and removed from the zircon population of interest, thus providing the true 'residence time'.

Many volcanic units from around the world, encompassing most large igneous provinces active today (Taupo Volcanic Zone, Toba, Yellowstone, Long Valley Caldera, Aegean Arc, Coso Volcanic Field) have now been dated using both Ar-Ar (on multiple mineral phases) and U-Th-Pb on zircons.

These latter show a surprisingly large range of crystallization ages, several hundreds of thousands to millions of years older than the eruption age (Simon *et al.*, 2008; Bindeman *et al.*, 2006; Costa, 2008; Reid, 2008; Table 10.3). Some single zircon crystals even show >200,000 years of growth history (Brown & Fletcher; 1999; Figure 10.7). These co-genetic zircon crystals would be typically called 'phenocrysts', but as some of them are several hundred of thousands of years older than the eruption age, a new term to defining them has appeared in the literature; antecryst (= old, recycled phenocryst; Charlier *et al.*, 2005; Miller *et al.*, 2007).

Interestingly, small volume volcanic units that have both U-Th-Pb ages on zircons and known eruption ages (either through Ar-Ar dating or historical records; Mt Unzen; Sano *et al.*, 2002, Coso Volcanic Field; Miller & Wooden, 2004, Mt St Helens; Claiborne *et al.*, 2008) also show several hundreds of thousands of years of magma residence time. As stressed by Costa (2008), there is no clear positive correlation between erupted volume and magma residence time on the basis of zircon U-Th-Pb and Ar-Ar dating; small and large

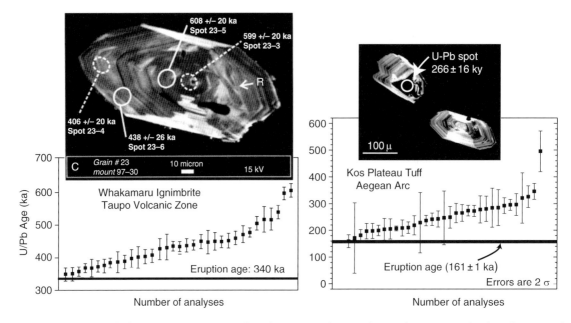

Fig. 10.7 *In-situ* U-Pb dating on zircons. Note that the zircons show evidence of continuous (within the error of individual analysis) crystallization over several hundred thousands years right up to the time of eruption. Modified from Brown & Fletcher, 1999 and Bachmann *et al.*, 2007a.

units carry zircons with approximately the same age range. However, whilst residence times and recurrence intervals are of similar magnitude for very large eruptions (on the order of 100,000 years), recurrence intervals are much shorter for small eruptions; for example, Mt St Helens has erupted over 50 times in the last 4,000 years (Mullineaux & Crandell, 1981), leading to an average of one eruption every ~100 years and a fairly significant volume of erupted magma over this millennium timescale. In fact, when comparing time-averaged volumetric volcanic output rates between small and large eruptions, the number are actually strikingly similar (~4–5 × 10⁻³ km³/yr; see Lipman, 2007 for supervolcanoes and White *et al.*, 2006 for rhyolites in arcs). Therefore, zircon crystallization ages date the growth of the magmatic plumbing system, but are poor indicators to estimate recurrence intervals for small eruptions. Parameters other than magma output rates must play a role on whether a volcanic eruption will be small or large (Jellinek & DePaolo, 2003),

but our understanding of eruption triggers is still limited, and more work is needed in this area.

Large silicic magma bodies: the plutonic perspective

Many silicic plutonic suites have also been dated using the U-Pb and Ar-Ar techniques. As for volcanic rocks, both sets of ages can be compared, although the interpretation of K-Ar and Ar-Ar results on plutons differs from those in volcanic rocks. Plutons crystallize to completion and cool to surface temperatures in millions of years instead of in a few days to months for erupted units. Cooling rate obviously proceeds at a much slower pace in plutons than in volcanic rocks and closure temperatures of the different minerals are reached sequentially (Figure 10.5). By dating feldspars, biotite and hornblende using the Ar-Ar technique, we can actually do *thermochronology* (= determine at what age the pluton reached the closure temperature of the different minerals), but the

Ar-Ar technique does not record a specific crystal-lization event in the lifetime of the pluton.

Despite the fact that the Ar-Ar dates do not have exactly the same meaning in plutonic and vol-canic rocks (cooling vs eruption ages), magma resi-dence times can also be estimated for the plutonic case. A major observation emerging from these studies is that pluton construction is much longer (≥10 Ma in some cases; Coleman *et al.*, 2004; Matzel *et al.*, 2006; Miller *et al.*, 2007; Walker *et al.*, 2007) than the life of a single 'eruptible' body. This 10^7 years timescale is comparable with the lifespan of large volcanic provinces (Oligocene Southern Rocky Mountain Volcanic field, Altiplano Puna Ridge Complex, Table 10.2 and de Silva *et al.*, 2006), but longer than most residence times obtained for a single volcanic unit. Plutonic zircon age distribution clusters in pulses (as do volcanic ones), indicating that they experienced periods of intense zircon crystallization (perhaps immediately following periods of magma replenishment), separated by stages of quiescence. It is also likely that plutonic rocks significantly outweigh their volcanic counterparts (high plutonic:volcanic ratios) during the waning stages of any magmatic provinces; warm crustal condi-tions and/or lower magma fluxes would favor preservation of plutonic bodies (Jellinek & DePaolo 2003; Coleman *et al.*, 2004; de Silva *et al.*, 2006; Bachmann *et al.*, 2007b; Lipman, 2007). Therefore, we should expect the plutonic record to show much longer residence times, particularly biased towards late-stage magmatic activity, whilst volcanic rocks mostly carry information about the high flux period of a given magmatic episode.

MODELING DIFFUSIVE RE-EQUILIBRATION OF CHEMICAL ELEMENTS IN MINERALS AND GLASSES

Many crystals growing from magmas act as 'tape recorders' of changing magmatic conditions. They add layers at their periphery that lock in the composition of their environment. When the environment rapidly changes (in P, T and/or com-position), for example following a magma mixing event, any crystals that were growing in the ini-tial environment have to re-adapt to the new con-ditions. Very commonly, diffusion profiles defined by element concentrations will be preserved in the minerals. They will carry time information that can be quantified if the diffusion coefficient of the element of interest in accurately known.

By analysing compositional variations along core-to-rim traverses in single crystals by *in-situ* geochemical analyses (electron microprobe, SIMS or LA-ICPMS), we can obtain high-resolution pro-files of how different element changes as a func-tion of distance to the rim (Figure 10.8, Panel A). Once such profiles are gathered, modeling tech-niques can be used to estimate the time since the crystal was emplaced into its new environment (Figure 10.8, Panel A). Diffusion timescales vary by several orders of magnitude, depending on the element used in the modeling (e.g. Li diffuse very fast, while Ba is much slower; Turner & Costa, 2007). This technique has also been applied to supervolcanic deposits, and while some processes still happen fast in large chambers (pre-eruptive reheating, Figure 10.8 Panel B; Wark *et al.*, 2007), the longest diffusive timescales (Ba in sanidine; >100,000 years; Morgan & Blake, 2006) are in agreement with crystal residence times in mag-mas measured with radiometric techniques.

MODELING TIME-DEPENDANT PHYSICAL PROCESSES

Melting (within the mantle or the crust), cooling of igneous bodies, magma extraction and trans-port within the crust all take time and constrain the rate of accumulation of large silicic magma bodies. Can we estimate timescales for these dif-ferent processes that occur in the lifetime of magmas?

As laid out by Petford *et al.* (2000), the lifespan of magmas can be divided into four stages:
1 formation (melting);
2 segregation from their source;
3 transport to their storage area; and
4 emplacement into their reservoir.

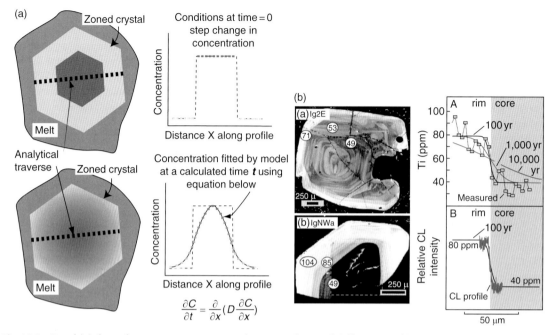

Fig. 10.8   Panel (a) shows how to estimate timescales using chemical diffusion profiles in zoned crystals. D stands for the diffusion coefficient of a given element. Panel (b) is an example of such measurements of timescales in quartz crystals from the Bishop Tuff. CL stands for cathodoluminescence. Numbers in circles are Ti content of different zones in each crystal (in ppm). Scale bars = 250 μm in each. Darkened area in (b), enclosed by dashed line, is due to beam damage (modified from Wark *et al.*, 2007).

Each of these events has a timescale inferred from knowledge of the underlying physics and the assumptions involved. As discussed for diffusion above, physical processes are time-dependant: some occur rapidly (gas-driven overturn in magma chambers; Ruprecht *et al.*, 2008), whilst others are much slower (compaction in silicic magma at low permeability; McKenzie, 1985). By identifying the main parameters of the system (e.g. viscosity, density, temperature gradients, pressure, etc.), we can calculate a range of realistic timescales for all these stages (Figure 10.9).

Whilst magmas are being transported and stored in the crust, time-dependant heat exchange occurs with the surrounding environment. Assuming a closed system (no external addition of heat), a cooling time can be estimated for a given size, magma composition and geothermal gradient. For typical large silicic magma bodies, these cooling times vary between 10^5 and 10^6 years (Spera, 1980). However, we know that magma bodies are also periodically reheated by intrusions of new, hotter magma batches into the base of the magma chamber. This episodic heating, which will ultimately depend on the mass flux of hotter magma re-injected from below (Dufek & Berganz, 2005; Annen, 2009), can significantly prolong the life of a magma body.

The timescales associated with these physical processes are generally poorly known, as evidenced by the large ranges displayed on Figure 10.9. All of them depend on many variables, such as magma temperature, pressure, density, viscosity and crystallinity, as well as system properties (e.g. geometry, volumes of the different bodies, geothermal gradient, magma fluxes), which remain inadequately constrained. Most

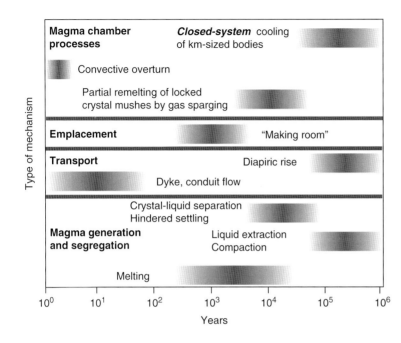

Fig. 10.9 Range of timescales for different physical processes that occur in large silicic magma bodies. See Spera, 1980; Sparks *et al.*, 1984; Bergantz & Ni, 1999; Petford *et al.*, 2000; Bachmann & Bergantz, 2004, 2006; Ruprecht *et al.*, 2008 for references.

of these processes also involve some complex coupling (variables are not independent of each others), inducing either positive or negative feedback and non-linear responses. A magma rising up a volcanic conduit is an example of this complex suite of coupling. Degassing and exsolution of bubbles will lower the bulk magma density, increasing its buoyancy and fostering mobility. However, bubble exsolution can also trigger crystallization (Sparks *et al.*, 1994; Blundy & Cashman, 2008) and consequently a significant viscosity increase of the liquid-crystal mixture, reducing its mobility. This highlights one of the biggest challenges for the future in understanding the inner workings of our planet: how to improve our knowledge on magma behavior from their source zones to the surface by better constraining the physical processes at play, and the system parameters that control the rates at which these processes occur. In particular, we need to work on how multiphase systems, such as magmas (containing multiple dense crystalline phases, high viscosity liquids and low density, low viscosity bub-

bles) evolve chemically and mechanically as they form, ascend, mix and erupt.

CONCLUDING REMARKS: HOW SOON WILL THE NEXT SUPERERUPTION OCCUR? THE CASE OF YELLOWSTONE

As mentioned in this chapter's introduction, a VEI 7 or 8 event could seriously impact life on Earth. Despite the effort of a large community of scientists, the question of when the next eruption will happen remains unsolved. Let us use Yellowstone as an example to see what we know and what we do not know. We obviously have a good documentation and understanding of its recent history (see multiple scientific reports, including Christiansen, 1984, 2001; Hildreth *et al.*, 1991; Gansecki *et al.*, 1996; Miller & Smith, 1999; Vazquez & Reid, 2002; Werner & Brantley, 2003; Brantley *et al.*, 2004; Husen *et al.*, 2004; Lowenstern, 2005; Nash *et al.*, 2006; Bindeman *et al.*, 2008; Christiansen & McCurry, 2008). The

Plio-Pleistocene focus of activity in northwestern Wyoming produced three large caldera-forming events since 2.1 Ma, with a total erupted rhyolite for these three units on the order of 3,500 to 4,000 km³. The last caldera-forming eruption, around 640,000 years ago (Lava Creek Tuff) outpoured ~1,000 km³ of crystal-poor rhyolite. It has since been erupting >500 km³ of rhyolitic lavas (the youngest being ~70,000 years old). Focusing on the last 640,000 years, the area has produced on average 2.5×10^{-3} km³/yr of rhyolite, which is barely lower than the worldwide average for rhyolites (~$4 \pm 1.4 \times 10^{-3}$ km³/yr). On the basis of the last three caldera-forming eruptions, Yellowstone has averaged ~700,000 years between major eruptive events. We could therefore expect one within 60,000 years, if this average was meaningful. Unfortunately though, this number is much too imprecise compared to a typical human timescale, and we should rely on other arguments to determine the state of the magma chamber.

Several observations leave little doubts that Yellowstone will erupt again some time in the future. Recent basaltic eruptions have only appeared at the periphery of the Yellowstone volcanic area (Hildreth et al., 1991), suggesting that a large, low density magma body in the upper crust blocks the ascent of the more dense, basaltic magma within the footprint of the caldera. The caldera floor is still 'breathing' (moving up and down; Husen et al., 2004), while massive degassing is underway (~45,000 tons/day of CO_2; Werner & Brantley, 2003). We know it is active, that there is magma beneath the surface, and that it can generate very large volume eruptions. However, determining the timing of the next cataclysmic volcanic eruption still defies our predicting capabilities.

Since no silicic magma has reached the surface for a long time (>70,000 years), and assuming that the productivity has not waned since the last caldera eruption, it is likely that there is a significant pool of rhyolitic magma in the upper crust at this time. Unfortunately, these bodies are notoriously difficult to delineate precisely with seismic imaging systems (Miller & Smith, 1999; Zandt et al., 2003; Lees, 2007), and we remain poorly informed on the present amount and geometry of the eruptible pool of magma beneath in the area.

Closely monitoring the whole Yellowstone caldera area (ground deformation, degassing rate, hydrothermal activity, etc.) is of prime importance, a task that the Yellowstone Volcano Observatory is rigorously fulfilling (http://volcanoes.usgs.gov/yvo/). Future research should also focus on studying units that predates other large caldera-forming eruptions anywhere on Earth, and compare the results with the most recent eruptions at Yellowstone. For example, knowing whether the magmatic column is on a heating or cooling trend might be one of the most important pieces of information scientists could gather about the future state of a given system. Adding heat and juvenile volatile-rich magmas into the upper crustal reservoirs will tend to make the reservoirs larger (Jellinek & DePaolo, 2003) and more eruptible (Keller, 1969; Bachmann et al., 2002; Wark et al., 2007; Molloy et al., 2008).

ACKNOWLEDGMENTS

Writing this chapter after many years of immersion in this fascinating topic of 'timescales of large silicic magma bodies' made me realize how much we have learned, but also highlighted some of major holes in my understanding of the processes involved. I am deeply indebted to numerous colleagues, who accompanied me through this scientific journey and include (in alphabetical order): George Bergantz, Bruce Charlier, Fidel Costa, Joe Dufek, Mike Dungan, Christian Huber, Peter Lipman, Jake Lowenstern, Felix Oberli. I also thank Tony Dosseto and two reviewers for their help on this manuscript. I was supported by NSF-EAR grant 08098208 during the completion of this chapter.

FURTHER READING

February 2008 issue of *Elements* (http://elements.geo scienceworld.org/content/vol4/issue1/) about many aspects of supervolcanoes.

Reviews articles of Fidel Costa and Mary Reid, both published in 2008 (see references list) for more coverage on timescales of supereruptions.

Lowenstern JB. 2005. Truth, fiction and everything in between at Yellowstone. *Geotimes* (http://www.agiweb.org/geotimes/june05/feature_supervolcano.html)

Bindeman I. 2006. *The Secrets of Supervolcanoes*. Scientific American (http://www.scientificamerican.com/article.cfm?id=the-secrets-of-supervolca#comments)

See the following websites on restless calderas (potential sites for future supereruptions): Yellowstone (http://volcanoes.usgs.gov/yvo/); Long Valley, CA (http://lvo.wr.usgs.gov/); Campi Flegrei caldera, Naples (http://www.wovo.org/0101_02.html).

REFERENCES

Ambrose SH. 1998. Late Pleistocene human population bottlenecks, volcanic winter, and the differentiation of modern humans. *Journal of Human Evolution* **34**: 623–651.

Annen C. 2009. From plutons to magma chambers: Thermal constraints on the accumulation of eruptible silicic magma in the upper crust. *Earth and Planetary Science Letters* **284**(3–4): 409–416.

Bachmann O, Bergantz GW. 2004. On the origin of crystal-poor rhyolites: extracted from batholithic crystal mushes. *Journal of Petrology* **45**: 1565–1582.

Bachmann O, Bergantz GW. 2006. Gas percolation in upper-crustal silicic crystal mushes as a mechanism for upward heat advection and rejuvenation of near-solidus magma bodies. *Journal of Volcanology and Geothermal Research* **149**(1–2): 85–102.

Bachmann O, Charlier BLA, Lowenstern JB. 2007a. Zircon crystallization and recycling in the magma chamber of the rhyolitic Kos Plateau Tuff (Aegean Arc). *Geology* **35**(1): 73–76.

Bachmann O, Dungan MA, Lipman PW. 2002. The Fish Canyon magma body, San Juan volcanic field, Colorado: Rejuvenation and eruption of an upper crustal batholith. *Journal of Petrology* **43**(8): 1469–1503.

Bachmann O, Miller CF, de Silva S. 2007b. The volcanic-plutonic connection as a stage for understanding crustal magmatism. *Journal of Volcanology and Geothermal Research* **167**: 1–23.

Bergantz GW, Ni J. 1999. A numerical study of sedimentation by dripping instabilities in viscous fluids. *International Journal of Multiphase Flow* **25**(2): 307–320.

Bindeman I, Schmitt A, Valley J. 2006. U–Pb zircon geochronology of silicic tuffs from the Timber Mountain/Oasis Valley caldera complex, Nevada: Rapid generation of large volume magmas by shallow-level remelting. *Contributions to Mineralogy and Petrology* **152**(6): 649–665.

Bindeman IN, Fu B, Kita NT, Valley JW. 2008. Origin and evolution of silicic magmatism at Yellowstone Based on Ion microprobe analysis of isotopically zoned zircons. *Journal of Petrology* **49**(1): 163–193.

Blundy J, Cashman K. 2008. Petrologic reconstruction of magmatic system variables and processes. In: Minerals, inclusions and volcanic processes. *Reviews in Mineralogy and Geochemistry* **69**: 179–239.

Bourdon B. Elliott T. 2010. Melt production in the mantle: constraints from U-series. In: *Timescales of Magmatic Processes: From Core to Atmosphere*, Dosseto A, Turner S, Van Orman JA (eds), Wiley-Blackwell, Oxford.

Brantley SR, Lowenstern JB, Christiansen RL *et al.* 2004. *Tracking Changes in Yellowstone's Restless Volcanic System*. USGS Fact Sheet 100-03.

Brown SJA, Fletcher IR. 1999. SHRIMP U-Pb dating of the preeruption growth history of zircons from the 340 ka Whakamaru Ignimbrite, New Zealand: Evidence for >250 ky magma residence times. *Geology* **27**(11): 1035–1038.

Charlier BLA, Wilson CJN, Lowenstern JB, Blake S, Van Calsteren PW, Davidson JP. 2005. Magma generation at a large, hyperactive silicic volcano (Taupo, New Zealand) revealed by U–Th and U–Pb systematics in zircons. *Journal of Petrology* **46**(1): 3–32.

Christiansen EN, McCurry M. 2008. Contrasting origins of Cenozoic silicic volcanic rocks from the western Cordillera of the United States. *Bulletin of Volcanology* **70**: 251–267.

Christiansen RL. 1984. Yellowstone magmatic evolution: Its bearing on understanding large-volume explosive volcanism. Explosive Volcanism: Its inception, evolution, and hazards. *National Research Council Studies in Geophysics*. National Academy Press, Washington DC, 84–95.

Christiansen RL. 2001. The Quaternary and Pliocene Yellowstone Plateau Volcanic Field of Wyoming, Idaho, and Montana. *US Geological Survey Professional Paper* 729-G: 145.

Claiborne LE, Miller CF, Clynne MA, Wooden J, Mazdab FK. 2008. Zircon from Mt St Helens: Implications for magma storage. In: *Goldschmidt Conference*, Vancouver, BC.

Coleman DS, Gray W, Glazner AF. 2004. Rethinking the emplacement and evolution of zoned plutons; geochronologic evidence for incremental assembly of the Tuolumne Intrusive Suite, California. *Geology* **32**: 433–436.

Costa F. 2008. Residence times of silicic magmas associated with calderas. In: *Dev elopment in Volcanology*, Marti J, Gottsmann J (eds), vol 10. Elsevier, Amsterdam, 1–55.

Costa F. Morgan D. 2010. Time constraints from chemical equilibration in magmatic crystals. In: *Timescales of Magmatic Processes: From Core to Atmosphere*, Dosseto A, Turner S, Van Orman JA (eds), Wiley-Blackwell, Oxford.

de Silva SL, Zandt G, Trumbull R, Viramonte J. 2006. Large-scale silicic volcanism – the result of thermal maturation of the crust. In: *Advances in Geosciences*, Chen Y-T (ed.), Vol. 1. World Scientific Press, Singapore, 215–230.

Dickins AP. 1997. *Radiogenic Isotope Geology*, Cambridge University Press, Cambridge, 471.

Dodson MH. 1973. Closure temperature in cooling geochronological and petrological systems. *Contributions to Mineralogy and Petrology* **40**: 259–274.

Dosseto A. Turner S. 2010. Magma cooling and differentiation – U-series isotopes. In: *Timescales of Magmatic Processes: From Core to Atmosphere*, Dosseto A, Turner S, Van Orman JA (eds), Wiley-Blackwell, Oxford.

Dufek J, Bergantz GW. 2005. Lower crustal magma genesis and preservation: a stochastic framework for the evaluation of basalt–crust interaction. *Journal of Petrology* **46**: 2167–2195.

Gansecki CA, Mahood GA, McWilliams MO. 1996. $^{40}Ar/$ ^{39}Ar geochronology of rhyolites erupted following collapse of the Yellowstone Caldera, Yellowstone Plateau volcanic field; implications for crustal contamination. *Earth and Planetary Science Letters* **142**(1–2): 91–108.

Hildreth W, Fierstein J. 2000. Katmai volcanic cluster and the great eruption of 1912. *Geological Society of America Bulletin* **112**(10): 1594–1620.

Hildreth W, Halliday AN, Christiansen RL. 1991. Isotopic and chemical evidence concerning the genesis and contamination of basaltic and rhyolitic magma beneath the Yellowstone Plateau volcanic field. *Journal of Petrology* **32**(1): 63–137.

Hughes GR, Mahood GA. 2008. Tectonic controls on the nature of large silicic calderas in volcanic arcs. *Geology* **36**(8): 627–630.

Husen S, Smith RB, Waite GP. 2004. Evidence for gas and magmatic sources beneath the Yellowstone volcanic field from seismic tomographic imaging. *Journal*

of Volcanology and Geothermal Research **131**(3–4): 397–410.

Jellinek AM, DePaolo DJ. 2003. A model for the origin of large silicic magma chambers: precursors of caldera-forming eruptions. *Bulletin of Volcanology* **65**: 363–381.

Keller J. 1969. Origin of rhyolites by anatectic melting of granitic crustal rocks; the example of rhyolitic pumice from the island of Kos (Aegean Sea). *Bulletin Volcanologique* **33**(3): 942–959.

LaMoreaux PE. 1995. Worldwide environmental impacts from the eruption of Thera. *Environmental Geology* **26**: 172–181.

Lee JKW, Williams IS, Ellis DJ. 1997. Pb, U, and Th diffusion in natural zircon. *Nature* **390**: 159–162.

Lees JM. 2007. Seismic tomography of magmatic systems. *Journal of Volcanology and Geothermal Research* **167**(1–4): 37–56.

Lipman PW. 1984. The roots of ash-flow calderas in western North America: windows into the tops of granitic batholiths. *Journal of Geophysical Research* **89**(B10): 8801–8841.

Lipman PW. 2000. The central San Juan caldera cluster: Regional volcanic framework. In: *Ancient Lake Creede: Its Volcano-Tectonic Setting, History of Sedimentation, and Relation of Mineralization in the Creede Mining District*, Bethke PM, Hay RL (eds), Geological Society of America Special Paper **346**: 9–69.

Lipman PW. 2007. Incremental assembly and prolonged consolidation of Cordilleran magma chambers: Evidence from the Southern Rocky Mountain volcanic field. *Geosphere* **3**(1): 1–29.

Lipman PW, McIntosh WC. 2008. Eruptive and noneruptive calderas, northeastern San Juan Mountains, Colorado: Where did the ignimbrites come from? *Geological Society of America Bulletin* **120**(7–8): 771–795.

Lowenstern JB. 2005. Truth, fiction and everything in between at Yellowstone. *Geotimes*. On-line article, access: http://www.geotime.org/june05/feature_supervolcano.html

Mason BG, Pyle DM, Oppenheimer C. 2004. The size and frequency of the largest explosive eruptions on Earth. *Bulletin of Volcanology* **66**: 735–748.

Matzel JEP, Bowring SA, Miller RB. 2006. Timescales of pluton construction at differing crustal levels: Examples from the Mt Stuart and Tenpeak intrusions, North Cascades, Washington. *Geological Society of America Bulletin* **118**(11/12): 1412–1430.

McDougall I, Harrison TM. 1999. *Geochronology and Thermochronology by the $^{40}Ar/^{39}Ar$ method*. Oxford University Press, Oxford, 269.

McKenzie DP. 1985. The extraction of magma from the crust and mantle. *Earth and Planetary Science Letters* **74**: 81–91.

Miller DS, Smith RB. 1999. P and S velocity structure of the Yellowstone volcanic field from local earthquake and controlled source tomography. *Journal of Geophysical Research* **104**: 15,105–115,121.

Miller J, Wooden J. 2004. Residence, resorption, and recycling of zircons in the Devils Kitchen rhyolite, Coso Volcanic Field, California. *Journal of Petrology* **45**(11): 2155–2170.

Miller JS, Matzel JEP, Miller CF, Burgess SD, Miller RB. 2007. Zircon growth and recycling during the assembly of large, composite arc plutons. *Journal of Volcanology and Geothermal Research* **167**(1–4): 282–299.

Molloy C, Shane P, Nairn I. 2008. Pre-eruption thermal rejuvenation and stirring of a partly crystalline rhyolite pluton revealed by the Earthquake Flat Pyroclastics deposits, New Zealand. *Journal of the Geological Society, London* **165**: 435–447.

Morgan DJ, Blake S. 2006. Magmatic residence times of zoned phenocrysts: introduction and application of the binary element diffusion modelling (BEDM) technique. *Contributions to Mineralogy and Petrology* **151**(1): 58–70.

Mullineaux DR, Crandell DR. 1981. The Eruptive History of Mt St Helens. In: *The 1980 Eruptions of Mt St Helens, Washington*, Lipman PW, Mullineaux DR (eds), US Geological Survey Professional Paper 1250, 3–15.

Nash BP, Perkins ME, Christensen JN, Lee D-C, Halliday AN. 2006. The Yellowstone hotspot in space and time: Nd and Hf isotopes in silicic magmas. *Earth and Planetary Science Letters* **247**(1–2): 143–156.

Newhall CR, Self S. 1982. The Volcanic Explosivity Index (VEI): An estimate of explosive magnitude for historical volcanism. *Journal of Geophysical Research* **87**: 1231–1238.

Petford N, Cruden AR, McCaffrey KJW, Vigneresse J-L. 2000. Granite magma formation, transport and emplacement in the Earth's crust. *Nature* **408**: 669–673.

Pyle DM. 2000. The sizes of volcanic eruptions. In: *Encyclopedia of Volcanoes*, Sigurdsson H, Houghton BF, McNutt SR, Rymer H, Stix J. (eds), Vol 1. Academic Press, London, 263–269.

Rampino MR, Ambrose SH. 2000. Volcanic winter in the Garden of Eden; the Toba supereruption and the late Pleistocene human population crash. In: *Volcanic Hazards and Disasters in Human Antiquity*, McCoy FW, Heiken G (eds), Vol 345. Geological Society of America, Special Paper, Boulder CO, 71–82.

Rampino MR, Self S. 1992. Volcanic winter and accelerated glaciation following the Toba supereruption. *Nature* **359**: 50–52.

Reid MR. 2008. How long to achieve supersize? *Elements* **4**: 23–28.

Robock A. 2000. Volcanic eruptions and climate. *Reviews in Geophysics* **38**: 191–219.

Ruprecht P, Bergantz GW, Dufek J. 2008. Modeling of Gas-driven magmatic overturn: tracking of phenocryst dispersal and gathering during magma mixing. *Geochemistry, Geophysics, Geosystems* **9**(7): doi:10.1029/2008GC002022.

Sano Y, Tsutsumi Y, Terada K, Kaneoka I. 2002. Ion microprobe U-Pb dating of Quaternary zircon: implication for magma cooling and residence time. *Journal of Volcanology and Geothermal Research* **117**(3–4): 285–296.

Self S. 2006. The effects and consequences of very large explosive volcanic eruptions. *Philosophical Transactions of the Royal Society* **A364**: 2073–2097.

Simkin T. 1993. Terrestrial volcanism in space and time. *Annual Review of Earth and Planetary Sciences* **21**: 427–452.

Simon JI, Renne PR, Mundil R. 2008. Implications of pre-eruptive magmatic histories of zircons for U-Pb geochronology of silicic extrusions. *Earth and Planetary Science Letters* **266**(1–2): 182–194.

Smith RL. 1960. Ash flows. *Geological Society of America Bulletin* **71**: 795–842.

Smith RL. 1979. Ash-flow magmatism. *Geological Society of America Special Paper* **180**: 5–25.

Soden BJ, Wetherald RT, Stenchikov GL, Robock A. 2002. Global cooling after the eruption of mount pinatubo: a test of climate feedback by water vapor. *Science* **296**(5568): 727–730.

Sparks RSJ, Barclay J, Jaupart C, Mader HM, Phillips JC. 1994. Physical aspects of magma degassing I. Experimental and theoretical constraints on vesiculation. In: *Reviews in Mineralogy. Volatiles in Magmas*, Carroll MR, Holloway JR (eds), **30**: 413–445.

Sparks RSJ, Huppert HE, Turner JS. 1984. The fluid dynamics of evolving magma chambers. *Philosophical Transactions of the Royal Society of London*, **310**: 511–534.

Sparks RSJ, Self S, Grattan J, Oppenheimer C, Pyle DM, Rymer H. 2005. *Super-eruptions: Global effects and future threats*. The Geological Society of London.

Spera FJ. 1980. Thermal evolution of plutons: A parametrized approach. *Science* **207**: 299–301.

Tamura Y, Gill JB, Tollstrup D, *et al.* 2009. Silicic Magmas in the Izu-Bonin Oceanic arc and implications for crustal evolution. *Journal of Petrology* **50**(4): 685–723.

Turner S, Costa F. 2007. Measuring timescales of magmatic evolution. *Elements* **3**: 267–272.

Turner S, Bourdon B. 2010. Melt transport from the mantle to the crust – U-series isotopes. In: *Timescales of Magmatic Pprocesses: from Core to Atmosphere*, Dosseto A, Turner S, Van Orman JA (eds), Wiley-Blackwell, Oxford.

Vazquez JA, Reid MR 2002. Timescales of magma storage and differentiation of voluminous high-silica rhyolites at Yellowstone caldera, Wyoming. *Contributions to Mineralogy and Petrology* **144**(3): 274–285.

Walker BJ, Miller CF, Lowery LE, Wooden JL, Miller JS. 2007. Geology and geochronology of the Spirit Mountain batholith, southern Nevada: implications for timescales and physical processes of batholith construction. *Journal of Volcanology and Geothermal Research* **167**: 239–262.

Wallace PJ, Gerlach TM. 1994. Magmatic vapor source for sulfur dioxide released during volcanic eruptions: evidence from Mt Pinatubo. *Science* **265**: 497–499.

Wark DA, Hildreth W, Spear FS, Cherniak DJ, Watson EB. 2007. Pre-eruption recharge of the Bishop magma system. *Geology* **35**(3): 235–238.

Werner C, Brantley S. 2003. CO_2 emissions from the Yellowstone volcanic system. *Geochemistry Geophysics Geosystems* **4**(27): doi:10.1029/2002GC000473.

White SM, Crisp JA, Spera FA. (2006) Long-term volumetric eruption rates and magma budgets. *Geochemistry Geophysics Geosystems* 7: doi:10.1029/2005GC001002.

Wilde SA, Valley JW, Peck WH, Graham CM. 2001. Evidence from detrital zircons for the existence of continental crust and oceans on the Earth 4.4 Gyr ago. *Nature* **409**(6817):175–178.

Woo G. 1999. *The Mathematics of Natural Catastrophes*. Imperial College Press, London, 292.

Zandt G, Leidig M, Chmielowski J, Baumont D, Yuan X. 2003. Seismic detection and characterization of the Altiplano-Puna magma body, central Andes. *Pure and Applied Geophysics* **160**: 789–807.

11 Timescales of Magma Degassing

KIM BERLO[1], JAMES E. GARDNER[2] AND JONATHAN D. BLUNDY[3]

[1]Department of Earth and Planetary Sciences, McGill University, Montreal, Canada
[2]Department of Geological Sciences, University of Texas at Austin, Austin, TX, USA
[3]Department of Earth Sciences, University of Bristol, Bristol, UK

SUMMARY

Magma degassing exerts a major control on the nature of volcanic eruptions. As volatile-rich magma ascends through the crust (or crystallizes), gas bubbles form by exsolution of volatile components, such as H_2O and CO_2 from the melt. These gas bubbles lend buoyancy to magma, but the exsolution of volatiles also results in increases in both density and viscosity of the melt, as well as in crystallization. Volatile exsolution and gas segregation of exsolved gas thus have profound consequences for magma ascent and eruption. Rapid magma ascent may result in limited volatile exsolution and segregation and potentially explosive eruption; more efficient gas loss, during slow magma ascent essentially defuses a pending eruption. Segregation occurs by a variety of mechanisms and over a range of timescales, from seconds to (in excess of) a thousand years. This chapter addresses the rates of magma degassing from volatile exsolution, through segregation to emission.

Timescales of Magmatic Processes: From Core to Atmosphere, 1st edition. Edited by Anthony Dosseto, Simon P. Turner and James A. Van Orman.
© 2011 by Blackwell Publishing Ltd.

THE IMPORTANCE OF DEGASSING

Volcanic eruptions take many forms, from quietly effusive lava flows (e.g. Hawaii) to violent, explosive events that affect the whole planet (e.g. Pinatubo). Despite the tremendous effort in understanding volcanic eruptions and close monitoring of a number of volcanoes, in many cases eruptions still take us by surprise. However, a number of key parameters that influence the how and when of volcanic eruptions have been identified; namely, the amount of dissolved gas that a volcanic magma carries, the rate at which the magma ascends through the crust and the interactions between processes such as degassing and crystallization that occur throughout ascent.

During magma ascent, dissolved volatile species, notably H_2O and CO_2, exsolve to form gas bubbles. Gas bubbles lend buoyancy to the magma and can speed up its ascent, but escape of gas results in a less buoyant and more viscous residual magma. Moreover, volatile exsolution can promote crystallization of anhydrous phases, which not only further increases its viscosity, but drives additional exsolution by concentrating volatile species in the melt phase. The extent to which magma is able to lose its volatiles during ascent largely determines eruptive style. If gas loss during ascent is inefficient, ascending magma will become progressively more bubble rich until it fragments explosively. The rate of gas loss from the magma in relation to the ascent rate is thus crucial to volcanic hazard assessment.

Magma degassing is not solely associated with eruption however, and there are many examples of degassing without eruption. Etna and Stromboli are classic examples of 'persistently degassing volcanoes' that discharge considerable quantities of gas into the atmosphere with little or no associated magma. At many volcanoes, gas plumes can be observed preceding eruption and in between eruptions, suggesting that volcanic gasses can travel through and escape from the magma (Stix *et al.*, 1993; Edmonds *et al.*, 2003). In other words, sub-surface magma may not only contain bubbles, but can also be permeable to gas flow. Observations of passively degassing volcanoes and radionuclide data, discussed in this chapter, suggest that degassing can occur decades or longer prior to eruption.

The continuing release of volcanic gasses over the lifetime of a volcano is in part responsible for the flux of metals to hydrothermal systems. Longevity of hydrothermal systems can result in the formation of ore deposits by a combination of leaching and deposition by heated fluids. The age of hydrothermal circulation cells is dependent upon the heat provided by the cooling intrusion of magma. Although single intrusions will cool rapidly (e.g. in hundreds of years), periodic recharge with new magma can keep large regions partially molten for millions of years, developing crustal hotzones (Annen & Sparks, 2002). The extent to which hydrothermal ore bodies form, and their composition, depends on the nature of gases being exsolved from magmas and the ability of these gases to dissolve and transport metals as well as an efficient precipitation mechanism.

THE FIRST STEP: MAKING A BUBBLE

Solubility

Volcanoes located in subduction-related settings are generally richer in volatiles, particularly H_2O, than those in other settings, because of fluid contributions from the subducting slab. This is

the main reason for the more common explosive nature of volcanoes in subduction zone settings. Estimates of the volatile concentrations in mantle-derived magma form the starting point of degassing studies. Fischer & Marty (2005) constrained the concentrations of H_2O and CO_2 in mantle-derived undegassed arc magma by using the unique origin of ^3He in the asthenospheric mantle. By comparing the ^3He flux at arc volcanoes with ^3He in the solid mantle and the volatile fluxes, they estimate that mantle-derived melts have ~8 to 16 wt.% H_2O and ~3,500 to 7,600 ppm CO_2 prior to differentiation and degassing. Volatiles, in particular H_2O, can thus be major constituents of pre-eruptive magma. Any crystallization of magma within the crust will only serve to increase volatile contents further. However, a significant fraction of the initial volatile budget is lost during ascent and cooling, because their volatile solubility, the maximum amount of that component that can be dissolved within the melt under a certain set of P, T, and compositional conditions, is much lower at the Earth's surface than at depth.

The solubility of H_2O and CO_2, as well as for a range of other species (i.e. S, Cl and F), has been determined experimentally for various compositions. The main factors controlling solubility are pressure and to a lesser extent oxygen fugacity, temperature and composition of the melt. The strong dependence of volatile solubility on pressure occurs because the molar volume of the gaseous form is so much larger than the molar volume of the dissolved species. As a consequence, magma ascent induces saturation of one or more species and formation of gas bubbles. The depth at which degassing starts depends on the initial concentration of volatiles and on the solubility of the species. Some species, such as the noble gases, have very low solubilities and could thus be expected to exsolve at a very early stage in the degassing process. However, such species are usually not present in sufficient quantities to form bubbles of their own and require another species to saturate, before migrating into the gas phase. Once magma reaches saturation in one of the more abundant

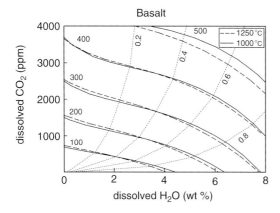

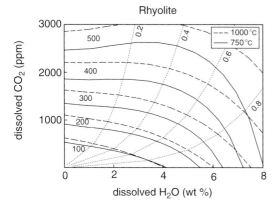

Fig. 11.1 Isobars for dissolved H_2O and CO_2 concentrations within volatile-saturated rhyolitic and basaltic melt at two different temperatures (Figure after Papale *et al.*, 2006).

volatile species, usually H_2O, CO_2 or S, gas bubbles form and the magma is referred to as being fluid- or vapor-saturated. The gas phase is however always a mixture of volatile species and never just pure H_2O or CO_2.

H_2O is dissolved in magma as OH^- and molecular H_2O, CO_2 as carbonate anions and molecular CO_2 (Silver *et al.*, 1990; Fine & Stolper, 1986). H_2O solubility correlates positively with pressure and negatively with temperature, with the dependence on pressure being stronger than that on temperature. The solubilities of H_2O and CO_2 also have significant compositional dependence (Figure 11.1). For a recent review of experimental

data and modeling of the solubility of H_2O and CO_2, see Moore (2008). Because H_2O and CO_2 are the main volatile constituents and the partial pressures of other components are relatively minor, it is possible to calculate at what pressure magma with specific H_2O and CO_2 concentrations would reach saturation with a multi-component gas phase using solubility models. The most widely used models for basaltic and rhyolitic magmas are by Newman & Lowenstern (2002) and by Papale *et al.* (2006). These take account of the compositional dependence of H_2O and CO_2 solubility and the non-ideal mixing of these species in the vapor phase. The latter factor accounts for the curvature of the saturation curves shown in Figure 11.1.

Bubble formation

Bubble nucleation

Once a melt reaches saturation in one or more volatile components, by a change in pressure, temperature or composition, these will exsolve to form a separate gas phase, such as a gas bubble. Other volatile species will then partition between the melt and the exsolved gas bubble. Bubble formation is the critical first step for most degassing processes; however, it is far from a straightforward, instantaneous appearance.

A gas phase is formed when enough volatile molecules have clustered together, which lowers the Helmholtz free energy of the system (*F*). Formation of a new phase requires an interface, separating the gas and melt. This interface is a structural mismatch and requires energy, often referred to as surface tension (σ), to form. The increased energy from surface tension can overwhelm the lowered free energy of clustered gas molecules when such clusters are small, leading to destruction of the cluster, preventing nucleation. If enough gas molecules are in the cluster however, the lowered free energy of the cluster, which varies with the volume of the cluster ($\sim r^3$, where r is cluster radius), can overcome surface tension, which increases only with the surface area of the cluster ($\sim r^2$), and bubble nucleation results.

If nuclei are assumed to form randomly and obey statistical laws, that is, nucleate homogeneously, then the probability of formation is proportional to $\exp(-\Delta F/k\mathrm{T})$, where ΔF is the resulting change in free energy from nucleation, k is the Boltzmann constant, and T is temperature. Assuming that nuclei are spherical, the rate of bubble nucleation (J) then equals:

$$J_o\exp[(-16\pi\sigma^3)/(3k\mathrm{T}\Delta P^2)]$$

where the driving force for (super)saturation is the change in pressure (ΔP). The pre-exponential term (J_o) takes into account kinetic considerations, and is often related in part to the number of molecules present, their diffusivity, and their separation (Hurwitz & Navon, 1994). Melt viscosity may also play a role in nucleation by inhibiting melt deformation to form an interface (Gardner, 2007). Importantly, even orders of magnitude differences in J_o have minor impact of J, compared to relatively small changes in σ. For example, supersaturations needed to produce a given nucleation rate differ by only $\sim$10 MPa when J_o changes by ten orders of magnitude from 10^{28} to 10^{38}, whereas a five-fold increase in σ from 0.02 to 0.1 N m^{-1}, increases the critical ΔP from 10 to 120 MPa (Navon & Lyakhovsky, 1998). For $\sigma > 0.1$ Nm^{-1}, supersaturations in excess of 150 MPa are needed to trigger homogeneous nucleation. Despite the impact that σ has on nucleation kinetics, measured values for σ exist only for high-silica rhyolite, dacite and phonolite. These indicate that less siliceous and/or more hydrous melts have lower values of σ. Interestingly, the relatively high surface tensions of hydrous rhyolites mean that water bubbles cannot nucleate homogeneously unless ΔP exceeds 120 to 150 MPa, implying that explosive eruptions of rhyolites with less $\sim$5 wt.% water are impossible.

In some cases, gas bubbles can nucleate at lower supersaturation in the presence of impurities or particles. Bubbles can nucleate on, for example, crystals because the presence of the crystal-bubble surface lowers the overall bulk surface tension. Geometrically, such so-called heterogeneous nucleation can be described by a wetting angle, which is the angle between the tangent of the bubble and the substrate. In the case of silicate melts, some crystals appear to act as better nucleation substrates than others. Magnetite is an especially good substrate, lowering critical supersaturation to below 5 MPa (Hurwitz & Navon, 1994; Gardner, 2007). Hematite can also wet water bubbles, more so at their ends than along their lengths, although greater supersaturations are needed than those for magnetite (Gardner, 2007). Plagioclase, on the other hand, appears to not wet water bubbles. In addition to the presence of crystals, heterogeneous nucleation may be more common, because supersaturation is likely to increase gradually with cooling and crystallization.

The number of bubbles that actually nucleate is often limited because once a bubble nucleates, more gas molecules can diffuse into that bubble, lowering the supersaturation of the melt and suppressing further nucleation (Toramaru, 1989). Indeed, the number (and size distribution) of bubbles recorded in volcanic pumice as vesicles is often significantly different than would be generated by either pure homogeneous or heterogeneous nucleation (Mourtada-Bonnefoi & Laporte, 2004). One potential way that bubble numbers could increase is if a second (or multiple) nucleation event is triggered in the magma. Such secondary events could result from magma ascending fast enough that supersaturation increases faster than gas molecules are able to diffuse to existing bubbles (Massol & Koyaguchi, 2005).

Bubble growth

Once a bubble nucleus is formed it can continue to grow by addition of more gas molecules. Gas molecules are added to the bubble by evaporation from the surface of the bubble. Evaporation locally depletes the melt, setting up a concentration gradient that drives further diffusion of gas towards the bubble. Bubble growth is thus controlled by gas diffusivity through the melt, which is favored by high gas concentrations, high temperatures and low-silica melts (Watson, 1994). Bubbles can additionally increase in size by expansion in response to a drop in confining pressure. Growth by expansion alone is only important at relatively low pressures, and hence in the final stages of magma degassing (Proussevitch *et al.*, 1993). All bubble growth requires the surrounding melt to deform, hence the viscosity of that melt can limit the rate of bubble growth. Very stiff melts can in fact viscously quench bubble growth, depending on the rate at which pressure changes (Thomas *et al.*, 1993). Given reasonable estimates for ascent rates during explosive eruptions, such viscous quenching is only important when viscosity exceeds ~10^9 Pa s (Gardner *et al.*, 2000). Because lower water contents reduce diffusivity and stiffen silicate melts, bubble growth should slow as degassing progresses. Mitigating those retarding effects however, is the decreasing distance between bubbles as they grow. Hence, decompression rates must increase as degassing progresses in order for silicate melt to supersaturate in gas during eruptions. Finally, when pressure changes, a finite amount of gas exsolves from the melt, which then diffuses towards the nearest bubbles. The number of bubbles present thus plays an important, and probably dominant, role in determining the resulting bubble sizes (Gardner *et al.*, 1999).

OPEN AND CLOSED SYSTEM DEGASSING

Degassing paths

Once gas is exsolved it can either stay with the magma, 'closed system degassing', or escape from the magma, 'open system degassing'. During closed system degassing, the gas maintains equilibrium with the melt during ascent, whereas during open system degassing, gas is constantly extracted from the melt. In most cases, however, degassing occurs by a combination of closed- and open-system processes. Initial degassing occurs essentially as a closed system, but once a critical gas volume (or threshold permeability) is reached, gas is able to segregate. For

example, gas bubbles need to have either a sufficient volume or sufficient time to separate from the melt. Open system degassing may result in fluxing gases through overlying magma, where this gas buffers the melt composition (Rust *et al.*, 2004).

The evolution of volatile concentrations within the melt can be tracked through melt inclusions (for recent reviews, see Blundy & Cashman, 2008; Métrich & Wallace, 2008). Melt inclusions are quenched droplets of silicate melt, preserved as glassy inclusions within volcanic phenocrysts (Figure 11.2). Each melt inclusion records the CO_2 and H_2O concentration of the melt at the time it was sealed off physically from the groundmass melt. Thus, comparing volatile concentrations in melt inclusions trapped at different depths and times yields an insight into the degassing history of a magma batch. Figure 11.3 shows how H_2O and CO_2 concentrations change with different degassing scenarios. During open system degassing, the CO_2 concentration of the melt decreases while the H_2O concentration remains essentially constant until the melt reaches saturation pressure for pure H_2O. Closed-system degassing results in greater H_2O exsolution at any given pressure. During closed system degassing, the magma coexists with an exsolved gas phase, which buffers the melt concentration. Examples of both open and closed system degassing pathways have been observed, but most melt inclusion datasets do not follow a single degassing trend. Blundy & Cashman (2008) demonstrate the potential variation in H_2O and CO_2 within a magma chamber, taking into account variations in temperature and crystallinity.

Magma storage pressures can be estimated from the volatile contents of melt inclusions (Wallace *et al.*, 1999; Schmitt, 2001; Liu *et al.*, 2006) provided there is evidence that the melt was volatile-saturated, i.e. it had bubbles, at the time the melt inclusions became sealed. The only unequivocal evidence for volatile saturation in a magma containing H_2O and CO_2 is an inverse correlation between CO_2 content of melt

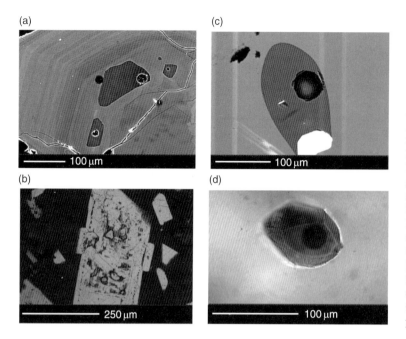

(a)

(b)

(c)

(d)

Fig. 11.2 Droplets of quenched silicate melt ('melt inclusions') trapped inside plagioclase phenocrysts from Mt St Helens, USA (a and b) and inside olivine phenocrysts from Fuego, Guatemala (c and d). In back-scattered electron micrographs (a and c) the melt inclusions appear as dark rounded blobs; the host plagioclase (a) is lighter grey and displays complex zoning in anorthite content from Ca-rich (lighter) to Na-rich (darker). Black circles inside some inclusions are bubbles. Bright crystal in c is a Fe-Ti oxide crystal. b and d are transmitted light photomicrographs.

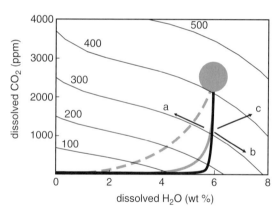

Fig 11.3 Different degassing paths for a basaltic magma from a source with 6 wt.% H_2O and 3,000 ppm CO_2 situated at >400 MPa: open system degassing (black), closed system degassing (gray) and closed system degassing with 30% exsolved bubbles (dashed). Also indicated are the effects of gas fluxing (a), isobaric crystallization (b) and fluid absent crystallization (c) isobars after Papale et al. (2006).

inclusions and some index of crystallization, for example incompatible trace elements (Wallace et al., 1999). This is because gas loss, whether driven by isobaric volatile-saturated crystalliza-

tion or by decompression of volatile-saturated magma, will always drive CO_2 preferentially into the vapor phase, leading to its preferential depletion relative to H_2O, in the melt. In cases where the melt is not volatile-saturated, the estimated pressure based on volatile saturation will underestimate the true storage pressure. Note that the presence of gas bubbles in a melt inclusion does not necessarily indicate that the magma was gas-saturated as bubbles can also form by shrinkage during cooling.

Interpreting melt inclusion data does present several challenges: do melt inclusions represent the surrounding melt? Or has their composition been changed during or after entrapment, for example through crystallization, leakage or diffusion? Some minerals are more likely to fracture and thus volatiles can leak out. In most cases this is easily verified as these melt inclusions tend to crystallize. Another way in which the composition of melt inclusions may be altered is by diffusion through the crystal lattice. This will be discussed below, as diffusion of volatile elements provides a time constraint on devolatilization of the melt. A third way is by partial crystallization of the melt inclusion.

Many melt inclusions contain crystals, some of which may have been entrapped with the melt, but others have crystallized later from the enclosed inclusion. Crystallization of anhydrous phases in particular will increase the H_2O and CO_2 concentrations of the melt.

Another consideration when interpreting inclusion data is that for melt inclusions to be entrapped, crystallization has to take place. Estimates of volatile concentrations in arc mantle for example are higher than maximum concentrations of volatiles observed in melt inclusions of primitive olivine crystals. Thus substantial loss of volatiles is likely to occur prior to the onset of crystallization. The next section shows that degassing itself promotes crystallization. Magma degassing and crystallization are intimately linked.

Degassing-induced crystallization

Dissolved H_2O has a dramatic effect on crystallization temperatures of magmas, reducing the liquidus temperature considerably, even at low concentrations. H_2O similarly has a pronounced effect on phase relations, inhibiting the saturation of some minerals, such as plagioclase, while favoring the crystallization of others, such as amphibole or biotite. Figure 11.4 shows the H_2O-saturated phase relations of Mt St Helens dacite,

compiled from published experimental data by Blundy & Cashman (2001).

As H_2O-saturated magma undergoes degassing due to decompression, it will crystallize due to the increase in liquidus temperature with decreasing pressure. This crystallization can be accomplished without any drop in temperature and therefore can operate on much faster timescales than cooling-driven crystallization, which requires the relative slow loss of heat through the wall rocks of the magma body (Blundy & Cashman, 2005). In fact, Blundy *et al.* (2006) show that in a system undergoing decompression-driven crystallization, the magma temperature can rise due to the release of latent heat of crystallization, leading to the apparent paradox that a magma undergoes both crystallization and heating as it ascends.

The tendency of ascending magma to crystallize has several important dynamic consequences. The growth of crystals leads to a sharp increase in the viscosity of the magma (crystals + melt), whilst the loss of H_2O from the melt leads to an increase in the viscosity of the melt phase itself. The net result is that ascending magma is slowed down dramatically once it encounters its H_2O-saturated liquidus, as long as it is moving slowly enough for crystals to form. The depth at which this occurs depends on the initial H_2O content of the magma and its tem-

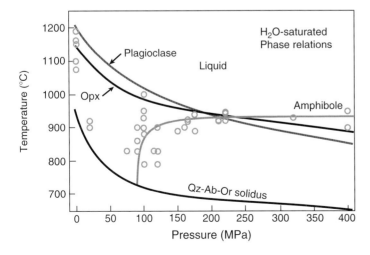

Fig. 11.4 Phase relations of H_2O-saturated dacite from Mt St Helens, constructed by Blundy & Cashman (2001) from the experimental datapoints (shown as open circles) of Merzbacher & Eggler (1984), Rutherford *et al.* (1985), Rutherford & Devine (1988), Rutherford & Hill (1993). Appearance of different mineral phases is shown by the solid lines. The iron-titanium oxides have been omitted for clarity. The solidus is taken from the haplogranitic quartz-albite-orthoclase system (Johannes & Holtz, 1996). See Plate 11.4 for a colour version of this image.

perature. Annen *et al.* (2006) describe this process
as 'viscous death' and suggest that it leads to the
stalling of magmas in the shallow crust and their
subsequent coalescence and cooling to form granitic plutons.

In contrast to H_2O, dissolved CO_2 has an almost
negligible effect on liquidus temperatures and
phase relations of typical silicate magmas. In fact,
the preferential loss of CO_2 from magmas initially

containing both CO_2 and H_2O serves to depress
liquidus temperatures by depressing the liquidus
as H_2O becomes enriched relative to CO_2 in the
residual melt. During the early stages of degassing, therefore, when CO_2 is the dominant
degassed species, crystallization does not occur.
Crystallization only begins in earnest when the
magma reaches sufficiently shallow depths that
almost all of the CO_2 has been lost.

Basaltic volcanism

Silicic volcanism

Fig. 11.5 Classification of volcanic
eruptions according to the magma
ascent rate and the degree of melt-
vapor segregation, essentially the
difference between ascent rates of
magma and gas. Where gas segregation is low and magma ascent rate
high, the system behaves largely as
a closed system and eruptions are
likely to be explosive. In contrast,
where the timescale of degassing is
faster than that of magma ascent,
there is a high degree of gas segregation and the system behaves like
an open system and only gas
emerges at the surface (Figure used
by permission of the Royal Society,
from Edmonds (2008), *Phil. Trans.
R. Soc.* A, Vol. 366, Fig. 7, p. 14).

Importance of degassing on eruption style

Gas exsolution has major consequences for the properties of magma, such as its density and viscosity. Hydrous magma has a lower density and viscosity, thus both viscosity and density increase during degassing unless gas bubbles are retained, which decreases the overall density. Degassing also promotes crystallization, adding to the increase in viscosity. If gas bubbles are able to segregate, magma thus becomes stiffer and harder to erupt as it ascends. However, if gas cannot segregate, the magma chamber may become over-pressurized and explosive eruptions can ensue. The interaction and rates of magma ascent, degassing and crystallization are thus crucial in whether and how magma is erupted (Figure 11.5). In very rapidly ascending magma, there may not be enough time for gas to segregate, resulting in violent eruptions of bubble-rich magma, for example, Plinian eruptions. Effusively and explosively erupted magma need not start with very different volatile concentrations. Rather, effusive eruptions appear to have lower magma ascent rates (Rutherford, 2008) allowing for more effective segregation of gas. The tendency of hydrous magma to crystallize upon reaching volatile saturation results in much hydrous magma never reaching the surface, forming plutons instead. Thus it is crucial to understand the interaction and kinetics of magma ascent, degassing and crystallization (see also Chapters 4 (O'Neill & Spiegelman, 2010), 5 (Turner & Bourdon, 2010), 7 (Costa & Morgan, 2010) and 8 (Dosseto & Turner, 2010)).

GAS SEGREGATION

Evidence for pre-eruptive bubbles

Volcanoes emit large quantities of gases, regardless of whether they are erupting. Some volcanoes, such as Sakurajima in Japan, have been persistently degassing for 50 years, which requires large amounts of magma. Such degassing suggests that gas segregation and escape (open-system degassing) can occur over timescales much longer than those of magma ascent during eruption.

Here we review evidence for the presence of gas bubbles in subsurface magma.

In 1991, huge amounts of gas were injected into the atmosphere by the eruption of Mt Pinatubo, but most of that gas could not have been dissolved in the magma at any reasonable sub-volcanic magma chamber depth. Since then, numerous other eruptions have been documented as having released 'excess' gas, suggesting that magmas often contain a separate gas phase at depth (Wallace, 2001). Estimating the composition and volume of that separate gas is a challenge, because some gas released during eruptions does in fact come from the melt. The amount lost from the melt can be approximated by the difference between gas concentrations in melt inclusions minus those in degassed groundmass glass, scaled to the mass of magma erupted, but this 'petrological' method overlooks any gas present in the separate gas phase. Often, the amount of sulfur released from the melt appears to be negligible, because of its very slow diffusivity, and so many workers focus on 'excess' sulfur to serve as a proxy for the separate gas phase. Although some excess sulfur could come from the degassing of un-erupted magma or the decomposition of sulfur-bearing minerals, such as anhydrite and pyrrhotite, the more likely source is gas bubbles that were in the erupted magma, which contained a mixed fluid that included sulfur. Following that hypothesis, a recent overview of eruptions suggests that 5 to 30 vol.% bubbles are commonly present in magma prior to ascent, accounting for 1 to 6 wt.% exsolved H_2O (Wallace, 2001). This implies that magma reaches volatile saturation well before it is erupted, and must have nucleated bubbles at depth.

Before the kinetics of bubble formation, both at depth and during ascent, can be properly addressed, we first need to know how many and how big those bubbles are. Volcanic pumices preserve an array of bubble sizes, often numerous smaller ones and a few large ones. Which of those are pre-eruptive bubbles? It has long been speculated that the relatively few large vesicles were formed in the magma before it erupted (Sparks & Brazier, 1982), although it cannot be ruled out that such bubbles result from coalescence of many smaller ones during magma ascent. Recently, Gualda & Anderson (2007) concluded that a relatively large vesicle in a pumice from the Bishop Tuff was a relic of a pre-eruptive bubble, that was in the order of 300 to 850 μm in diameter before it was erupted. That bubble may have been held within the magma by being attached to many dense magnetite crystals, making it neutrally buoyant. If such a bubble is typical of pre-eruptive bubbles, there would be hundreds to thousands of these per cubic centimeter of a magma containing 5 to 30 vol.% gas. If pre-eruptive bubbles are instead typically much smaller, in the order of 10 μm in size, then that same magma would have hundreds of millions of bubbles per cubic centimeter. Identifying the numbers and sizes of pre-eruptive bubbles is made especially difficult because any such bubbles will grow and potentially coalesce during ascent to the surface. Recent experimental work on decompressing bubbly melt found that small bubbles grow substantially, even during decompressions as fast as ~2 MPa s^{-1} (Gardner, 2009). That implies that only the relatively large vesicles seen in volcanic pumice may be relics of pre-eruptive bubbles, even if small bubbles were originally present at depth. Because vesicle distributions in pumice are used to ground truth numerical models of explosive eruptions and kinetics of syn-eruptive bubble formation, deciphering the numbers and sizes of any pre-eruptive bubbles is critical but remains essentially unexplored.

Open-system degassing

Ascent of bubbles – a basaltic affair

Gas bubbles are more buoyant than the surrounding melt and once a gas bubble forms it will rise through the melt (Figure 11.6). However, this process is hampered by the viscosity of the melt. The rise of a gas bubble can be described by Stokes' Law:

$$v = \left[\frac{(\rho_m - \rho_g)g}{18\mu}\right] \times d^2 \approx \frac{\rho_m g d^2}{18\mu}$$

where v is the velocity of the bubble, d is the bubble diameter, ρ_m and ρ_g the density of the

melt and gas, respectively, and μ is the dynamic viscosity of the melt. For a typical basaltic melt, a 1 mm bubble would rise at 4 km/yr, whilst in a water-rich rhyolitic melt, the same bubble would rise at only 0.4 m/yr. In viscous magma, gas bubbles are therefore likely to ascend with the magma and continue to interact with it, unless the magma remains stationary for prolonged periods. However, for some basaltic magma, ascent of bubbles does overlap or surpass magma ascent velocity. Thus ascent of bubbles is unlikely to result in significant segregation in silicic magma, but could result in some segregation in basaltic magma, particularly in shallow magma.

One way to enhance gas segregation, in particular in basaltic magma, is to grow larger bubbles, for example when bubbles form clusters or coalesce. Large gas bubbles (meters in diameter!) have been observed just before they explode upon reaching the surface. There are currently two main models for explosive basaltic eruptions, such as Strombolian and Hawaiian eruptions. Vergniolle & Jaupart (1990) show how such large 'gas slugs' could form. In their experiments, small gas bubbles ascend in a magma reservoir until they encounter a bottleneck, whereupon the bubbles slow and accumulate until a critical 'foam' thickness is reached. Bubbles then coalesce and escape as gas slugs up the conduit. Parfitt (2004) suggested that gas slugs can also form by differential ascent rates of bubbles of different sizes. Since large bubbles ascend faster they will scavenge slower, smaller bubbles and grow at their expense. This process relies critically on the magma ascent rate: in a slowly ascending magma the bubbles will travel further and there is more time to create large gas slugs. Clearly, accumulation and interaction between bubbles is a key process in efficient gas segregation, as will be discussed in detail in the next section.

Convection

Convective overturn can also aid the ascent of individual bubbles and has been suggested as an effective way to generate large gas emissions, such as during passive (non-eruptive) degassing (Stevenson

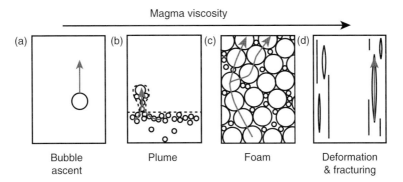

Fig. 11.6 Schematic illustration of different ways in which gas can segregate from magma. (a) ascent of individual bubbles through low viscosity magma (e.g. basalt, carbonatite); (b) accumulation and coalescence of bubbles either at an interface (compositional, temporal) or by larger bubbles catching up with smaller ones; (c) gas flow through an interconnected network of bubbles; (d) deformation of bubbles and fracturing of magma in high stress regions such as along the conduit walls (high viscosity magma only). See Plate 11.6 for a colour version of these images.

& Blake, 1998). Bubble-rich or volatile-rich magma is less dense than surrounding volatile-poor magma and will thus ascend faster. At some depth, bubbles will have expanded sufficiently to either segregate from the melt or accumulate against a boundary (e.g. encounter an increase in crystal content), forming a bubble-rich layer that caps the convection cell. The melt, having lost both its dissolved volatiles, because of the decrease in pressure during ascent, and its gas bubbles, is now denser and can descend again. Such convection cells can result in the degassing of large volumes of magma, which may never reach the surface, simply because it has lost its buoyancy.

Bubble suspensions

Highly vesicular magma forms not only by ascent of individual bubbles, as discussed in the previous section, but also by expansion of bubbles during magma ascent (Figure 11.6). Bubble suspensions (often termed 'foam', though few have the 74 vol.% bubbles required to qualify as such) can thus occur both in basaltic and silicic magma.

Interacting bubbles allow gas to migrate from one bubble to the next, establishing a pathway for gases. Permeability will eventually lead to gas escape and bubble collapse. The development of open system degassing requires a degree of per-

meability within the bubble-rich magma. The relationship between porosity (Φ, bubble volume fraction) and the permeability (k) can be described using percolation theory, which requires a certain critical porosity Φ_{cr} for gas flow:

$$k(\Phi) = c(\Phi\text{-}\Phi_{cr})^n$$

when $\Phi \geq \Phi_{cr}$

$$k(\Phi) = 0$$

when $\Phi < \Phi_{cr}$

in which c is a proportionality constant and n is dependent upon the pathway geometry and typically is between 2 and 4. Below the percolation threshold Φ_{cr}, no continuously connected network of bubbles exists and thus the system is impermeable. In a system with ideal spheres, permeability is reached at around 30 vol.% of bubbles, the percolation threshold. Figure 11.7 shows that in explosively erupted volcanic rocks, permeability is reached at around this same percolation threshold; the threshold appears to occur at much lower porosity in effusively erupted volcanic rocks.

However, many volcanic rocks have porosities of up to 70 vol.%. In particular, such high porosities are observed in pumices from explosive eruptions (Figure 11.7), which presents a conundrum,

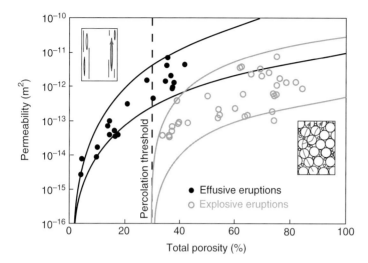

Fig. 11.7 Porosity (vesicularity)-permeability relation such as measured in volcanic rocks (modified after Mueller et al., 2005). In bubble foams connectivity is reached at ~30% bubble volume, the percolation threshold. Thus gas loss during explosive eruptions is late and bubbles can expand until they reach permeability (closed system degassing). Effusively erupted magma is permeable at much lower vesicularity and the nature of the connectivity differs, fracturing and deformation provide efficient pathways for gases to escape (open system degassing). See Plate 11.7 for a colour version of this image.

because percolation starts at 30 vol.% resulting in gas loss and explosive eruptions are thought to result partly from the inability of the magma to lose its volatiles. The explanation lies in the relative rates of percolation and degassing (Burgisser & Gardner, 2005). At high decompression rates, such as during explosive eruptions, gas escape through a bubble network is limited by the time required to establish permeability.

Bubble deformation and fracturing- the silicic case

In systems where foams do not form, permeable gas flow can occur through the volcanic conduit and its margins. In very viscous magma, high yield stresses, especially along the conduit margins, cause bubbles to deform and can eventually lead to shear-fracturing of the magma itself. In addition, the host rock surrounding the magma reservoir can be fractured and porous, leading to gas loss through the walls as well as up the magma-filled conduit (Figure 11.6). Both elongate bubbles, deformed by shear stresses, such as in tubular pumices, and fractures will influence the porosity-permeability relationship. In the previous section, it was shown that in an ideal, spherical close-packed structure, permeability is reached only at around 30 vol.% bubbles. However, particularly in

extrusive eruptions, permeability is observed at much lower porosity (Figure 11.7, Mueller et al., 2005), in such cases, gases escape through deformed bubbles and fractures. Microfractures have been observed in pumice, but it is unclear whether they formed whilst the groundmass was molten or when it quenched to glass.

Magma itself can fracture in response to shear stresses induced by flow of highly viscous magma (Tuffen et al., 2008). Veins of pyroclastic material and banding in obsidian have been suggested to represent healed fractures, which once provided gas escape routes (Stasiuk et al., 1996; Rust et al., 2004; Tuffen et al., 2003). Such fractures are transient and the crystal and glass fragments are re-deposited by the gas-particle mixture, which flows along the fractures. The transient fracture networks provide highly efficient pathways to transport gases. Rust et al. (2004) suggest that brecciated conduit margins act as highly permeable channels for continuous gas flow.

Apart from the loss of gas through a permeable network of bubbles and/or cracks, the unstable nature of bubbly magma can be resolved another way: by fragmentation of the surrounding magma. Fragmentation occurs when the gas is exposed to a sudden pressure gradient, such as occurred at Mt St Helens on May 18, 1980, when a landslide on the northern flank of the volcanic cone removed part of the overburden. In such cases the gas loss from the magma is insufficiently fast compared to the ascent rate of the magma and the expanding bubbly magma will fragment.

TIMESCALES OF DEGASSING

During closed-system degassing, the timescales of gas exsolution are governed by the rates of the processes that cause exsolution, for example, magma ascent or cooling. Estimates of magma ascent rates can be inferred from diverse methods, such as seismic data or petrologic observations. A recent review of magma ascent rates can be found in Rutherford (2008) (see also Chapters 4 (O'Neill & Spiegelman, 2010) and 5 (Turner & Bourdon, 2010)). Estimates range from 0.001 to 0.015 m/s for extrusive eruptions and in excess of meters per second for explosive eruptions. During such very rapid ascent rates, volatile exsolution may be hampered by the ability of gases to diffuse into bubbles.

The previous section has shown that most magma will at some point become permeable to gas, either by differential ascent of gas and magma or by expansion of the gas phase, resulting in open-system degassing and gas loss. In such cases the rate of degassing exceeds that of magma ascent. Determining the timescales of degassing can then provide important clues to the nature of the gas segregation mechanism and vice versa. Timescales of magma ascent and crystallization have been discussed in the previous chapters (Costa & Morgan, 2010; Dosseto & Turner, 2010; O'Neill & Spiegelman, 2010; Turner & Bourdon, 2010); here the timescales of gas exsolution and segregation will be discussed.

Diffusive timescales

Diffusive timescales

Not only do different volatile species have different solubilities, they also have different diffusivities. H_2O, the most abundant volatile species, is also the fastest at diffusing. When bubbles form during decompression or magma ascent, segregation can occur too fast to enable equilibrium partitioning of all species between the gas and melt phases. Thus, not all H_2O or other volatile species are completely exsolved from the melt upon eruption, because the melt does not have enough time to fully re-equilibrate with the gas phase at the new pressure. Because each species diffuses over different timescales, it is possible to use the extent of degassing of each species to assess the magma ascent rate as well as the timescale of the incomplete exsolution. A study by Humphreys *et al.* (2008) shows that there are H_2O gradients in tubular melt inclusions, which were not yet fully closed off at the time of eruption. They used these diffusion profiles to estimate a magma ascent time during the 1980 Plinian eruption of Mt St Helens of 40 to 60 m/s. Castro *et al.* (2005) measured H_2O concentrations in obsidian, which decrease towards bubbles indicating incomplete degassing. By fitting the measured profiles to a model of H_2O diffusion, they established that the time between bubble nucleation and magma quenching was between 0.4 and 15 days.

Diffusion profiles of H_2O have also been observed within crystals. Even anhydrous minerals, such as olivine, can contain ppm levels of H_2O. Thus H_2O, and probably other volatiles as well, can diffuse through the crystal lattice, implying that even enclosed melt inclusions will respond to devolatilization of the groundmass melt. This response is, however, dependent upon diffusion kinetics, which provides a timescale of the crystal's ascent in a degassing magma. There are several ways in which H_2O can diffuse through a crystal: as molecular H_2O via dislocations within the lattice, or by hydrogen diffusion. Portnyagin *et al.* (2008) showed experimentally that relatively dry enclosed melt inclusions in olivines increase their H_2O within hours when subjected to a H_2O-rich environment. Such rapid readjustment would imply that H_2O-rich melt inclusions have spent less than a few hours at low pressures; that is, they ascended very rapidly. Hydrogen diffusion profiles have also been observed in xenocrystic mantle olivines from kimberlites and alkali basalts (Peslier et al., 2008). These diffusion profiles suggest ascent rates in excess of 5 to 37 m/s to capture the diffusive release of H_2O from the olivines en route to the surface.

Diffusive loss of volatiles during eruption

One of the fastest rates of cooling for magma is that of pumice in the atmosphere. When an eruption plume is injected into the atmosphere, it mixes with air and cools quickly in a matter of minutes. Hort & Gardner (2000) found through numerical modeling, however, that hot pumice clasts in the plume will also cool rapidly by losing heat to the plume, but not uniformly. Instead, the rim of the pumice will cool almost as fast as the plume, whereas its interior can stay relatively hot for many minutes, with that time being longer for larger pumices. Ash shards, however, are expected to cool essentially as rapidly as the plume itself, because of the very short distance over which heat must be conducted.

Based on cooling timescales estimated for pumices of various sizes, Hort & Gardner (2000) predicted that the amount of water that will remain dissolved in the glass should be low, but not as low as expected from the solubility of water at atmospheric pressure. Slightly elevated water contents are, in fact, often found in volcanic glass, although these are much lower than would be expected if the vesicle content of pumice recorded the total amount of degassing. For example, magmas erupted in 1991 at Mt Pinatubo and in 1883 from Krakatau initially contained ~5 to 6 wt.% water, and their resulting pumices typically have 60 to 80 vol.% vesicles (Gerlach et al., 1996; Mandeville et al., 1998). If those pumice were to represent the end-product of closed-system degassing, then their glass should still have ~1 to 2 wt.% water dissolved within them; instead they contain only ~0.3 to 0.8 wt.%. Hence, 10 to 30% of the water released from melt occurs after the network of bubbles becomes permeable. Hydrogen isotopic data from volcanic glass supports the loss of water through open-system degassing of the pumice (Westrich et al., 1991).

Both chlorine and sulfur are often found in low concentrations in volcanic glass, with the differences between un-degassed glass inclusions and matrix glass tending to increase as the melt becomes hotter and more basic in composition (Devine et al., 1984; Palais & Sigurdsson, 1989). Those differences are usually much smaller than for water and can, at least partly, be attributed to open-system degassing of cooling pumice, because both sulfur and chlorine diffuse more slowly than water, but both diffuse faster as temperature increases (Watson, 1994). In fact, the diffusivity of sulfur becomes slow enough at temperatures of ≤800°C that essentially no sulfur will escape from the melt. Indeed, the differences between sulfur contents in glass inclusions and matrix glass from cool rhyolite eruptions is essentially negligible. In that case, all sulfur released to the atmosphere must come from gas bubbles already present before eruption.

Breakdown of hydrous minerals

Hydrous minerals, such as hornblende and biotite, are commonly stable in water-rich melts at high water pressures, because water in their structures occupies a smaller molar volume than it does as a fluid in water-saturated melt. Conversely, those hydrous minerals are not stable at low water pressures. Hence, when magma loses water, for example during ascent to the surface, that loss causes hydrous minerals to break down, and the rate at which they break down can reflect the rate of magma ascent. The reaction of hornblende in intermediate and evolved magmas has been used the most to determine magma ascent, mainly because hornblende is a common mafic mineral in such magmas. Other volatile-rich minerals, such as biotite and anhydrite, are also known to react with silicate melt.

When hornblende is taken to low water pressures outside its stability field, it reacts with its surrounding melt to produce is a rind composed mainly of plagioclase + pyroxenes + Fe-Ti oxides. The growth rate of that reaction rind has been calibrated experimentally in a limited number of magmas, to allow the rind thickness to be used to

$$^{226}\text{Ra} \xrightarrow{\text{1599 yrs}} {}^{222}\text{Rn} \xrightarrow{\text{3.82 days}} {}^{210}\text{Pb} \xrightarrow{\text{22.6 yrs}} {}^{210}\text{Bi} \xrightarrow{\text{5.0 days}} {}^{210}\text{Po} \xrightarrow{\text{138 days}} {}^{206}\text{Pb}$$

Fig. 11.8 The part of the ^{238}U decay series, which is of particular interest to degassing, showing the half-lives of the parent in each step. The extremely short-lived nuclides that form by decay of ^{222}Rn: ^{218}Po, ^{214}Pb, ^{214}Bi and ^{214}Po, with half-lives of minutes to milliseconds, have been omitted.

estimate ascent rate. Experiments show that the reaction is slow, and no rind is seen in the first few days after decompression, which limits the use of the 'hornblende geospeedometer' to slowly rising magma in lava and dome extrusions. Its usefulness is also limited by the overall size of the hornblende. Regardless, variations in hornblende decomposition have been used to understand how fast magma can rise, for example, during dome extrusions at Mt St Helens (Rutherford & Hill, 1993), Redoubt (Wolf & Eichelberger, 1997) and Montserrat (Devine *et al.*, 1998).

Because use of the hornblende reaction is restricted to relatively slowly rising magma, water diffusion and bubble growth are fast enough that they do not limit the rate of the reaction. Water loss from the melt does raise the viscosity of the melt and slows crystal growth, both of which impact on the growth of the reaction rind. The reaction rate should thus vary as magma rises to the surface. Indeed, experimental studies show that the reaction is fastest at water pressures just below that needed for hornblende to be stable, and slows dramatically as pressure lowers (Browne & Gardner, 2006). Using the total thickness of the reaction rind as a proxy for ascent along the entire, integrated path to the surface is thus an over-simplification. In fact, Browne & Gardner's (2006) experiments found that the texture, mineralogy and thickness of the reaction are all functions of decompression path. Once those complexities are taken into account, it becomes clear that extruding magma can follow complex ascent paths, sometimes stalling one or more times during ascent before reaching the surface.

Radionuclides

Radionuclides – magma

The decay of ^{238}U, which is naturally present in magma in trace amounts, results in the produc-

tion of daughter isotopes, each of which is radioactive itself (Figure 11.8). The decay chain involves isotopes of different elements, each with a different half-life and physio-chemical properties. Thus fractionation between parent and daughter isotopes will occur during processes such as crystallization and degassing. The decay of ^{226}Ra is particularly interesting for degassing studies as it produces ^{222}Rn, which, as a noble gas, prefers the gas phase, whereas ^{226}Ra does not. Gas loss, and thus radon loss, can result in a relative shortage (or 'deficit') of its 'granddaughter' ^{210}Pb. Note that the intermediate nuclides (^{218}Po, ^{214}Pb, ^{214}Bi, ^{214}Po) are usually ignored, because of their extremely short half-lives (minutes or less). Fractionations between nuclides of decay chains are transient and their magnitude depends on the initial fractionation and the time that passed since. Thus, measurement of ^{226}Ra and ^{210}Pb in magma shortly after its eruption can be used to study timescales of degassing. Measurement of ^{222}Rn and other volatile nuclides in the gas phase at the surface provides another useful constraint on the gas escape time, which will be discussed in the next section.

Gauthier & Condomines (1999) modeled the deficit of ^{210}Pb in the melt with respect to ^{226}Ra due to degassing of ^{222}Rn:

$$\left(\frac{^{210}\text{Pb}}{^{226}\text{Ra}}\right)_m = 1 - f + f \times \exp(-\lambda_{210}t)$$

in which the parentheses stand for activity ($A = N_{210}\lambda_{210}$; the number of particles times the decay constant), f is the fraction of radon lost from the system (estimated as being close to 1, e.g. Gill *et al.*, 1985), λ is the decay constant of ^{210}Pb ($\lambda = \ln(2)/\tau_{1/2}$, where $\tau_{1/2}$ is the half-life) and t is time. Because of the very different half-lives of ^{222}Rn and ^{210}Pb, at equal activity there are many more

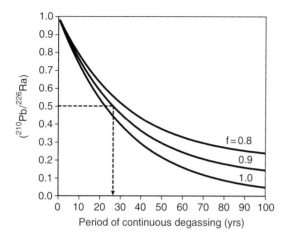

Fig. 11.9 Change in $(^{210}Pb/^{226}Ra)$ in magma with continuous degassing of a non-renewed magma reservoir (Figure after Gauthier & Condomines, 1999). f is the fraction of ^{222}Rn lost from the magma.

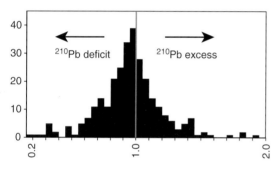

Fig. 11.10 Histogram of $(^{210}Pb/^{226}Ra)$ in volcanic rocks (from Compilation in Berlo & Turner, 2010). To compare magma erupted at different times; all ratios have been recalculated to their eruption date. Hence only analyses of samples with known eruption age are shown (n = 301). Most ratios lie between 0.5 and 1.5, but values as low as 0.2 and as high as 13 have been documented.

was continuous and no recharge occurred (Figure 11.9). This implies that magma can be fluid-saturated and permeable for at least 25 years prior to eruption. Moreover, to create ^{210}Pb deficits at all, gas escape has to be fast compared to the half-life of ^{222}Rn (3.82 days). However, there are other ways in which ^{210}Pb can be fractionated from ^{226}Ra, including dynamic melting of the mantle (Rubin et al., 2005), fractional crystallization (Reagan et al., 2008), precipitation of sulfides, or interaction with cumulates of plagioclase and amphibole (Van Orman & Saal, 2009). Thus it is important to check that there are no correlations between fractionation indices and $(^{210}Pb/^{226}Ra)$ before considering degassing as the culprit.

In symmetry with the hypothesis for ^{210}Pb deficit formation by gas loss, ^{210}Pb excesses (Figure 11.10), which are mainly observed in arc volcanoes, can be explained by streaming gas through overlying melt. Whilst traversing the overlying magma, the ^{222}Rn in the gas phase decays to less volatile ^{210}Pb, which is redissolved in the melt. The build-up of ^{210}Pb excesses in the magma that is being traversed can occur on very short timescales, if the relative volume of Rn-supplying magma to Rn-receiving magma is large. ^{210}Pb deficits and ^{210}Pb excesses are coupled and have been observed in the same eruption when magma from a range of depths is erupted, such as during the 1980 eruption of Mt St Helens (Berlo et al., 2006). Unlike ^{210}Pb deficits, it appears to be difficult to create ^{210}Pb excesses without involvement of a gas phase. However, Reagan et al. (2008) show that in evolved magma it may be possible to create excesses using amphibole and K-feldspar fractionation.

Radionuclides – gases

Besides ^{222}Rn, the decay of ^{226}Ra produces a number of other short-lived nuclides, which partition into the vapor phase during magma degassing. ^{222}Rn-^{210}Pb-^{210}Bi-^{210}Po (in order of decay, Figure 11.8), are highly enriched in volcanic plumes relative to standard atmosphere (Lambert et al., 1985/86). These nuclides owe their enrichment to the volatility of their metals, as well as their affinity for halide and sulfide compounds at high

atoms of ^{210}Pb than ^{222}Rn, requiring continued withdrawal of ^{222}Rn over years in order to see the effect on the $(^{210}Pb/^{226}Ra)$ ratio (Figure 11.9). Figure 11.10 shows a compilation of $(^{210}Pb/^{226}Ra)$ as measured within volcanic rocks all over the world. Most analyses plot between 0.5 and 1.5. The lower extent of this range would suggest a minimum degassing time of 25 years, if degassing temperatures. However, they are not equally volatile and their volatility depends on temperature and the compositions of the melt and gas. Emanation coefficients are used to characterize the ratio of atoms released, to those originally present within the magma (Lambert *et al.*, 1985/86). Polonium, with an emanation coefficient of close to 1, is almost completely lost during degassing, whereas bismuth and lead are less volatile, with emanation coefficients of 0.1 to 0.3 and 0.99 to 0.98, respectively (Gill *et al.*, 1985; Lambert *et al.*, 1985/86; Pennisi *et al.*, 1988; Rubin, 1997; Gauthier *et al.*, 2000; Le Cloarec & Gauthier, 2003). ^{226}Ra, their relatively long-lived parent isotope ($\tau_{1/2} \sim 1,599$ yrs), is not volatile and remains with the magma. Thus these radionuclides are fractionated from their parent and from each other during degassing. Note that their abundance in magma is so small that they do not degas on their own, but require a carrier gas such as H_2O or CO_2.

Gas emitted directly from magma should have relative abundances of short-lived radionuclides based upon their half-life and their emanation coefficients, thus $(^{210}Po/^{210}Pb) > (^{210}Bi/^{210}Pb) > 1$ (Le Cloarec & Pennisi, 2001; Gauthier *et al.*, 2000). Such ratios have been observed at, for example, Etna (Lambert *et al.*, 1985/86; Le Cloarec & Pennisi, 2001), Stromboli (Gauthier *et al.*, 2000) and Merapi (Le Cloarec & Gauthier, 2003), suggesting that the time between partitioning of nuclides into the gas phase and its measurement at the surface is in the order of a few hours to tens of days. However, where escape routes are longer, for example, a deeper magma source, non-eruptive degassing, cooling, deposition and decay can modify the short-lived radionuclide ratios. At Vulcano for example, Le Cloarec *et al.* (1994) suggest that the emitted ^{210}Po is derived from both magmatic degassing and from Pb-rich sublimates. Gases can condense upon fracture walls, resulting in Pb-enriched encrustations. Such sublimates are both a sink for ^{210}Pb in the gas and a source for ^{210}Po and ^{210}Bi from the decay of ^{210}Pb.

The most stringent constraint on gas escape timescales comes from ^{222}Rn, with a half-life of only 3.82 days. Radon is inert and remains gaseous even at low temperature. It does, however, dissolve in groundwater. Radon is constantly produced by decay in minerals, but only the radon along grain boundaries is free to be entrained in water or gas. Rock fracturing increases surface area and thus liberates radon. In some cases peaks in radon emissions from soil and groundwater have been observed prior to earthquakes (Igarashi *et al.*, 1995). The whole volcanic edifice, rocks, soil and aquifers, is thus a potential source of ^{222}Rn emissions and it is difficult to distinguish between ^{222}Rn derived from degassing magma at depth or anything in between there and the surface. At Stromboli, spatial and temporary fluctuations in ^{222}Rn activity have been interpreted as precursors to volcanic activity (Cigolini *et al.*, 2005). However, the uncertain origin of the ^{222}Rn makes the interpretation of the time constraints provided by its short half-life difficult. A recent study by Giammanco *et al.* (2007) at Etna uses correlations with CO_2 derived from degassing magma and an even shorter-lived radon isotope, ^{220}Rn, which is derived from soil interaction, to distinguish between magma and soil degassing.

In summary, short-lived radionuclides in volcanic gases present a unique opportunity to derive timescales of degassing. Observations of excess radionuclides, ^{222}Rn, ^{210}Po and ^{210}Bi, with half-lives of days in volcanic plumes (Lambert, 1983) and their absence in rock samples collected immediately after eruption (Sato & Sato, 1977; Bennett *et al.*, 1982; Gill *et al.*, 1985; Reagan *et al.*, 2006), testify to degassing timescales of days. Fumaroles provide another source of information, which is more complicated to interpret, because of potential contributions from soil, groundwater and sublimates. At Stromboli and Etna, gas escape can occur in as little as hours (Gauthier *et al.*, 2000), but most estimates are on the order of a few days to tens of days, with longer estimates for non-eruptive periods (Lambert *et al.*, 1985/86; Le Cloarec & Pennisi, 2001). The observation of $(^{210}Bi/^{210}Pb)>1$ at Merapi (Le Cloarec & Gauthier, 2003) suggests that in more evolved systems gas escape time is similar.

LONG-TERM CONSEQUENCES
OF DEGASSING

The preceding sections dealt with degassing timescales of seconds to decades – eruptive timescales – where degassing is largely driven by magma decompression. However, many volcanoes are underlain by magma bodies that remain partially molten for thousands or even millions of years (see Chapters 8 (Dosseto & Turner, 2010) and 10 (Bachmann, 2010)). Many such magma bodies are unlikely to be volatile saturated when emplaced, but cooling and crystallization will lead to saturation and gas exsolution. Degassing caused by subsurface cooling of magma bodies is likely to be slower than by magma ascent, especially during eruption. Periodic recharge provides a continued flux of heat and volatile-rich magma to this environment. The effects of long-term or periodic gas release from magma are evident in the form of ore deposits.

Potential for non-eruptive magmatic gas release on the millennial timescale

The dependence of the solubility of H_2O and CO_2 on temperature is smaller than that of pressure. In fact, the solubility of H_2O actually increases during isobaric cooling. Thus cooling alone is a much less efficient driving force for degassing compared to decompression. However, cooling causes crystallization and it is this that drives degassing. The timescales of crystallization are therefore highly relevant to those of degassing and are discussed in detail in Chapters 7 (Costa & Morgan, 2010) and 8 (Dosseto & Turner, 2010). Here we focus on the cooling rates of subsurface magma bodies that can be estimated independently.

A simple estimate of cooling rates of a magma body can be obtained by looking at the conductive cooling of a sheet-like body. In a sheet, with a smaller vertical than lateral extent, heat dissipation can be estimated using a one-dimensional approach. In this simple approach we are ignoring convection within the magma body. The thickness of the sheet is $2w$ and the initial temperature is $T_0 > T_{cr}$, with T_{cr} being the host rock temperature:

$$\partial T/\partial t = \alpha \partial^2 T/\partial x^2 \qquad -\infty < x < \infty$$

where $\alpha = \kappa/\rho c_p$ is the thermal diffusivity. For boundary conditions $\partial T/\partial x = 0$ at $x = 0$ and $x \Rightarrow \infty$, the solution to this equation is (Carslaw & Jaeger, 1959):

$$\frac{(T - T_{cr})}{T_0 - T_{cr}} = \frac{1}{2}\left[erf\left[\frac{w - x}{2\sqrt{\alpha t}}\right] + erf\left[\frac{w + x}{2\sqrt{\alpha t}}\right]\right] -\infty < x < \infty$$

which gives the temperature distribution within the sheet and surrounding rock at any time. A single dyke of 100 m thickness with initial temperature of 1,200°C will cool to a maximum temperature of 900°C within ~400 years. Observations of plutons suggest that magma chambers form by additions of multiple intrusions over extended time periods (Glazner *et al.*, 2004). Repeated intrusion of magma provides not only a direct influx of volatiles, but also extends the lifespan of magmatic systems.

Petford & Gallagher (2001) modeled the effects of periodic intrusion of sills in the lower crust. They showed that single intrusions, regardless of size, will lead to short heat pulses and little melting of the host rock, whereas multiple intrusions, particularly when the interval between intrusions is short, can lead to substantial melting. Their model considers only the process of melt generation in the surrounding crust that has become heated by the sills. The ability of rocks to generate melt once heated depends critically on the amount of H_2O available. Solid rocks are limited in the amount of H_2O they can store by the abundance of hydrous minerals, such as amphibole and micas. A typical amphibolite, with 40% amphibole can only contain 0.8 wt.% H_2O, which severely limits its ability to generate melt once heated by a process known as dehydration melting.

Annen & Sparks (2002) and Annen *et al.* (2006) extended the models of Petford & Gallagher (2001) to consider the generation of evolved melts both from heating country rock and from crystallizing basalt within the sills. In many settings, notably above subduction zones, basalts forming lower crustal sills contain high initial contents of

H_2O, typically much higher than the 0.8 wt.% in amphibolite. This vastly increases their ability to generate evolved melts during crystallization. Annen *et al.* (2006) show that melt productivity in the lower crust is vastly enhanced when we consider both potential sources of melt. There is also a significant 'incubation time' after injection of the first sill, during which the crust temperature becomes elevated above the initial crustal geotherm, by advection of heat from the sills. Slow magma intrusion rates lengthen the incubation time; fast rates shorten it.

An important aspect of melt generation in the Deep Crustal Hot Zone concept of Annen *et al.* (2006) is that the residual melt concentrates H_2O present in the initial basalt, because the minerals that crystallize close to the liquidus are H_2O-poor. Thus for an initial basalt H_2O content of 2.5 wt.%, 75% crystallization, a typical value for the generation of silicic andesite melt, leads to a four-fold increase in H_2O. Melts generated in this way can have very high initial H_2O contents before they begin their ascent. At lower crustal pressures, H_2O solubility is sufficiently high that all of this volatile burden is fully dissolved within the melt and is only released at shallow depths, leading, as described above, to copious decompression crystallization once the H_2O-saturated liquidus is reached. CO_2 present in the initial sill-forming basalts will also be concentrated in a similar way. However, the lower solubility of CO_2 relative to H_2O means that CO_2 will start to degas at much greater depths in the crust, and will be significantly reduced at the onset of decompression crystallization. In summary, Deep Crustal Hot Zones are likely to play an important part in modifying mantle-derived volatile-bearing basalts beneath many arcs, and provide a means to generate significant volumes of volatile-charged evolved melts that can then ascend and degas within the crust over a considerable range of depths.

Ore deposits as a result of long-term degassing

The existence of large, long-lived bodies of solidifying magma within the crust has profound consequences for the surrounding rocks. Not only do these bodies provide heat, they also supply volatile elements expelled from the magma to the country rock. Heating may cause the surrounding host rock to partially melt and some hybridization between intruding magma and (partially) molten host is likely to occur. The combined effects of the temperature gradient and the flux of volatile elements from the magma drives hydrothermal circulation of fluids through both solidified parts of the magma body and the host. Such hydrothermal circulation cells are the basis for the formation of ore deposits surrounding long-lived magmatic systems. The extent to which the ore elements originate from the magma or are leached from the country rock is variable (Hedenquist & Lowenstern, 1994). Porphyry deposits (Au, Cu, Mo) are closely associated with magmatic activity and derive much of the ore-forming fluid directly from magma. They are commonly found in orogenic arcs, such as the Andes, where magma accumulation in the lower to middle crust has occurred over millions of years. Other types of ore deposits, which have a high input of magmatic fluids, are high-sulfidation epithermal deposits, which are thought to be shallower extensions of the porphyry deposits based upon their spatial relation (Sillitoe, 1973; Arribas *et al.*, 1995). There are many volcanoes with well-established surficial hydrothermal systems, for example White Island in New Zealand, which may be surface manifestations of ongoing ore deposition at depth.

Observations at potential sites of active ore deposition at volcanoes raise the question of how long it takes to form an entire ore body. One way to find out is by measuring the fluxes of metals coming out of volcanoes, for example in those with well established hydrothermal systems, and compare this to fluid chemistry in established ore deposits. Simmons & Brown (2006) measured the Au-flux at one of the world's largest gold deposits; the Ladolam hydrothermal system hosted by Luise Volcano in Papua New Guinea. Based on their analyses, they concluded that the current deposits could have formed in several 10^4 years. Using a similar approach, Simmons & Brown (2007) showed that in the Taupo Volcanic Zone, metal fluxes are actually even higher, yet so far these high fluxes may not have resulted in ore formation as there are no known deposits. This suggests that not only the element flux, but also the efficiency of the precipitation mechanism is important in the formation of ore deposits. Detailed studies also show that epithermal ore deposits are the result of multiple episodes of magmatic activity (Bendezu *et al.*, 2008). A recent review by Tosdal *et al.* (2009) suggests that ore deposits can form in a few thousand to a few hundred thousand years.

Porphyry deposits are associated with large intrusive complexes, which provide heat to drive regional-scale hydrothermal systems. As shown in the previous section, such systems can remain a significant source of heat for millions of years, especially if the system experiences regular influxes of new magma. Dating of ore minerals has shown that most porphyry deposits formed over extended periods of time (Harris *et al.*, 2008; Marsh *et al.*, 1997). However, as with epithermal systems, the formation occurs episodically rather than continuously. The Bajo de la Alumbrera porphyry copper-gold deposit in Argentina, for example, formed over a three to four million-year period (Harris *et al.*, 2008). However, without recharge, the magmatic system could not have provided the heat flux over this period and would have solidified within <10,000 years.

The formation of an ore deposit may take as little as several 10^3 years, or grow incrementally over millions of years. Each such increment may be linked to magma recharge, providing a new influx of volatile rich magma, which is gradually expelled. It remains challenging to distinguish such episodes with the available chronometers.

SYNTHESIS

Degassing is an all-encompassing term that has been applied to a wide variety of processes that result in the loss of volatiles from magma. Consequently this chapter has discussed degassing timescales of seconds to thousands of years.

It should be clear now that it is impossible to visualize degassing as a single, isolated process. Degassing is driven by magma decompression and cooling and thus intimately linked to magma ascent, crystallization and recharge. Here we attempt to integrate the timescales of the different processes that lead to volatile loss. In a conceptual model (Figure 11.11), the first step of degassing is exsolution of volatiles. The first volatile species to exsolve is, in most cases, CO_2 and this can reach saturation at depths of 15 km or more, depending on the initial concentration. Rates of exsolution are dependent upon the diffusivity as well as the bubble number density. This is essentially closed-system degassing and will continue until equilibrium partitioning is reached and there are no changes in pressure

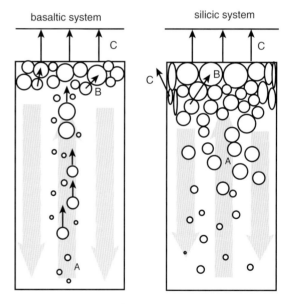

Fig. 11.11 Conceptual model to illustrate degassing stages. Degassing starts as a closed-system degassing of variable duration depending on decompression and cooling rates (a). The transition to open system degassing (b) can be brought about in a variety of ways, for illustrative purposes the basaltic case is shown as bubble ascent aided by convection. Alternatively, accumulated bubbles may become permeable to gas, leading to gas escape (c).

and temperature to drive further exsolution. Maximum estimates for the duration of closed-system degassing are thus commensurate with timescales of magma ascent and cooling.

The transition from closed-system degassing to open-system degassing can be effected in a number of different ways. In low viscosity systems (e.g. basaltic, Figure 11.11), bubbles may be able to ascend through the magma to some extent, causing a gradual change in concentration. In Figure 11.11, the transition is visualized as a bubble 'foam'. Foams are transient features, but can be continuously or periodically recreated. Destruction of foam can lead to resorption, fragmentation or coalescence and gas escape. However when interrupted, for example by eruption, exsolution may be incomplete and the sudden transition from closed system to open system can take place in seconds, as shown by Burgisser & Gardner (2005). Consequently, the onset of open-system degassing can be sudden or more gradual, where gas bubbles grow to sufficient size to escape the melt, for example.

Once gas is able to escape from the melt, it will continue to do so until it reaches equilibrium under the magmatic conditions. Rates derived from diffusive profiles in melt and minerals, as well as breakdown rims on hydrous minerals, are essentially decompression rates, whereby volatile elements attempt to readjust to changing pressure conditions. For magma to continue to exsolved volatiles pressure, temperature or composition conditions need to change. Timescales of decades, such as derived from $(^{210}Pb/^{226}Ra)$, are thus likely related to convection or continuing ascent of magma. Where volatiles are periodically replenished by additions of new magma over timescales of 10^3 years, ore deposits can form. Short-lived radionuclides measured in volcanic gases give estimates of the time it takes for gas to travel from magma to surface and can yield information on the depth of degassing magma and the permeability of the intermediate rock and soil. How fast gases can move through magma itself will depend on the physical mechanism by which it is transferred.

REFERENCES

Annen C, Sparks RSJ. 2002. Effects of repetitive emplacement of basaltic intrusions on thermal evolution and melt generation in the crust. *Earth and Planetary Science Letters* **203**: 937–955.

Annen, C, Blundy JD, Sparks RSJ. 2006. The genesis of intermediate and silicic magmas in deep crustal hot zones. *Journal of Petrology* **47**: 505–539.

Arribas A, Hedenquist JW, Itaya T, Okada T, Concepcion RA, Garcia JS. 1995. Contemporaneous formation of adjacent porphyry and epithermal Cu-Au deposits over 300 ka in northern Luzon, Philippines. *Geology* **23**: 337–340.

Bachmann O. 2010. Timescales associated with large silicic magma bodies. In: *Timescales Of Magmatic Processes: From Core to Atmosphere*. Dosseto, A, Turner S, and J. A. Van Orman JA (eds), Wiley-Blackwell, Oxford.

Bendezu R, Page L, Spikings R, Pecskay Z, Fontbote L. 2008. New $^{40}Ar/^{39}Ar$ alunite ages from the Colquijirca district, Peru: evidence of a long period of magmatic SO_2 degassing during formation of epithermal Au-Ag and Cordilleran polymetallic ores. *Mineralium Deposita* **43**: 777–789.

Bennett JT, Krishnaswami S, Turekian KK, Melson, Hopson WG. 1982. The uranium and thorium decay series nuclides in Mt. St. Helens effusives. *Earth and Planetary Science Letters* **60**: 61–69.

Berlo K, Turner S, Blundy J, Black S, Hawkesworth C. 2006. Tracking pre-eruptive magma degassing using $(^{210}Pb/^{226}Ra)$ disequilibria in volcanic deposits of the 1980–1986 eruption of Mt St Helens. *Earth and Planetary Science Letters* **249**: 337–349.

Berlo K, Turner S. 2010. ^{210}Pb-^{226}Ra disequilibria in volcanic rocks. *Earth and Planetary Science Letters* **296**: 155–164.

Blundy J, Cashman KV. 2001. Ascent-driven crystallisation of dacite magmas at Mt St Helens, 1980–1986. *Contributions to Mineralogy and Petrology* **140**: 631–650.

Blundy J, Cashman K. 2005. Rapid decompression-driven crystallization recorded by melt inclusions from Mt St Helens volcano. *Geology* **33**: 793–796.

Blundy J, Cashman K, Humphreys MCS. 2006. Magma heating by decompression-driven crystallisation beneath andesite volcanoes. *Nature* **443**: 76–80.

Blundy J, Cashman K. 2008. Petrologic constraints on magma system variables and processes. *Reviews in Mineralogy and Geochemistry* **69**: 179–239.

Browne BL, Gardner JE. 2006. The influence of magma ascent path on the texture, mineralogy, and formation of hornblende reaction rims. *Earth and Planetary Science Letters* **246**: 161–176.

Burgisser A, Gardner JE. 2005. Experimental constraints on degassing and permeability in volcanic conduit flow. *Bulletin of Volcanology* **67**: 42–56.

Castro JM, Manga M, Martin MC. 2005. Vesiculation rates of obsidian domes inferred from H_2O concentration profiles. *Geophysical Research Letters* **32**: L21307.

Cigolini C, Gervino G, Bonetti R, *et al.* 2005. Tracking precursors and degassing by radon monitoring during major eruptions at Stromboli Volcano (Aeolian Islands, Italy). *Geophysical Research Letters* **32**: L12308.

Costa F Morgan D. 2010. Time constraints from chemical equilibration in magmatic crystals. In: *Timescales of Magmatic Processes: From Core to Atmosphere.* Dosseto A, Turner S, Van Orman JA (eds), Wiley-Blackwell, Oxford.

Carslaw HS, Jaeger JC. 1959. *Conduction of heat in solids*, 2nd edn. Clarendon Press, Oxford.

Devine JD, Sigurdsson H, Davis AN, Self S. 1984. Estimates of sulfur and chlorine yield to the atmosphere from volcanic eruptions and potential climatic effects. *Journal of Geophysical Research* **89**: 6309–6325.

Devine JD, Rutherford MJ, Gardner JE. 1998. Petrologic determination of ascent rates for the 1995–1997 Soufriere Hills Volcano andesitic magma. Geophysical Research Letters **25**: 3673–3676.

Dosseto A, Turner S. 2010. Magma cooling and differentiation – U-series isotopes. In: *Timescales of Magmatic Processes: From Core to Atmosphere.* Dosseto A, Turner S, Van Orman JA (eds), Wiley-Blackwell, Oxford.

Edmonds M, Oppenheimer C, Pyle DM, Herd RA, Thompson G. 2003. SO_2 emissions from Soufrière Hills Volcano and their relationship to conduit permeability, hydrothermal interaction and degassing regime. *Journal of Volcanology and Geothermal Research* **124**: 23–43.

Edmonds M. 2008. New Geochemical insights into volcanic degassing. *Philosophical Transactions of the Royal Society* **A388**: 4559–4579.

Fine GJ, Stolper EM. 1986. Carbon dioxide in basaltic glasses: concentrations and speciation. *Earth and Planetary Science Letters* **76**: 263–278.

Fischer TP, Marty B. 2005. Volatile abundances in the sub-arc mantle: insights from volcanic and hydro-thermal gas discharges. *Journal of Volcanology and Geothermal Research* **140**: 205–216.

Gardner JE. 2007. Heterogeneous bubble nucleation in highly viscous silicate melts during instantaneous decompression from high pressure. *Chemical Geology* **236**: 1–12.

Gardner JE. 2009. The impact of pre-existing gas on the ascent of explosively erupted magma. *Bulletin of Volcanology* **71**: 835–844.

Gardner JE, Hilton M, Carroll MR. 1999. Experimental constraints on degassing of magma: isothermal bubble growth during continuous decompression from high temperature. *Earth and Planetary Science Letters* **168**: 201–218.

Gardner JE, Hilton M, Carroll MR. 2000. Bubble growth in highly viscous silicate melts during continuous decompression from high pressure. *Geochimica et Cosmochimica Acta* **64**: 1473–1483.

Gauthier P-J, Condomines M. 1999. [210]Pb-[226]Ra radioactive disequilibria in recent lavas and radon degassing: inferences on the magma chamber dynamics at Stromboli and Merapi volcanoes. *Earth and Planetary Science Letters* **172**: 111–126.

Gauthier PJ, Le Cloarec M-F, Condomines M. 2000. Degassing processes at Stromboli Volcano inferred from short-lived disequilibria ([210]Pb-[210]Bi-[210]Po) in volcanic gases. *Journal of Volcanology and Geothermal Research* **102**: 1–19.

Gerlach TM, Westrich HF, Symonds RB. 1996. Pre-eruption vapor in magma of the climactic Mount Pinatubo eruption: Source of the giant stratospheric sulfur dioxide cloud. In: *Fire and Mud: Eruptions and Lahars of Mt Pinatubo, Philippines*, Newall CG, Punongbayan RS (eds.), Quezon City, Philippine Institute of Volcanology and Seismology, and Seattle, University of Washington Press, Washingtom DC, 415–431.

Giammanco S, Sims KWW, Neri M. 2007. Measurement of [220]Rn and [222]Rn and CO_2 emissions in soil and fumarole gases on Mt. Etna volcano (Italy): Implications for gas transport and shallow ground fracture. *Geochemistry Geophysics Geosystems* **8**: Q10001.

Gill J, Williams R, Bruland K. 1985. Eruption of basalt and andesite lava degasses [222]Rn and [210]Po. *Geophysical Research Letters* **12**: 17–20.

Glazner AF, Bartley JM, Coleman DS, Gray W, Taylor Z. 2004. Are plutons assembled over millions of years by amalgamation from small magma chambers? *GSA Today* **14**: 4–11.

Gualda GAR, Anderson AT. 2007. Magnetite scavenging and the buoyancy of bubbles in magmas. Part1:

Discovery of a pre-eruptive bubble in Bishop rhyolite. *Contributions to Mineralogy and Petrology* **153**: 733–742.

Harris AC, Dunlap WJ, Reiners PW, *et al.* 2008. Multimillion year thermal history of a porphyry copper deposit: application of U-Pb, $^{40}Ar/^{39}Ar$ and (U-Th)/He chronometers, Bajo de la Alumbrera copper-gold deposit, Argentina. *Mineralium Deposita* **43**: 295–314.

Harris AC, Dunlap WJ, Reiners PW, *et al.* 2008. Multimillion year thermal history of a porphyry copper deposit: Application of U-Pb, $^{40}Ar/^{39}Ar$ and (U-Th)/He chronometers, Bajo de la Alumbrera copper-gold deposit, Argentina. *Mineralium Deposita* **43**: 295–314.

Hedenquist JW, Lowenstern JB. 1994. The role of magmas in the formation of hydrothermal ore deposits. *Nature* **370**: 519–527.

Hort M, Gardner J. 2000. Constraints on cooling and degassing of pumice during Plinian volcanic eruptions based on model calculations. *Journal of Geophysical Research* **105**: 25,981–26,001.

Humphreys MCS, Menand T, Blundy JD, Klimm K. 2008. Magma ascent rates in explosive eruptions: constraints from H_2O diffusion in melt inclusions. *Earth and Planetary Science Letters* **270**: 25–40.

Hurwitz S, Navon O. 1994. Bubble nucleation in rhyolitic melts- experiments at high pressure, temperature and water-content. *Earth and Planetary Science Letters* **122**: 267–280.

Igarashi G, Saeki S, Takahata K, *et al.* 1995. Groundwater radon anomaly before the Kobe earthquake in Japan. *Science* **269**: 60–61.

Johannes W, Holtz F. 1996. Petrogenesis and experimental petrology of granitic rocks. Springer-Verlag, Berlin, 335 pp.

Lambert G, Le Cloarec MF, Ardouin B, Le Roulley JC. 1985/86. Volcanic emission of radionuclides and magma dynamics. *Earth and Planetary Science Letters* **76**: 185–192.

Lambert G. 1983. Volcanic emission of radon daughters. In: *Forecasting Volcanic Events*, Tazieff H, Sabroux JC (eds), Elsevier Science, New York, 475–484.

Le Cloarec M-F, Gauthier P-J. 2003. Merapi volcano, Central Java, Indonesia: A case study of radionuclide behavior in volcanic gases and its implications for magma dynamics at andesitic volcanoes. *Journal of Geophysical Research* **108**(B5): 2243.

Le Cloarec MF, Pennisi M. 2001. Radionuclides and sulfur content in Mount Etna plume in 1983–1995: New constraints on the magma feeding system. *Journal of Volcanology and Geothermal Research* **108**: 141–155.

Le Cloarec M-F, Pennisi M, Corazza E, Lambert G. 1994. Origin of fumarolic fluids emitted from a nonerupting volcano: Radionuclide constraints at Vulcano (Aeolian Islands, Italy). *Geochimica et Cosmochimica Acta* **58**: 4401–4410.

Liu Y, Anderson AT, Wilson CJN, Davis AM, Steele IM. 2006. Mixing and differentiation in the Oruanui rhyolitic magma, Taupo, New Zealand: Evidence from volatiles and trace elements in melt inclusions. *Contributions to Mineralogy and Petrology* **151**: 71–87.

Mandeville CW, Sasaki A, Saito G, Faure K, King R, Hauri E. 1998. Open-system degassing of sulfur from Krakatau 1883 magma. *Earth and Planetary Science Letters* **160**: 709–722.

Massol H, Koyaguchi T. 2005. The effect of magma flow on nucleation of gas bubbles in a volcanic conduit. *Journal of Volcanology and Geothermal Research* **143**: 69–88.

Marsh TM, Einaudi MT, McWilliams M. 1997. $^{40}Ar/^{39}Ar$ Geochronology of Cu-Au and Au-Ag Mineralization in the Potrerillos district, Chile. *Economic Geology* **92**: 784–806.

Merzbacher C, Eggler DH. 1984. A magmatic geohygrometer: application to Mt St Helens. *Geology* **12**: 587–590.

Métrich N, Wallace PJ. 2008. Volatile abundances in basaltic magmas and their degassing paths tracked by melt inclusions. *Reviews in Mineralogy and Geochemistry* **69**: 363–402.

Moore G. 2008. Interpreting H_2O and CO_2 contents in melt inclusions: constraints from solubility experiments and modeling. *Reviews in Mineralogy and Geochemistry* **69**: 333–361.

Mourtada-Bonnefoi CC, Laporte D. 2004. Kinetics of bubble nucleation in a rhyolitic melt: an experimental study of the effect of ascent rate. *Earth and Planetary Science Letters* **218**: 521–537.

Mueller S, Melnik O, Spieler O, Scheu B, Dingwell DB. 2005. Permeability and degassing of dome lavas undergoing rapid decompression: an experimental determination. *Bulletin of Volcanology* **67**: 526–538.

Navon O, Lyakhovsky V. 1998. Vesiculation processes in silicic magmas. In: *The Physics of Explosive Volcanic Eruptions*, Gilbert JS, Sparks RSJ. (eds), Geological Society, London, Special Publications **145**: 27–50.

Newman S, Lowenstern JB. 2002. VolatileCalc: A silicate melt-H_2O-CO_2 solution model written in visual basic for excel. *Computers and Geosciences* **28**: 597–604.

O'Neill C, Spiegelman M. 2010. Formulations for simulating the multiscale physics of magma ascent.

In: *Timescales of Magmatic Processes: From Core to Atmosphere*. Dosseto A, Turner S, Van Orman JA (eds), Wiley-Blackwell, Oxford.

Palais JM, Sigurdsson H. 1989. Petrologic evidence of volatile emissions from major historic and pre-historic volcanic eruptions. In: *Understanding Climate Change*, J.W. Kidson, JW (ed.), American Geophysical Union, Geophysics Monograph **52**: 31–53.

Papale P, Moretti R, Barbato D. 2006. The compositional dependence of the saturation surface of $H_2O + CO_2$ fluids in silicate melts. *Chemical Geology* **229**: 78–95.

Parfitt EA. 2004. A discussion of the mechanisms of explosive basaltic eruptions. *Journal of Volcanology and Geothermal Research* **134**: 77–107.

Pennisi M, Le Cloarec M-F, Lambert G, Le Roulley JC. 1988. Fractionation of metals in volcanic emissions. *Earth and Planetary Science Letters* **88**: 284–288.

Peslier AH. Woodland AB. Wolff JA. 2008. Fast kimberlite ascent rates estimated from hydrogen diffusion profiles in xenolithic mantle olivines from southern Africa. *Geochimica et Cosmochimica Acta* **72**: 2711–2722.

Petford N, Gallagher K. 2001. Partial melting of mafic (amphibolitic) lower crust by periodic influx of basaltic magma. *Earth and Planetary Science Letters* **193**: 483–499.

Portnyagin M, Almeev R, Matveev S, Holtz F. 2008. Experimental evidence for rapid water exchange between melt inclusions in olivine and host magma. *Earth and Planetary Science Letters* **272**: 541–552.

Proussevitch AA, Sahagian DL, Kutolin VA. 1993. Stability of foams in silicate melts. *Journal of Volcanology and Geothermal Research* **59**: 161–178.

Reagan MK, Tepley FJ, Gill JB, Wortel M, Garrison J. 2006. Timescales of degassing and crystallization implied by [210]Po–[210]Pb–[226]Ra disequilibria for andesitic lavas erupted from Arenal volcano. *Journal of Volcanology and Geothermal Research* **157**: 135–146.

Reagan MK, Turner S, Legg M, Sims KWW, Hards VL. 2008.[238]U-and [232]Th-decay series constraints on the timescales of crystal fractionation to produce the phonolite erupted in 2004 near Tristan da Cunha, South Atlantic Ocean. *Geochimica et Cosmochimica Acta* **72**: 4367–4378.

Rubin K. 1997. Degassing of metals and metalloids from erupting seamount and mid-ocean ridge volcanoes: Observations and predictions. *Geochimica et Cosmochimica Acta* **61**: 3525–3542.

Rubin KH, van der Zander I, Smith MC, Bergmanis EC. 2005. Minimum speed limit for ocean ridge magmatism from [210]Pb–[226]Ra–[230]Th disequilibria. *Nature* **437**: 534–538.

Rust AC, Cashman KV, Wallace PJ. 2004. Magma degassing buffered by vapor flow through brecciated conduit margins. *Geology* **32**: 349–352.

Rutherford MJ. 2008. Magma ascent rates. *Reviews in Mineralogy and Geochemistry* **69**: 241–271.

Rutherford MJ, Devine JD. 1988. The May 18, 1980 eruption of Mt St Helens 3. Stability and chemistry of amphibole in the magma chamber. *Journal of Geophysical Research* **93**: 11,949–11,959.

Rutherford MJ, Hill PM. 1993. Magma ascent rates from amphibole breakdown: An experimental study applied to the 1980–1986 Mt St. Helens eruption. *Journal of Geophysical Research* **98**: 19,667–19,686.

Rutherford MJ, Sigurdsson H, Carey S, Davis A. 1985. The May 18, 1980 eruption of Mount St. Helens 1. Melt composition and experimental phase equilibria. *Journal of Geophysical Research* **90**: 2929–2947.

Sato K, Sato J. 1977. Estimation of gas-releasing efficiency of erupting magma from [226]Ra-[222]Rn disequilibrium. *Nature* **266**: 439–440.

Schmitt AK. 2001. Gas-saturated crystallization and degassing in large-volume, crystal-rich dacitic magmas from the Altiplano-Puna, northern Chile. *Journal of Geophysical Research* **106**: 30,561–30578.

Sillitoe RH. 1973. The tops and bottoms of porphyry copper deposits. *Economic Geology* **68**: 799–815.

Simmons SF, Brown KL. 2006. Gold in magmatic hydrothermal solutions and the rapid formation of a giant ore deposit. *Science* **314**: 288–291.

Simmons SF, Brown KL. 2007. The flux of gold and related metals through a volcanic arc, Taupo Volcanic Zone, New Zealand. *Geology* **35**: 1099–1102.

Silver LA, Ihinger PD, Stolper E. 1990. The influence of bulk composition on the speciation of water in silicate glasses. *Contributions to Mineralogy and Petrology* **104**: 142–162.

Sparks RSJ, Brazier S. 1982. New evidence for degassing processes during explosive eruptions. *Nature* **295**: 218–220.

Stasiuk MV, Barclay J, Carroll MR, *et al.* 1996. Degassing during magma ascent in the Mule Creek vent (USA). *Bulletin of Volcanology* **58**: 117–130.

Stevenson DS, Blake S. 1998. Modelling the dynamics and thermodynamics of volcanic degassing. *Bulletin of Volcanology* **60**: 307–317.

Stix J, Zapata GJA, Calvache VM, *et al.* 1993. A model of degassing at Galeras Volcano, Colombia, 1988–1993. *Geology* **21**: 963–967.

Thomas N, Tait S, Koyaguchi T. 1993. Mixing of stratified liquids by the motion of gas bubbles: application to magma mixing. *Earth and Planetary Science Letters* **115**: 161–175.

Toramaru A. 1989. Vesiculation process and bubble-size distributions in ascending magmas with constant velocities. *Journal of Geophysical Research* **94**: 17,523–17,542.

Tosdal RM, Dilles JH, Cooke DR. 2009. From source to sinks in auriferous magmatic-hydrothermal porphyry and epithermal deposits. *Elements* **5**: 289–295.

Tuffen H, Dingwell DB, Pinkerton H. 2003. Repeated fracture and healing of silicic magma generate flow banding and earthquakes? *Geology* **31**: 1089–1092.

Tuffen H, Smith R, Sammonds PR. 2008. Evidence for seismogenic fracture of silicic magma. *Nature* **453**: 511–514.

Turner S, Bourdon B. 2010. Melt transport from the mantle to the crust – U-series isotopes. In: *Timescales of Magmatic Processes: From Core to Atmosphere*. Dosseto A, Turner S, Van Orman JA (eds), Wiley-Blackwell, Oxford.

Van Orman JA, Saal AE. 2009. Influence of crustal cumulates on 210Pb disequilibria in basalts. *Earth and Planetary Science Letters* **284**: 284–291.

Vergniolle S, Jaupart C. 1990. Dynamics of degassing at Kilauea volcano, Hawaii. *Journal of Geophysical Research* **95**: 2793–2809.

Wallace PJ, Anderson AT, Davis AM. 1999. Gradients in H_2O, CO_2 and exsolved gas in a large-volume silicic magma system: Interpreting the record preserved in melt inclusions from the Bishop Tuff. *Journal of Geophysical Research* **104**: 20,097–20,122.

Wallace P. 2001. volcanic SO_2 emissions and the abundance and distribution of exsolved gas is magma bodies. *Journal of Volcanology and Geothermal Research* **108**: 85–106.

Watson EB. 1994. Diffusion in volatile-bearing magmas. *Reviews in Mineralogy and Geochemistry* **30**: 371–411.

Westrich HR, Eichelberger JC, Hervig RL. 1991. Degassing of the 1912 Katmai magmas. *Geophysical Research Letters* **18**: 1561–1564.

Wolf KJ, Eichelberger JC. 1997. Syneruptive mixing, degassing, and crystallization at Redoubt Volcano, eruption of December, 1989 to May 1990. *Journal of Volcanology and Geothermal Research* **75**: 19–37.

Index

Timescales of Magmatic Processes: From Core to Atmosphere, 1st edition. Edited by Anthony Dosseto, Simon P. Turner and James A. Van Orman.
© 2011 by Blackwell Publishing Ltd.